Cognitive Morphodynamics

European Semiotics: *Language, Cognition, and Culture*
Sémiotiques Européennes: *langage, cognition et culture*

Edited by / Série dirigée par
Per Aage Brandt (Cleveland), Wolfgang Wildgen (Bremen/Brême),
and/et Barend van Heusden (Groningen/Groningue)

Volume 11

PETER LANG
Bern · Berlin · Bruxelles · Frankfurt am Main · New York · Wien

Jean Petitot
In collaboration with René Doursat

Cognitive Morphodynamics

Dynamical Morphological Models of Constituency in Perception and Syntax

PETER LANG
Bern · Berlin · Bruxelles · Frankfurt am Main · New York · Wien

Bibliographic information published by Die Deutsche Bibliothek
Die Deutsche Bibliothek lists this publication in the Deutsche Nationalbibliografie; detailed bibliographic data is available on the Internet at ‹http://dnb.ddb.de›.

A catalogue record for this book is available from *The British Library*, Great Britain.

Library of Congress Cataloging-in-Publication Data

Petitot, Jean, 1944-
Cognitive morphodynamics : dynamical morphological models of constituency in perception and syntax / Jean Petitot ; in collaboration with René Doursat.
p. cm. – (European semiotics: language, cognition, and culture sémiotiques européennes: langage, cognition et culture; v. 11)
Includes bibliographical references and index.
ISBN 978-3-0343-0475-7
1. Mathematical linguistics. 2. Computational linguistics. 3. Grammar, Comparative and general–Grammar categories. 4. Cognitive grammar. I. Doursat, René. II. Title.
P123.P448 2011
415`.9–dc22

2011006638

ISSN 1423-5587
ISBN 978-3-0343-0475-7

Hochfeldstrasse 32, CH-3012 Bern
info@peterlang.com, www.peterlang.com, www.peterlang.net

Printed in Switzerland

Grammatically specified structuring appears to be similar, in certain of its characteristics and functions, to the structuring in other cognitive domains, notably that of visual perception.

Len Talmy

By the same author

- *Les Catastrophes de la Parole. De Roman Jakobson à René Thom.* Paris: Maloine,1985.
- *Morphogenesis of Meaning.* Bern: Peter Lang, 2004.
- *Physique du Sens.* Paris: Editions du CNRS, 1992.
- *Neurogéométrie de la vision. Modèles mathématiques et physiques des architectures fonctionnelles.* Paris: Les Éditions de l'École Polytechnique, Distribution Ellipses, 2008.

******

- (ed., with F. Varela, J.-M. Roy & B. Pachoud) *Naturalizing Phenomenology: Issues in Contemporary Phenomenology and Cognitive Science*, Stanford University Press, 1999.
- (ed.) "Linguistique cognitive et Modèles Dynamiques", *Sémiotiques*, 6-7, 1995.
- (ed. with J. Lorenceau) "Neurogeometry and Visual Perception", *Journal of Physiology – Paris*, 97, 2003.
- (ed. with A. Sarti & G. Citti) "Neuromathematics of vision", *Journal of Physiology – Paris*, 103, 2009.

******

- e-mail : jean.petitot@polytechnique.edu
- URL : http://www.crea.polytechnique.fr/JeanPetitot/home.html

Contents

Introduction

1. Purpose and scope of this book

The purpose of this book is to present mathematical models of the relations between perception and language. If we had to condense it into one formula, we could say that it tries to show that *syntax is to perception what algebraic topology is to differentiable manifolds*: spanning several levels of categorization, it identifies in the visual geometry of perceptual scenes abstract invariants that can be reformatted and redescribed as syntactic constituent-structures. Algebraic topology examines the universe of differentiable manifolds and makes explicit rough information about their global structure. This is made possible by categorizing these structures into algebraic structures such as homotopy, homology, and cohomology groups. In this book we will look at the universe of images and try to make explicit rough information concerning their *morphological* structure. We will show that this is possible if we use specific mathematical theories for categorizing the structures into *non-symbolic* syntactic scripts or frames, which can then be translated into symbolic syntactic structures.

Our investigation takes place in the context of a *naturalist* approach to structures conceived in the *structuralist* sense. Claude Lévi-Strauss famously claimed that "social sciences will be structural sciences or will not be" ("*les sciences humaines seront structurales ou ne seront pas*"). We would like to add that "social sciences will be *natural* sciences or will not be". Of course, this statement can have some plausibility only if we broaden the classical concept of natural sciences to the point where structural phenomena, too, can be construed as natural phenomena.

From the outset, this was one of the main purposes of the research program of *Morphodynamics* initiated in the 1960's by René Thom, on the mathematical basis of the theories of singularities and dynamical systems. During the 1970's and the 1980's, we applied morphodynamical models to structural phonetics, categorical perception, and visual perception, and, with a few colleagues such as Wolfgang Wildgen and Per Aage Brandt, to structural syntax and structural semiotics.[1]

1 Perhaps the reader will allow us a few bibliographical indications. In what concerns categorization and categorical perception in phonetics, see for instance our texts [261],

At that time, the use of topological and dynamical models in semiolinguistics was completely new and raised a lot of questions since it disrupted the dominant formalist epistemology. The very idea that abstract structures of meaning could be *natural* structures susceptible of being modeled as a kind of physical and biological phenomena sounded rather provocative. To emphasize the significance of such a "naturalistic" and "morphodynamical" turn, we coined in [279] the neologism "physics of meaning". In reference to it, René Thom later introduced the term "semiophysics".

If one's goal is to "naturalize" semiolinguistics structures, one has to account for them as a special kind of emerging Gestalts. A key consequence of this conversion of paradigm is to abandon the requirement that models of *natural* syntactic structures be *formal* (algebraic, combinatorial, etc.). Indeed, in natural sciences, the mathematical structures used for modeling an empirical phenomenal realm have nothing to do with any "ontology" of this realm. Their scope is to provide appropriate computational tools for *reconstructing* phenomena. It is therefore a deep epistemological mistake to believe that natural languages have necessarily to be modeled using formal languages.

During the 1980's, the morphodynamical approach to semiolinguistics became more easily and widely accepted due to the tremendous development of connectionist neurocognitive models, which are typical examples of morphodynamical models.[2] It also deeply interacted with the new trends in *cognitive grammars*—in the sense of Len Talmy, Ron Langacker, Ray Jackendoff, George Lakoff and Terry Regier—focused on the perceptual grounding of linguistic structures.

The core of this work is constituted by the development of this theoretical perspective—structural semiolinguistics, morphodynamics, connectionism, cognitive grammars—during the 1990's. *One of our main goals is to offer a rigorous and operational mathematical basis to the intuitive "image-schemata" of cognitive grammars.*

2. Acknowledgements

This book advances and expands upon our previous works *Morphogenesis of Meaning* and *Physics of Meaning*, which owed much to René Thom's seminal ideas. It relied highly on Per Aage Brandt's support and was devised during two stays at the *Center for Semiotic Research* (Aahrus University), where he was the director at that time. It is for me a great personal pleasure to thank Per Aage who made so many fundamental contributions, whether theoretical

[269], [293]. In what concerns a topological and dynamical approach to structural syntax and semiolinguistics, we began to work on the subject since [258]. We connected this morphodynamical setting with case grammars, relational grammars, and cognitive grammars in [260], [261], [262], [265], [266], [267], [268]. For a critical presentation see Ouellet [250].

2 See our papers [276] and [275]. See also Visetti [395] and [396].

or institutional, to dynamical cognitive semiotics. I am also grateful to all the friends of the CSR. I also want to thank the other prominent specialist of Thom's linguistics, Wolfgang Wildgen, who worked out so many interesting applications of morphodynamical models.

At the outset, this work enjoyed many discussions with Daniel Andler, Elie Bienenstock, Yves Marie Visetti, and also Hugh Bellemare and René Doursat in the context of the DSCC (*Dynamical Systems, Connectionism and Cognition*) project of the CREA (*Centre de Recherche en Épistémologie Appliquée* at École Polytechnique). It also greatly benefited from the two Royaumont meetings about *Compositionality in Cognition and Neural Networks* organized by Daniel Andler, Elie Bienenstock and Bernard Laks (May 1991 and June 1992), and two other meetings, *Motivation in Language* organized by Umberto Eco and Patrizia Violi at the *International Center for Semiotic and Cognitive Studies* at the University of San Marino (December 1990), and *Le Continu en Sémantique linguistique* organized by Bernard Victorri and Catherine Fuchs at the University of Caen (June 1992). My joint researches with my colleague and friend Jean-Pierre Desclés were also essential.

All this technical material was elaborated in an already rather rich context. First, I had the privilege of discussing with eminent linguists such as Hansjakob Seiler and Bernard Pottier who supported Thom's perspective. Then, there was a dense network of colleagues interested in the new trends in dynamical structuralism: Jean-Claude Coquet, Franson Manjali, Pierre Ouellet, Bernard Victorri, Peter Gärdenfors, David Piotrowski, and many others. In visual perception, there was also a network of mathematicians interested in the geometry of vision: David Mumford, Jean-Michel Morel, Bernard Teissier, Giuseppe Longo, and also the psychologist Jan Koenderink (with his group at the University of Utrecht) who used singularity theory in vision.[3] Finally, I was also closely associated with specialists of Gestalt theory and phenomenology (in Husserl's and Merleau-Ponty's sense) such as Barry Smith and Kevin Mulligan, Jean-Michel Roy and Bernard Pachoud, Roberto Poli and Liliana Albertazzi.

In this supporting environment, I was very fortunate to have many opportunities to discuss with Len Talmy, Ron Langacker, Paul Smolensky, and George Lakoff. Particular thanks are due to Tim van Gelder and Bob Port for their idea of organizing the important conference on *Mind as Motion*.

Another meeting that played an important role in my work was the Conference *Topology and Dynamics in Cognition and Perception*, which I organized on 11-13 December 1995 at the *International Center for Semiotic and Cognitive Studies*. Many participants belonged to these scientific networks: L. Albertazzi (Univ. of Trento), P. Bozzi (Univ. of Trieste), P. A. Brandt (Aarhus Univ.),

3 For the actuality of these works, see my recent book (2008) *Neurogéométrie de la vision. Modèles mathématiques et physiques des architectures fonctionnelles* [304], and also [295], [298], [302].

R. Casati (CREA), R. Doursat (CREA), M. Gnerre (Univ. of Cassino), S. Gozzano (Univ. of Roma), R. Langacker (Univ. of Califonia, San Diego), M. Leyton (Rutgers Univ.), R. Poli (Univ. of Trento), T. Regier (Univ. of Chicago), B. Smith (SUNY, Buffalo), L. Talmy (SUNY, Buffalo).

Finally, I extend special thanks to two collaborators: René Doursat who played a fundamental role in the results presented in Chapter 3, which he co-authored for a large part, and in establishing the final version of the manuscript, and Franson Manjali, the translator of *Morphogenesis of Meaning*, who translated Chapter 1.

Jean Petitot

CHAPTER 1

The Cognitive and Morphodynamical Turns

1. Introduction

In the 1970's and early 1980's a number of works were devoted to the use of morphodynamical models—that is, dynamical mathematical models of forms, patterns and structures—in structural and semiolinguistic disciplines. We dedicated three books and several papers to their applications in phonetics (analysis of the relationships between audio-acoustics and phonological categorization, models of categorical perception), actantial theory[1] (case grammars, structural syntax), and structural semiotics (semantic categorization, models of Greimas' narrative schemes and Lévi-Strauss' canonical formula of myth). These models came from natural sciences and participated in the increasingly radical *naturalization* of mind undertaken by cognitive science. To emphasize this point, we qualified them to be part of a *Physics of Meaning* (see [279]).

In parallel, during the 1980's the development of cognitive grammars led to a complete reversal of the theoretical status of the syntactic-semantic structures of natural languages. The convergence of profound theoretical transformations resulted in a spectacular progress of dynamical approaches—first with connectionist models of neural networks, then with dynamical models proper, the latter being a natural generalization of the former. In fact, as Daniel Amit [11] has shown, introducing a hypothesis of full feedback and recurrence in a neural network allowed to reinterpret Hebbian reverberation as the stabilization of its dynamics into one of several attractors during a "psychological" time (a few hundred ms).[2] As Tim van Gelder [393] states:

> If connectionism was the most dramatic theoretical revolution of the 1980's, it appears that dynamics is the connectionism of the 1990's.

In fact, we observe an irresistible movement of *naturalization* of eidetic and structural descriptions, not only in the philosophy of mind and in linguistics, but also, for example, in Husserlian phenomenology.[3]

---

1 Throughout this book, we use the terms "actant", "actantial", and "actantiality" to refer to semantic roles in the sense of case grammars and narrative grammars.

2 For a discussion of these neuromimetic dynamical models, see the special issue [12] of *Behavioral and Brain Sciences* devoted to a "target article" by D. Amit.

3 On the naturalization of phenomenology, see Petitot [294] as well as the whole volume *Naturalizing Phenomenology* [245].

The naturalization of eidetic descriptions has of course raised the question of their implementation in physical and biological substrata. During the 1980's, neuromimetic connectionist models have considerably advanced our understanding of fundamental cognitive phenomena such as categorization, learning or inductive inference. The use of sophisticated models coming from statistical physics has led specialists to formulate them in a mathematical universe where the dynamical point of view was dominant. This is why the connectionist implementation of cognitive structures and processes has converged with the dynamical point of view and, more precisely, with structuralist morphodynamical models.

In this introductory chapter, we present (rather rhapsodically) a few elements of this conceptual debate.

2. Morphodynamics in cognitive semiolinguistics

As mentioned above, during the 1980's the conceptual basis of semiolinguistic disciplines was deeply transformed. A new sensitivity emerged and the focus shifted to problems that had been left in the shadow so far. New foundations were asked for and new tools of conceptualization and formalization were transferred from other disciplines so far considered alien to semantic and syntactic problems. Our goal is not to analyze here in a detailed way this paradigm shift, but to explain how it provided the contextual background of our reflections.

2.1. Characteristics of the cognitive turn

The most striking characteristics of this mutation were the following:

2.1.1. *Critique of formalism.* The first characteristic was the desire to do away with the deficiencies of the formalist conceptions of *natural* language, which privileged mathematical tools adapted only to the analysis of *formal* languages (formal logic, formal semantics, intensional logic, categorical grammars, category theory and topoï, etc). In particular, the generativist point of view, i.e., the mechanistic conception of grammars as algorithms that generate languages from finite sets of rules, was the most strongly criticized (often unjustly and unfairly). Henceforth, the *naturality* of natural languages was foregrounded, while the dogma of the centrality and *autonomy* of syntax was firmly questioned.

If we take the naturality of natural language seriously, the consequences that we can draw for the conception and modeling of linguistic structures are considerable. Indeed, we must then consider:

(i) language as resulting from a phylogenetic evolutionary process concerning human cognitive abilities;
(ii) the universals of language as universals of a cognitive nature correlated with the structures of perception and action;

(iii) these cognitive structures themselves as processing an information that is present in the environment (semantic realism);
(iv) linguistic structures as natural phenomena that are as transcendent to our consciousness as the physical, chemical or biological phenomena constituting our body: we "dwell" in our language as we dwell in our body, i.e., without being able to convert its intuition into a knowledge; our consciousness of our body bears only a false "naive" biology and no seed of any scientific biology; in the same way, our consciousness of language bears only a false "naive" linguistics and no seed of any naturalist scientific theory of language;
(v) the formal automatisms of competence as emerging from the natural underlying mechanisms of performance.

2.1.2. *Conceptual structures, embodiment and phenomenal world.* A second characteristic of the cognitive turn is the search for conceptual cognitive structures grounding natural language. The idea is that a conceptual structure underlying language can account for the compatibilities of language with perception and action. It is strongly supported by many works in cognitive science (see, e.g., Mandler [222] and [223]), which extensively show that there exists in infants a *preverbal conceptual thought* built from the perceptual categorization of objects, spatial relations, and events.

This leads to rejecting the classical thesis of the autonomy of syntax and insisting correlatively on the primacy of semantics, and the inseparability of meaning and grammar. As Ronald Langacker [205] claims:

> A pivotal theoretical issue is the relation between meaning and grammar. (....) The central claim of cognitive grammar [is] that meaning and grammar are indissociable.

The rejection of the autonomy of syntax leads not only to privileging *semantic* structures but also grounding them in a theory of cognitive acts (a noetics in the sense of Husserl), on the one hand, and a *phenomenology* or ecology of the natural world (in the sense of Gibson), on the other hand. The latter concerns the qualitative structuring of the sensory world in things, qualities, states of affairs, processes, events, which are *morphologically* structured, both objectively (i.e., on physical bases) and perceptually (see, e.g., neo-ecological theories of perception such as David Marr; but we shall return to this). The idea is that both a cognitive psychology and a phenomenology of the natural world constrain universally the syntactic-semantic structures of natural languages. This strong phylogenetic hypothesis about language's naturality puts the deep structures of language very far away from the surface linguistic level. But its influence has nevertheless been growing in strength everywhere in linguistics.

The opening of the conceptual structure onto the phenomenal world is also an opening onto the body. Mind is "embodied" and semiolinguistic structures and universals are fundamentally constrained by the compatibility between language, perception and action. Hence the spectacular renewal of *phenomenological* problematics (those of the later Husserl and Merleau-Ponty).

2.1.3. *The organic connections with theories of perception.* Given their critique against formalism in linguistics, the cognitive theories employed in the new paradigm are evidently not those relevant to the classical cognitivist paradigm. In the classical paradigm, the external physical information transduced into neural information (via sensory receptors and modular peripheral systems) is processed by means of a formal symbolic computation operating at successive levels of mental representations that share the structure of *formal* languages, with their symbols, expressions, rules, and inferences.

The theories used in this book are rather theories of perception, in particular those that admit the existence of *geometric-topological* and analogical mental representations (as in Shepard's and Kosslyn's works on mental images) as well as those which treat cognitive acts in terms of dynamical models of performance (as in connectionism), and not in terms of formal descriptions of competence.

These new linguistic orientations are of course related to converging achievements in neighboring scientific domains. Major advances in image analysis, both in neurobiology of vision and computational models of image processing, have helped us better understand the multifarious representational levels of perception—from the lowest (early vision: retina and primary visual areas) to the highest (face recognition, etc.) cognitive levels. These discoveries made possible a whole set of new technical studies bearing upon the links between visual scenes and the syntactic-semantic structure of the statements describing these scenes. Similarly, new insights into the fundamental relations between perception and action have led to thorough works on the embodiment of these conceptual structures. The important consequences of an "embodied cognition" were especially well exemplified in robotics, e.g., with the work of Rodney Brooks at the MIT Lab of *Computer Science and Artificial Intelligence.*

From the cognitive viewpoint, semantic issues in natural languages underwent a significant reformulation. Semantics here is no longer a matter of "distinctive features", "generative semantics", or "selection rules". The question becomes rather to explain how language can be *applied* to perceptual reality and actively structure it, as well as how this structuration is essential to further actions. Recent results let us imagine a near future in which robots will be able to trigger their motor behavior on the basis of the linguistic description of images acquired through their sensory devices, and communicate this description to other robots. Conducting this kind of research implies foraging into the deepest levels of motor and perceptual controls and their neural implementation. There lie many technical and difficult problems, whether neurobiological, algorithmic, computational, or mathematical, whose resolution is key to these cognitive approaches to semantics.

2.1.4. *Grammaticalization of Gestalts.* A fundamental non-formalist thesis proceeding from the non-autonomy of syntax is that *grammar specifies semantic contents.* As we shall see, this thesis is crucial for example for Ray Jackendoff, Leonard Talmy and Ronald Langacker. One starts from the observation that lexicon and grammar can be differentiated by distinguishing *open* lexical classes (parts of speech: verbs, nouns, adjectives, etc.), whose cardinal is large and indeterminate, from *closed* grammatical classes, whose cardinal is small and fixed.[4] The thesis (especially in Talmy) is that the closed classes grammatically specify certain very particular notions. These grammatically specified structures are *schematic* with respect to the states of affairs (the visual scenes) that they structure. They are idealized, abstract and topologically plastic.

2.1.5. *Iconicity and morphological structuration.* The importance of the connections between perception and language thus leads to the thesis that the latter is *anchored* into the former. Hence the problematic of *iconicity.* Iconicity of structures, particularly syntactic structures, does not mean that structures are concrete figures. It does not involve any "figurativity" in the classical sense but only an abstract iconicity of a schematic nature. Mental representations are construed as schemata, as generalized Gestalts, as a mental imagery that, as Kant had already explained in his theory of the *schematism* of empirical concepts, is a system of rules for the construction of referents. The image-schemata structuring mental representations are *types*, not tokens.

This gestaltic[5] conception of the structures of language became so influential in the 1990's that Herbert Simon himself, in a target paper *Bridging the Gap. Where Cognitive Science Meets Literary Criticism* of a special issue of the *Stanford Humanities Review* [346], defended the thesis that meanings are visualized as mental images and even claimed that:

> a mental picture formed by retrieving some information from memory or by visualizing the meaning of a spoken or written paragraph is stored in the same brain tissue and acted on by the same mental processes as the picture recorded by the eyes.

The schematic iconicity of mental structures is in fact a thesis about their *format.* It questions the *propositionalist* dogma (which is the cornerstone of all formalist conceptions), according to which mental contents must share a propositional format. The iconicity thesis is on the contrary that the format of mental contents is topological-dynamic. We could trace it back to Kant's schematism and Peirce's existential graphs. The idea is that the *spatio-temporal a priori* is deeper than the symbolic a priori: the human visual system is

4 This is a revival of the traditional opposition between "categorematic" and "syncategorematic"

5 We take the liberty of using the adjectival form 'gestaltic'.

inherited from a very long natural evolution, while ideography and writing are extremely recent cultural acquisitions.

2.1.6. *Dynamical Structuralism.* Overcoming the formalist point of view also led to a change in the concept of structure. Structures can no longer be conceived as formal assemblages of symbolic elements connected by means of formal relations. They are now conceived as natural, organic, qualitatively self-organized and dynamically regulated wholes, as forms, Gestalts, or patterns. The perspective is now organizational, dynamical, and emergential: structures emerge from substrata, be they internal (neuronal) or external, while symbolic, discrete and sequential structures formally described by the classical paradigm are now equated with qualitative, structurally stable and invariant structures emerging from an underlying dynamics.

2.1.7. *Schematicity and categorization.* Schematicity grounds semiolinguistic structures in a basic cognitive activity, namely *categorization.* In cognitive grammars, even the most abstract syntactic structures are construed as data typing by means of prototypes (e.g., categorization of events).

These points of view have been well summarized by Peter Gärdenfors in *Conceptual Spaces. The Geometry of Thought* [118]:

(i) meaning is defined by conceptualization within cognitive models (and not by truth conditions in possible worlds);
(ii) cognitive models are perceptually tailored: "a central hypothesis of cognitive semantics is that the way we store perceptions in our memories has the *same form* as the meanings of words";
(iii) "semantic elements are based on *spatial* or *topological* objects (not symbols)";
(iv) "cognitive models are primarily *image-schematic* (not propositional)";
(v) semantics is primarily in relation to syntax, the latter is not a formal computation;
(vi) "concepts show *prototype* effects."

2.2. The path-breaking point of view of Morphodynamics

On a number of essential points, the "cognitive turn" shows striking analogies with previous approaches and particularly with the "morphodynamical turn" operated by René Thom in the late 1960's (see Chapter 5).

Thom's and Zeeman's works were mainly concerned with models allowing the transition from neuronal dynamics to emerging cognitive structures, on the one hand, and from the dynamics of external substrata to emergent morphologies, on the other hand. In both cases, the basic problem is essentially the same: understanding how the interactions of a very large number of "micro" elementary units are able to generate "macro" morphological structures. This is the general problem of emergent structures in complex systems.

Let us briefly mention three key Thomian ideas that are still currently worked out:

1. The idea that a mental content can be identified with the topology of an attractor (i.e., a structurally stable asymptotic state) of an underlying neural dynamics, and that the syntactic trees of generative grammars are an abstraction of the bifurcations of such attractors into sub-attractors. This allows us to interpret the formal kinematics of linguistic competence and its logical-combinatorial structures as stable macroscopic regularities that emerge from the underlying microscopic dynamics. Hence a key analogy with physical models of *critical phenomena*, in particular with thermodynamical models of phase transitions (Thom [379], Petitot [271]). This idea was taken further in connectionist models within the framework of the subsymbolic connectionist paradigm (see, e.g., Smolensky [355]). According to this perspective, entities possessing a semantics are, on the "micro" subsymbolic level, global and complex patterns of activation of elementary local units mutually interconnected and computing in parallel. Their semantics is an emergent holistic property. The discrete and sequential symbolic structures of the "macro" symbolic level (symbols, expressions, rules, inferences, etc.) are qualitative, stable and invariant structures, emerging from the subsymbolic level through a cooperative process of aggregation. Here again, there is a key analogy with phase transitions. If we now introduce the Lyapunov functions of the attractors considered—what Paul Smolensky calls a "harmony" function (Smolensky [352], [355])—we are naturally brought back to Thom's morphodynamical models.
2. The idea that there exist objective qualitative morphological structures in the environment, which are of physical origin and emerge from the statistical physics of their substrata, and that it is therefore possible to develop mathematically a qualitative ontology of the phenomenal world (qualitative physics).
3. Finally, the idea—basic for what is called the *localist hypothesis* in linguistics (see Chapter 7.2 and Petitot [258], [261], [266])—that topological and qualitative *spatio-temporal* relations between the actants of a spatio-temporal scene are indistinguishably local and grammatical and, consequently, their *interactions* can be taken as general schemata for grammatical connections (in the sense of actantial relations). Hence, an iconic schematism of deep actantiality. This idea was echoed by cognitive linguists, cited above, and particularly by Ray Jackendoff.

There are however a few differences between Thom's and Zeeman's approaches and current cognitive connectionist models.

1. In contemporary connectionist models, the internal dynamics is explicitly specified while in Thomian models, it is only implicit. This difference changes nothing at the theoretical level since the bifurcation

schemata of Lyapunov functions of attractors are in some sense universal (i.e., independent of the fine structure of the dynamics). But it changes a lot at the level of numerical simulations and at the experimental level.

2. Thom construed phenomenology and ecology of the natural world in terms of a qualitative ontology, that is, in objective and emergential terms. Thus, for him, qualitative physics was a true mathematical physics. The problem was therefore to link it with cognition, in particular visual cognition, and understand how the retrieval of such qualitative morphological structures can be equivalent, on the part of the cognitive subject, to a certain type of information processing. In what is currently called qualitative physics, on the contrary, one usually thinks of qualitative and morphological structures in terms of Artificial Intelligence.
3. Finally, as far as language is concerned, Thom related his morphodynamical approach directly to linguistic surface structures, which raised subtle issues. At that time, the mediation via conceptual structures—that is, precisely, the cognitive turn—was missing.

3. Three main examples of a cognitive approach to language

We now take a closer look at three main specialists of cognitive grammars: Ray Jackendoff, Ron Langacker and Len Talmy, and wonder why they did not attempt to mathematize—rather than just schematize—their insightful theoretical concepts.[6]

3.1. Ray Jackendoff

Let us first consider Ray Jackendoff's theses, which exemplify the revival of phenomenology and Gestalt theory in cognitive science. In order to understand the relations between perception, action, language and thought, Jackendoff introduced in *Semantics and Cognition* [169] the hypothesis of a *conceptual structure* (*CS*), a deep cognitive level of mental representations where linguistic, sensory and motor information become compatible.

> There is a single level of mental representation, conceptual structure, at which linguistic, sensory, and motor information are compatible ([169], p. 17).

The intent of this hypothesis is to allow a better understanding of the structural constraints imposed on a theory of cognition, as well as the relations between universal grammar, cognitive abilities in general and the structure of thought. It assumes that language "reflects" thought and the world ("realist"

6 As we will see when discussing Terry Regier's works in Section 5.3, the case of George Lakoff is rather different.

and "ontological" thesis), that there is a psychological reality of linguistic information and, therefore, that there exist *semantic* constraints that condition syntax, where these constraints are themselves constrained by the structure of perception. This hypothesis is widely supported by cognitive pychology (Mandler [222]) and it reintroduces certain theses that were already developed by Charles Osgood and Alexander Luria:

> It seems perfectly reasonable to think that much, if not all, that is universal in human language is attributable to underlying cognitive structures and processes. Perceptual and linguistic sequences must, at some level, share a common representational (semantic) system and a common set of organizational (syntactic) rules, cognitive in nature. (Osgood [248])
>
> We must look for the roots of basic linguistic structures in the relations between the active subject and reality and not in the mind itself. (Luria [216])

A conceptual structure encodes external physical information. It processes (represents) this information and refers to the world, but only *projectively*, i.e., by transforming the real physical world *RW* into a *projected world PrW*, the qualitative sensory world of phenomenological experience.

The projected world is a cognitive construction—noetic-noematic, Husserl would say—without any other physical content than the information contained in the peripheral stimuli (light waves, sound waves, etc.). As Jackendoff has shown in *Consciousness and the Computational Mind* [170], it allows us to pursue many themes of Husserlian phenomenology if we note that "projectability" is a *property*: phenomenal consciousness must not be confused with mental computation. Mental information proceeds from computation by the constituents of the conceptual structure. But the greater part of the internal structure of these constituents is not projectable. Phenomenal experience does not manifest its internal computational structure. What is manifested are only what Husserl called the noematic correlates of noetic syntheses.[7]

Let us consider the standard example of *color* given by Jackendoff. In *RW*, there are electromagnetic waves. As for the sensory quality (the qualia) /color/, it belongs to *PrW* and results from the processing of physical information by a conceptual constituent [COLOR] belonging to *CS*. [COLOR] is the structure of /color/ formally expressed in the internal structure of the corresponding mental computation, the relation between [COLOR] and /color/ raising the classical "mind-body problem" (Jackendoff [169], pp. 31-34) (see Figure 1).

The projected world is not, by definition, the real world. But it is not just the perceived world either. Even if it is the world "for us" in the phenomenological sense of the term, the noematic part of phenomenal consciousness, it is not an imaginary set of appearances. It is a genetically constrained construction, adapted to the real world and universal for our species.

7 For some remarks on phenomenology in the context of semiolinguistics and cognitive science, see, e.g., Petitot [261], [282], and [294].

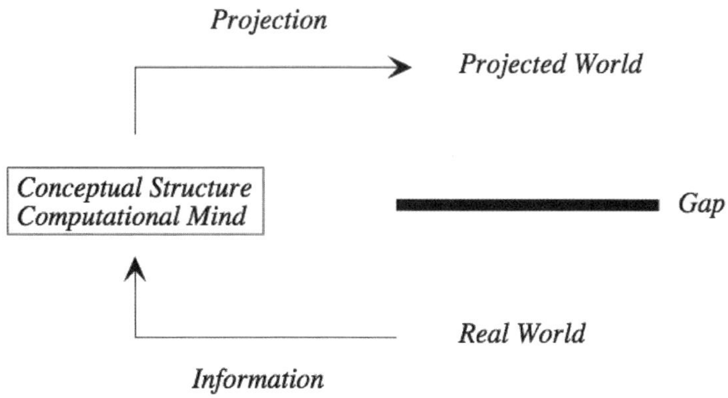

FIGURE 1. The transformation of the real world *RW* into the projected world *PrW* through the conceptual structure *CS* according to Jackendoff [169].

Language and perception possess an ontological content, but this concerns the ontology of *PrW* and not that of physical *RW*. Building on this basis, Jackendoff undertakes a cognitive analysis of *CS* in its projective relation to *PrW*. This leads him to identify the ontological categories of *PrW*, which consists of spatio-temporally delimited /things/, /forms/, /places/, /states/, /events/. These primitives are represented in *CS* by constituents [THING], [FORM], [PLACE], [STATE], [EVENT] that process sensory information, particularly visual information. Like Aristotelian categories, they are associated with different types of possible questions.

We must emphasize the fact that, for Jackendoff, the level of *CS*, "at which linguistic and non-linguistic information are mutually compatible", constitutes the same level as that of the semantic structure,

> the level at which semantic properties of sentences such that synonymy, anomaly, presupposition, and inference can be formally captured. (Jackendoff [169], p. 95)

The primitive categories [THING], [PLACE], [STATE], [EVENT] and [PATH] (which may be specialized and complexified using semantic rules of well-formedness) allow us to easily describe verbs as /be/, /go/, /stay/, /leave/, /cause/, etc., with their corresponding case (prepositional) system. The fundamental hypothesis is that (i) they come from perception and (ii) have nevertheless a universal semantic relevance. Jackendoff refers here to Gruber:

> In any semantic field of [EVENTS] and [STATES], the principal event-, state-, path- and place- functions are a subset of those used for the analysis of spatial location and motion. (Ibid, p. 188)

Jackendoff is aware of the powerful phylogenetic content of such a cognitive hypothesis:

> The psychological claim behind this methodology is that the mind does not manufacture abstract concepts out of thin air, aether. It adapts machinery that is already available, both in the development of the individual organism and in the evolutionary development of the species. (Ibid, p. 189)
>
> [The claim is] that all [EVENTS] and [STATES] in conceptual structure are organized according to a very limited set of principles, drawn primarily from the conceptualization of space. (Ibid, p. 209)

The hypothesis depends crucially, as we see, on the ontology of *PrW*. It expresses an

> evolutionary conservation in cognition—the adaptation of existing structure rather than the development of entirely new mechanisms,

and leads

> to highly structured hypotheses about the structure of thought. (...) It integrates linguistic theory and methodology fully into the fabric of cognitive psychology. (Ibid, p. 210)

Although seeking to define a qualitative ontology of the phenomenal world and criticizing the formalist approaches to semantics, Jackendoff still retains a clear-cut dualist division between physical objectivity and the projected world. At no time does he consider the hypothesis that part of this ontology could concern the emergence of qualitative structures out of physical reality by a self-organizing (non-cognitive, non-symbolic, non-computational) process of phenomenalization. For him, *PrW* is a purely cognitive construction. The external physical information—which is non-meaningful as such for the cognitive system—is transduced into meaningful information by the peripheral sensory organs (retina, cochlea, etc.), and then processed at several levels. But none of these levels possesses a cognitively relevant physical content.

3.2. Ronald Langacker

Let us say now a few words about Langacker's conception.

3.2.1. *The theses of cognitive grammars.*
◇ *Grammar is a high-level categorization process*

According to Langacker, cognitive grammar (CG) is a perceptually rooted "space grammar". It is also a "natural" (as opposed to "formal") grammar. This is the thesis of the inseparability of syntax and semantics, and moreover of the primacy of semantics:

> Grammar (or Syntax) does not constitute an autonomous formal level of representation. (I, p. 2)[8]
>
> There is no meaningful distinction between grammar and lexicon. (I, p.3)

This entails the centrality of conceptual structure:

---

8 The quotations of this section refer to *Foundations of Cognitive Grammar I, II* [203], [204].

> Semantic structure is conceptualization tailored to the specification of linguistic convention. (I, p. 98)

Syntactic structures are construed "organically rather than prosthetically" (I, p. 12). Constituency is not, as in logic, a matter of symbolic combinatorics but rather, as in biology, a matter of self-organization. As Langacker explains in the beginning of Volume II,

> a speaker's linguistic knowledge (his internal grammar) is not conceived as an algorithmic constructive device giving (all and only) well-formed expressions as "output". (II, p. 2)

It is rather a set of cognitive routines that allows the speaker to *schematize* and *categorize* the utterance events. Linguistic structures are essentially schematic and their correlated image-schemata are simultaneously iconic and abstract.

Every predication profiles entities on a background and organizes a scene, a state of affairs, in such a manner that some elements become salient. In particular, the grammatical relations implied in constituency and compositionality must not be defined only prototypically in terms of semantic roles but also schematically in terms of *profiled* relations linking salient entities ("trajectors") with other secondary entities ("landmarks").

The insistence on schematicity and iconicity of structures and on the centrality of meaning leads to the affirmation that grammar is not formal, but "conceptual"—it would be better to say "semiotic"—and that its approach has to be cognitive:

> Cognitive grammar equates meaning with conceptualization (explicated as cognitive processing). (II, p. 5)

Primacy is then given to concepts conceived as *schemata* and sets of *instructions* (routines) for structuring thought. These are not "containers for meaning" but "entrenched cognitive routines" possessing a hierarchy of subroutines (II, p. 162). Hence a deep and difficult problem: if placed in typical generic referential situations of linguistic descriptions of scenes, how would we describe the *figurative information* selected and expressed in the description? We cannot accept a *predicative* description since, in CG, predication is not a primitive fact but, on the contrary, a complex and sophisticated process which has to be accounted for. This is why grammar is not autonomous and of semiotic nature, a conventional symbolization of semantic structure (p. 2).

> Grammar itself, i.e., patterns for grouping morphemes into progressively larger configurations, is inherently symbolic and hence meaningful. (I, p. 12)[9]
>
> Grammar is simply the structuring and symbolization of semantic content. (p. 12, see below Talmy's similar theses).

Langacker insists a lot on the heterodoxy of this thesis (p.76):

---

9 Langacker's use of the term "symbolic" is rather confusing since the classical formalist paradigm is itself called "symbolic". For Langacker, "symbolic" is a semantic notion and not a syntactic one. It has to do with meaning and not at all with logical symbols.

> Counter to received wisdom, I claim that basic grammatical categories such as noun, verb, adjective and adverb are semantically definable. (p.119)

◇ *Scanning*

At the cognitive level, the most basic processing of CG is, according to Langacker, *scanning*, that is, a local operation of contrast detection. Scanning is:

> an ubiquitous process of comparison and registration of contrast that occurs continuously throughout the various domains of active cognitive functioning. (I, p. 116)

It picks up qualitative discontinuities and builds from them a schematic imagery which is at the same time linguistic and perceptual.

The image-schemata used for structuring visual scenes and conceptual situations vary according to differences in attention—what Langacker calls "focal adjustments"—which selectively increase the salience of certain aspects of the scene: perspective (figure/background, deixis, subject/object difference, etc.), point of view, level of abstraction, etc.

◇ *The geometrization of concepts*

An essential theoretical concept in Langacker's CG is the concept of *domain* and, above all, of *basic domain*. Since meaning consists of knowing how to *use* concepts as strategies of conceptual structuration, the semantic units of language are *contextual* (I, p. 155). The context that characterizes a semantic unit is its domain, where domains are cognitive entities (mental experiences, representational spaces, conceptual complexes). But there exist universal basic domains that can no longer be characterized in terms of more fundamental domains (space, time, sensory qualities such as color, etc.). They are all endowed with some sort of *geometrical* structure.

> By definition, basic domains occupy the lowest level in hierarchies of conceptual complexity: they furnish the primitive representational space necessary for the emergence of any specific conception. (I, p. 149)

This point is really crucial, because the non-primitive concepts will be *grounded* into the basic domains through chains of intermediary concepts. For instance, "space" is the basic domain for "body", which is itself an abstract domain for "parts of the body", etc.

The constitutive feature by which CG differs from the formalist traditions is therefore the following. The basic domains are endowed with geometric features (dimensions, distances, continuous /discrete topologies, boundaries, etc.) that express the fact that the basic domains are not conceptually definable in the classical sense (I, p. 150). This allows us to give a new definition of concepts as *positions* or *configurations* in basic domains. There will be "position" (that is, "location") in the "locational" basic domains (temperature, color, etc.), where the coordinates are *intrinsic* and the points non-equivalent (non-homogeneity). There will be "configuration" (that is, form) in the "configurational" basic domains (real space, etc.) where the points are equivalent (homogeneity, symmetry groups, relativity) (I, p. 152). Langacker rediscovers here the grand old philosophical opposition between "intensive" and "extensive" magnitudes.

The consequence of such a perspective is that concepts are now construed as "fillings" of extensive forms with intensive qualities. As this is indeed the exact definition of a *morphology*, we can call it a morphological approach to concepts. Cognitive scanning (detection of qualitative discontinuities in general qualitative spaces) opens onto a *geometrization of meaning.*

Let us now recall briefly how Langacker defines *Things*, *Relations* and *Processes.*

3.2.2. *Things.* Langacker begins by developing a morphological approach to things. He invokes the classical idea that nominal predicates denote things, but he defines the latter cognitively. His definition is purely morphological (morpho-cognitive).

Generally, a predication has a scope—the aspects of the scene that are specifically included in it. It selects from the scene a maximally salient referent. This corresponds to the old gestaltic figure/ground opposition, here referred to as profile/base. The *base* is the cognitive structure from which the referent of a semantic structure is detached as a figure on a background; the *profile* (the contour) is the entity denoted by the semantic structure; it is salient and functions as a focal point (I, p. 183). A thing is therefore identified with a contour—a profile, a sketch—on a background, in other words, a "region in some domain" (I, p. 189).

These regions can be bounded or not and can connect several domains (for example spatial extension and sensory qualities). Just as domains can be abstract, their definition is very general. It is important to understand that it involves the *cognitive universality* of morphological notions such as "region", "limit", "center", "periphery", "interior/exterior", "form", etc., i.e., notions concerning the *segmentation* of general spaces by qualitative *discontinuities.*

A thing is "a region in some domain". As in Gestalt theory, its scanning distinguishes a "profile" (shape) from a "base" (ground) via the detection of a boundary. A first scanning scans the regular points interior to the domain, another scans the exterior regular points, and a third one scans the singular points of the boundary (see Figure 2).

The elementary events are sensations $q(p)$ of a quality degree q (e.g., color) at some position p of the underlying space W of the thing. A scanning scans the spatial difference $\Delta p = p_2 - p_1$ between two neighboring positions, and the qualitative difference $\Delta q = q(p_2) - q(p_1)$ between the quality q at these two positions.

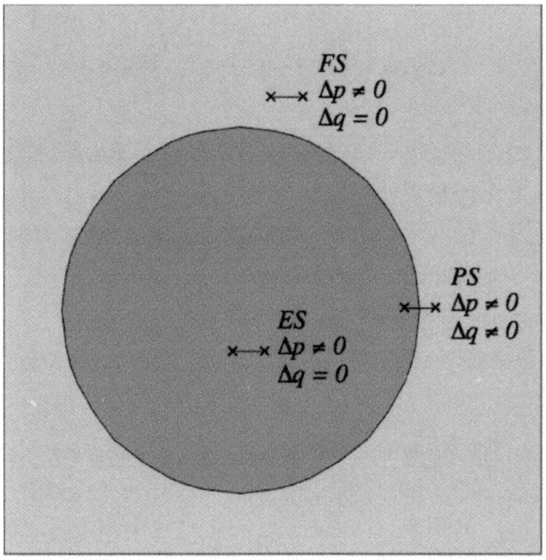

FIGURE 2. Field scanning (FS), expanse scanning (ES), and periphery scanning (PS) according to Langacker [203].

(i) For the *field scanning* (FS) of the ground (the base), $\Delta p \neq 0$ but $\Delta q = 0$. The judgment is therefore one of local homogeneity. Langacker then introduces a second basic and general cognitive principle according to which "the occurrence of a cognitive event facilitates its recurrence (I, p. 210)". This is a principle of globalization, by which the cognitive system scans the global homogeneity—"the qualitative uniformity"—of the ground.

(ii) For the *expanse scanning* (ES) of the shape (the profile), one also has $\Delta p \neq 0$ and $\Delta q = 0$ but with another value of q. ES scans the global homogeneity of the profile.

(iii) The scanning of the boundary constitutes a third sort, the *periphery scanning* (PS). "This is a higher-order comparison chain that constitutes the perception of a boundary, i.e., an interface between two regions (I, p. 210)".

PS detects boundary points, which are characterized by $\Delta p \neq 0$ and $\Delta q \neq 0$. The boundary B, is "a continuous, line-like entity extending through the visual field" (p. 211).[10]

One can therefore:

(i) scan locally the interior and the exterior of the shape separated by the boundary and

10 For the analysis of Langacker's scanning, see also Wildgen [409], p. 127.

(ii) execute the transition from the local to the global level through a *propagation process*—in agreement with the general cognitive principle "occurrence facilitates recurrence".

Mathematically, Langacker's description fits perfectly well with the Thomian definition of *a morphology* as a system of boundaries (of qualitative discontinuities). Let W be the "region" occupied by the thing in the basic domain E. W is filled by qualities $q_i = q_1, ..., q_n$ that vary with $w \in W$: $q_i = q_i(w)$. Phenomenologically, a fundamental opposition holds between :

(i) local homogeneity, i.e., a continuous variation of all the $q_i(w)$;
(ii) local heterogeneity, i.e., a discontinuous variation of some of the $q_i(w)$.

This opposition can be topologically formalized in the following manner. Let us call a point $w \in W$ *regular* if there exists a neighborhood U of w in W such that all the $q_i(w)$ are continuous functions in U (local homogeneity). By definition, the global set $\mathcal{U} = \{w \in W \mid w \text{ regular}\}$ is an open subset of W. Let $K = W - \mathcal{U}$ be its complementary set. K is a closed subset of W. The points $w \in K$ are called *singular*. In every neighborhood U of a singular point w, some of the $q_i(w)$ present a discontinuity (local heterogeneity). K is a boundary, a system of interfaces, which decomposes W in qualitatively homogeneous regions. In fact, the regular/singular and open/closed oppositions are deeply correlated with the "structurally stable/structurally unstable" opposition: in $\mathcal{U}$ the qualities $q_i(w)$ are stable, but at the crossing of K they become unstable.

3.2.3. *Relations.* As we have seen, one of the main challenges of CG is to develop a satisfactory theory of relations and processes.

What could therefore be the profiling of a relation? In a relational profile there exists an asymmetry between a salient figure—called a "trajector"—and the other parts of the profile, which operate as "landmarks". These constituents are linked by abstract spatial schemata of relations between *places*, which are filled by real objects. The morphological schematization of things as regions cut out in abstract qualitative spaces imply that relations are decomposed into "basic conceptual relations" of a spatial nature (still in an abstract qualitative sense).

According to Langacker, all static relations can be reduced to four basic conceptual relations: identity $[A_1 \text{ ID } A_2]$, inclusion $[A_1 \text{ IN } A_2]$, separation $[A_1 \text{ OUT } A_2]$, and association$[A_1 \text{ ASSOC } A_2]$ (see Figure 3). It must be stressed that these relations are positional. They are the possible basic spatial relations of co-location in a general space.

The relation of association deserves particular attention. In a relation of separation $[A_1 \text{ OUT } A_2]$, A_1 and A_2 are in some sense independent from one another. They do not constitute a whole on their own. On the contrary, in a relation of association $[A_1 \text{ ASSOC } A_2]$, one of the entities, the "trajector", is localized relative to the other, the "landmark". As we will see, this fact is essential.

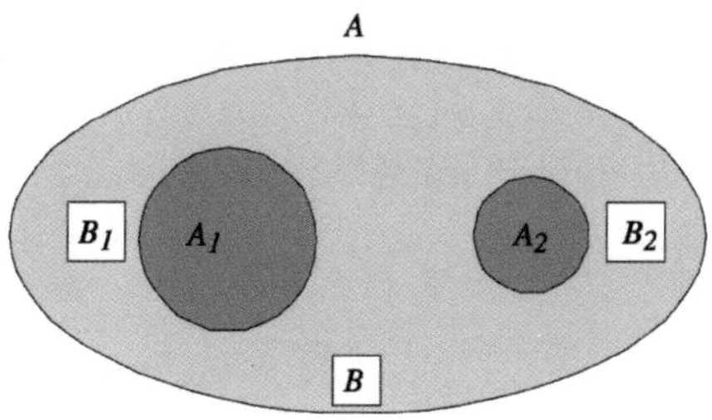

FIGURE 3. The ASSOCIATION relation [A_1 ASSOC A_2] according to Langacker [203]. A_1 and A_2 are objects with boundaries B_1 and B_2 associated through a superordinate object A of boundary B.

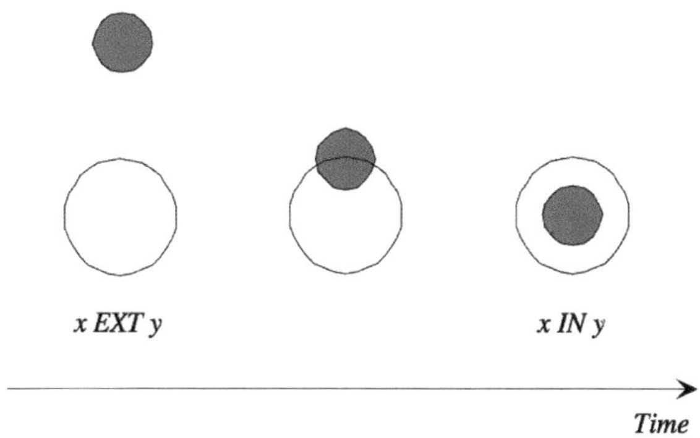

FIGURE 4. The schematization of a process like [ENTER] according to Langacker [203]. An initial state x EXT y evolves toward a final state x INT y through a chain of intermediary states.

The problem is then to scan these positional relations, which are global and informationally infinite (in the sense that they can vary in a continuous way), in agreement with the two fundamental processes of local scanning and local-to-global propagation. As we shall see in Chapter 3, it is a highly non-trivial problem.

Let us suppose, nevertheless, that this main problem is solved. Then, how should we analyse processes? The basic idea is a *temporal* profile, i.e., profiling a temporal sequence of profiled relations.

3.2.4. *Processes.* Processes profile temporal sequences of profiled relations. Let us consider as an example the schematization of a process like [ENTER] (see Figure 4).

It consists of a temporal sequence of states beginning with an initial state of outsideness x EXT y and ending with a final state of insideness x INT y (x and y are the places of the actants). It is important to note here that a process is more than a mere predication involving a temporal dimension. It is a temporal profiling of a series of states which are themselves individually profiled as relations. Hence the general definition:

> A process is a relationship scanned sequentially during its evolution through conceived time. (I, p. 254)

If the conceived time is the real time of the process, the scanning is "sequential". If the different states of the process are conceived "in parallel", then the scanning is "synchronous".

Here again, the cognitive approach in terms of scanning turns into a morphological analysis that is not only spatial but also temporal. It is easy from there, as Langacker shows, to develop a (rudimentary) cognitive theory of *aspectuality*. A process is *imperfective* when the sequentially scanned relations are qualitatively identical (no temporal accidents: the process is without event, without qualitative change). The conceived time is profiled consequently as homogeneous and non-limited (open interval) (I, p. 258). On the contrary, a perfective process is temporally limited: it contains accidents, events, qualitative discontinuities in time (I, p. 260).

3.2.5. *Geometric space grammar.* To summarize, at the most basic level concepts are construed in CG as positions—"locations"—or configurations in some geometrical (topological, differentiable, metric, linear, etc.) manifold, and CG leads to the following identifications:

(i) terms (or fillers) ≡ localized (possibly filled) domains in some concrete or abstract space;
(ii) relations ≡ positional relations between locations;
(iii) processes ≡ temporal deformations of positional relations;
(iv) events ≡ interactions between locations;
(v) agentivity ≡ causal control of the (inter)actions;
(vi) semantic roles ≡ types of transformations and controlled interactions (configurational definition).

3.3. Len Talmy

In an impressive set of works, Len Talmy has developed a cognitivist point of view of linguistics that introduces *cognitive representations* (CR) to play the role of deep semantic structures. CR have a topological-dynamical nature compatible with the structures of perception and action. In a fundamental paper "The Relation of Grammar to Cognition" (Talmy [369][11]), he resorts to three simple and powerful ideas.

(i) A sentence evokes—by way of complex cognitive processes of production and reception—a cognitive representation, or rather a "scene" in Fillmore's sense, conceived as a complex meaning that processes perceptual information. Lexical semantics provides the content of the CR, and grammar provides its syntactic structure.

(ii) The difference between lexicon and grammar can be characterized in a *non-semantic* manner using the opposition between open and closed classes. Indeed, as we have stressed in Section 2.1.4, lexical classes (verbs, nouns, adjectives, etc.) are open. Their number is large and easy to increase. On the contrary, the grammatical class elements (nominal and verbal inflections, conjunctions, demonstratives, types of syntactic constructions, grammatical relations, word-order, auxiliaries, etc.) are closed. Their number is small and fixed.[12]

(iii) With these premises, the key idea is that certain special semantic notions are grammatically specified by the closed classes.

Instead of accepting the simple opposition between syntax and semantics, Talmy duplicates semantics. First, there is the lexical semantics. Second, there are the syntactic forms that structure sentences. But some of these syntactic forms grammatically specify special semantic *contents*.

Hence the idea of enumerating the grammatically specified notions. Comparative studies show that they exhibit a great deal of universality (for example most of the languages grammatically specify the numeral opposition one/many by means of the grammatical category of number, while none of them grammatically specify color). They fall into classes forming a system: number and quantification; structuration of space, prepositions and deictics, localization procedures and landmarks; structuration of time and aspectuality; states of division (continuous/discrete); degrees of extensionality (point/limited extension/unlimited extension); gradients that structure qualitative dimensions involving degrees; procedures of selection of an element in a set, levels of typicality and exemplarity, etc. Having analyzed this rich material in a highly detailed way, Talmy concluded that the content of all these notions is essentially topological and dynamical:

11 All of Talmy's articles were gathered into the two volumes of his magnum opus *Toward a Cognitive Semantics*—I: *Concept Structuring Systems* [374], and II: *Typology and Process in Concept Structuring* [375].

12 Naturally, this dichotomy is not perfectly sharp and clear cut. As it was pointed out by Ron Langacker [207] for the case of prepositions, "If one considers not just the core set of fully grammaticized prepositions (*in*, *on*, *under*, *beside*, etc.), but the entire range of conventional prepositional locutions (including *by the side of*, *at the top of*, *in the bottom of*, etc.), it is not at all clear that the class is really closed."

> grammatically specified structuring appears to be similar, in certain of its characteristics and functions, to the structuring in other cognitive domains, notably that of visual perception. (Talmy [369])

In another work "How Language Structures Space" [370], Talmy analyzed in detail the morphological information specified by the prepositions of English. He concludes with the central idea that

> the closed-class forms of a language taken together represent a skeletal conceptual microcosm

and that

> this microcosm may have the fundamental role of acting as an organizing structure for further conceptual material. (Talmy [370], p. 228)

The morphological information has a strongly schematic nature, to the extent that it only retains a small portion (very qualitative, but nonetheless very rich) of all the available information (for example the metric information is in general not taken into account). According to Talmy, the four generic properties of such schematizations are the following:

(i) *Idealization.* In order to be applied projectively to various particular scenes, grammatical schemata require a preliminary strong idealization. This is the cognitive process through which external states of affairs are grasped according to schemata.

(ii) *Abstraction.* This is the partner process of idealization. If the informationally limited schemata can be applied to states of affairs, it is because many features (physical, metrical, etc.) are ruled out.

(iii) *Topological plasticity.* Grammatical schemata neglect scale and other metric features of forms.

(iv) *Alternative schematizations.* The semantic fields of grammatically specified notions are generally multidimensional (e.g., more than 20 dimensions for the prepositional system of English). Contrary to what one might believe, they are not categorized by schemata that are all at the same level because this would need too many schemata. Rather, schemata are distributed in an optimal way: some are general schemata that articulate large domains, other are more specialized and fine-grained, and refine the coarse-grained representations. As a consequence, there are alternatives for the choice of the schemata to be applied. Very often, the states of affairs will be simultaneously underdetermined with respect to overspecified schemata and overdetermined with respect to underspecified schemata. A recourse to lexicon will thus be necessary and this is why (Talmy [369]) "a major aim in cognitive linguistics must be to investigate the interactions between lexical and grammatical specifications arising in a single sentence."

On such a basis, Talmy identifies four major "imaging systems" that are grammatically specified and encoded in what he calls the "fine structure".

(i) A system specifying the qualitative "geometrical" features of objects and situations.
(ii) A system specifying the "point of view". It can be, for example, outside the scene, static and synoptic, or contrarily, mobile, local, and inside the scene, etc.
(iii) A system specifying the distribution of attention: focalization, thematization, gestalt-like complementarity figure-ground, etc.
(iv) Finally, a system specifying "force dynamics".

We see that one of Talmy's most compelling ideas is that in a configuration of objects the ground is not just a mere passive frame but its geometry is also grammatically specified. Thus we are challenged to find how this geometry can vary with linguistic circumstances, e.g., for the English or French case systems, how prepositions select only certain morphological features from the perceptual data, leaving out most of the others. Words as simple as "in", "above", "across", etc. can be considered true *shape-processing algorithms*, taking the perceptual image as input, processing it in a highly specific way, and outputting a semantic schema. This kind of *active* semantics is obviously very far from the classical lexical notion of semantics.

Finally, taking seriously the tenets of cognitive linguistics, one is led to the following postulate: given a perceptual scene (already resulting from low-level preprocessing), the semantics of natural language further organizes this material through a package of morphological routines that (a) enrich the scene with new virtual structures, which did not originally belong to it (e.g., invisible boundaries), and (b) at the same time select from the scene the right morphological features to process. Along these lines, the semantic schemata would trigger additional processing routines, whether object-centered or global.

4. Previous cognitive perspectives

There exist a few theoretical antecedents to the cognitive viewpoints, in particular Fillmore's and Chafe's case grammars that we have analyzed in length in our previous works *Morphogenesis of Meaning* [261] and *Physics of Meaning* [279]. There are also the works of Bernard Pottier[13] and Antoine Culioli. For example, Culioli has often emphasized the fact that

> there is nothing that allows the semantics of natural languages to be reduced to the interpretative semantics of formal systems. (Culioli [71], p. 7)

According to him, we can assume that

> at a very deep level (probably pre-lexical) there is a grammar of primitive relations where the distinction between syntax and semantics makes no sense. (Ibid., p. 8)

At this level, the main problem is to express what Gilles-Gaston Granger called the

---

13 See, e.g. Pottier [313], [314], [315].

> a priori conditions of the creation of linguistic forms. (Granger [128])

It is the problem of universals and as Granger emphasized:

> an essentially syntactic conception of these universals, suggested by a logicist theory of natural language, would screen the moment of content. In fact, it seems to us that the distinction between syntax, semantics and pragmatics, while absolutely necessary for the analysis of the state of a language, is subsequent to the universals. The latter are indissolubly, at one and the same time, acts of enunciation, "natural" categories of the objects of the world and abstract rules of symbolic sequences. (Granger [128])

But perhaps the main precursor of cognitive grammars is Lucien Tesnière. In his seminal paper for the Tesnière conference organized in 1992 by Françoise Madray-Lesigne, Ron Langacker offered evidence in favor of "structural syntax".

> Let me start on a personal note by indicating the considerable empathy I feel for Tesnière. (...) Like Tesnière's structural syntax, this theory [cognitive grammar] represents a detailed and coherent formulation that attempts to capture the essential nature of grammar while affording sufficient flexibility to accommodate the full range of grammatical phenomena encountered cross-linguistically.[14]

After having recalled that

> a pivotal theoretical issue is the relation between meaning and grammar,

Langacker observes that we could think at first glance that, with his concept of stemma,

> Tesnière shares with Chomsky the view that grammatical structure is distinct and autonomous vis-à-vis semantics.

But he immediately adds that

> the *structural connections* that constitute for Tesnière the very crux of grammar do not exist independently of the *semantic connections* they express. (...) Such statements are contrary to the original vision of an autonomous syntax, but perfectly compatible with the central claim of cognitive grammar, namely that meaning and grammar are indissociable.

We know that, concerned with the rational construction of a "pure" syntax, Tesnière reduced sentences to a set of connections (1.2, 1.3).[15] Now, these connections are "incorporeal" (immaterial, disembodied) (1.4) and do not have any marker (16.12). They can be grasped only by the mind's eye. As Jean Fourquet noted in his preface to the *Élements*, a sentence realizes with its connections

---

14 A shorter version of the paper appeared in the volume of *Sémiotiques* on *Modèles dynamiques et Sémiotique cognitive*, [205].

15 The references are to *Éléments de Syntaxe Structurale* (Tesnière [378]). For a discussion of Tesnière's theory, see Petitot [261] and [291].

> the expression of a *lived experience* with linguistic structure, the structuration of an event in view of its communication through language.

Connection is the principle of syntax (1.12) and thought is expressed by means of it (1.7): the production of a sentence by a speaker consists in connecting lexemes (1.9), and comprehension of the sentence by a receiver consists in grasping these connections.

The theoretical problem raised by the Tesnierian concept of connection is that *it is not* a relation in the logical sense of the term. And we do not see very well what alternative eidetic status can be conferred to it. Tesnière never ceased repeating that connections are not of a logical essence and involve a "subjective and unconscious" activity, a "deep, elementary" phenomenon (20.13), in fact a cognitive structuration. Connections are relations which are simultaneously ideal (abstract) and realized concretely (implemented) in a subject of performance. With the stemmas that hierarchically organize them, they

> express the *speaking activity* which is distinguished, under the name of "parole" from the result of that activity, that is the "langue". (3.11)[16]

Vital organic principles, as well Humboldtian *energeia* as Saussurian *parole*, connections are clearly on the side of performance as a principle of production of sentences, and not on the side of the formal rules of competence. Consequently, they are part of a theory of performance, which allows us to understand them as constraining, stable and self-regulated organic structures. The problem is rather similar to morphogenesis, embryogenesis, and organogenesis in biology, and certainly just as difficult.

It is not by accident that these two problems have been worked out together by Thom in the late 1960's, in a common theoretical framework for biology and linguistics.[17] The fundamental thesis was that Tesnierian connections were positional connections between actantial places located in a conveniently chosen space. Hence the following program of a topological syntax:

- Actants, whether as participants or as pure places, occupy positions.
- Relations correspond to positional relations.
- The system of connections (stemmas) creates positional configurations (positional Gestalts).
- Processes correspond to temporal transformations of such configurations (with the associated aspectuality).
- The lexicalized events (generally lexicalized by verbs) correspond to interactions between positions filled by actants.

---

16 "Langue" and "parole" come from Saussure.

17 I can testify to the deep admiration that René Thom had for Lucien Tesnière. He told me on numerous occasions that one of his biggest regrets was to have stayed several years in Strasbourg at the beginning of his career without ever finding an opportunity to meet the master, whose work became one of the major sources of his topological and dynamic conception of linguistics.

- Finally, the case semantic roles correspond to interaction types. They categorize possible actions.

We can appreciate to what extent Thom was close to the theses of cognitive grammars. In the early 1970's, his pioneering topological point of view was not well understood in spite of the efforts by masters like Roman Jakobson, Hansjakob Seiler[18] and Bernard Pottier. One had to wait until the development of cognitive grammars, and particularly the work on schematicity and syntactic iconicity, for the thesis of "topological ideality" (distinct from "logical ideality") of syntax to be fully accepted.

5. The problem of formalization and modeling

The developments of cognitive grammars bring to light the deep, difficult and fascinating problems of formalization. First of all, it must be observed that cognitive grammars are critically lacking formal models. This is a very striking and puzzling fact. Formalization should therefore provide them with more solid foundations and make them evolve from an intuitive and (richly) descriptive stage to a truly systematic and scientific stage. This lack of modeling can be explained by a certain suspicion against any "formalization" in the style that we have been accustomed to by theoretical linguistics and AI since the 1950's. Rejecting rule-based Chomskyan and AI-like theories, the tenets of cognitive linguistics were also led to reject formalization as a whole. As their opponents did, they implicitly took for granted that "formalization" equals "formal language" without considering that there existed *other types* of formal models involving other types of working tools than symbols. Today, by contrast, numerous studies have shown that fundamental cognitive processes such as categorization, learning, inference, or even rule-extraction, can be explained in terms of dynamical systems, which bear no resemblance whatsoever to logical-combinatorial systems. This involves a complete shift in the conception of formalization. Given the difficulty and the epistemological intricacies of the problem, a closer examination is in order.

5.1. The limits of formalism

During the second half of the 20th century, the dominant conception of modeling in linguistics has been formalist, inspired by logic and computer science. This was due to two basic factors. First, the considerable achievements of formal logic and axiomatics have led a majority of philosophers, epistemologists, and linguists (but not mathematicians!) to think that mathematical theories could be reduced to formal languages in the framework of the syntax/semantics opposition in model-theoretic logic.[19] The question of mathematizing linguistic structures has thus become an attempt to formalize natural languages in

18 See, e.g., Seiler [339] and [340].

19 For an introduction to model theory, see Petitot [259] and its bibliography.

terms of formal languages. The developments, even more considerable, of computer sciences have definitely reinforced this point of view. As we have seen, the main consequence has been that only the automatisms of competence, and their formal kinematics, have been formalized, and not at all the dynamical mechanisms of performance. Whether in the framework of such and such formal logic, or in theories of automata, generative grammars (Chomsky), categorical grammars (Montague), formal semantics, intensional logic (Hintikka-Kripke), or theories of categories and topoï, the same idea was gradually technically pursued and elaborated upon: linguistic representations are formal symbolic representations, upon which symbolic computation operates.

By definition, such formalisms "denaturalize" natural language. They equate competence with a system of rules and reduce the problem of concrete performance to mere problems of implementation.

It was not remarked very often that, in the standard transformational-generative conception of grammar, the reduction of syntax to a formal description of competence involves a fundamental constraint. As Chomsky claimed, transformations must be applied sequentially and must therefore be applied to objects of the *same formal type* as those they produce. Consequently, in the regression from surface structures to deep structures, one obtains abstract primitive structures ("atomic propositions" as in logic, "kernel sentences" as in Harris, etc.) which are of the same formal type as the surface structures (for instance syntactic trees). It is thus impossible to investigate their links with perception and action, on the one hand, and understand their emergence in terms of underlying dynamical mechanisms of performance, on the other hand. Now, the structures of perception and action, as well as the dynamical mechanisms of performance, undoubtedly impose certain *universal* constraints on grammatical structures. If we do not take them into account, we are committed to interpreting these constraints as genetically innate.

That is why, as early as 1975, we have argued against this "obvious" classical symbolism, in which our ignorance of the physical foundations of mental structures forces us to reduce ourselves to an abstract characterization of competence. The conclusion of this "bad syllogism" is that the structural properties of language that cannot be derived from such an abstract characterization must be explained in innatist terms. It seemed to us, on the contrary (see [258]), that the alternative "good syllogism" was the following:

(i) we do not as yet know the physical neurophysiological bases of language;
(ii) but we can nonetheless posit them and assume the existence of dynamical processes underlying performance, processes from which emerge the formal and abstract kinematic structures of competence;
(iii) formal grammars formalize only certain aspects of these emerging formal structures;

(iv) but there exists other aspects, linked with perception and action, which impose additional cognitive constraints on the "humanly accessible" grammars.

5.2. Computationalism: the symbolic/physical dualism

As far as cognition is concerned, the formalist point of view is inseparable from the classical symbolic cognitivist paradigm, according to which cognitive sciences are the sciences of mental representations expressed in an internal formal language that manipulates denoting symbols. It is postulated, as Daniel Andler has explained, that

> the contact with the world allows the cognitive system to equip its internal symbols with meaning. (Andler [15], p. 7)

In other words, it is assumed that

> the structural properties of the world are expressible, by means of a sufficiently rich formal language, in the form of representations and rules. (Ibid., p. 8)

5.2.1. *The symbolic level.* The classical paradigm is computational, symbolic and functionalist.

(i) First of all, it postulates the existence of neurophysiologically implemented mental representations, and differs on this point from purely eliminativist, reductionist, and physicalist conceptions which consider that mental representations are only artifacts of the psychological descriptions and do not possess as such any objective existence (see, e.g., Churchland [61]).

(ii) Then, it postulates that these representations are symbolic, i.e., pertain to an internal language of thought (Fodor's "mentalese") possessing the structure of a formal language (symbols, well-formed expressions, inference-rules, etc.). On this point, it differs from conceptions that assume that experimental results (for example, rotating mental images) argue in favor of non-propositional but geometric-topological mental representations (see, e.g., Kosslyn [190] and Shepard-Cooper [344]).

(iii) Finally, it postulates that, as in computer science, one can separate problems of hardware from problems of software and that symbolic mental representations are, as far as their formal structure and informational contents are concerned, independent of their implementation in physical substrata. It differs on this point from the emergential conceptions which, on the contrary, consider that these formal structures must be conceived as stable structures emerging from underlying dynamical and statistical processes (see Thom [379], [382], Zeeman [418], Rumelhart's and McClelland's P.D.P. [253], Smolensky [355], Petitot [262], [267], [271]). Initially developed by Hilary Putnam and Jerry Fodor (but later rejected by Putnam himself), functionalism is the classical solution to the problem of relations between mental and brain states. It rests on the observation that a computational program, which is a set of logical instructions, can be implemented in computers possessing very different physical structures. In other words, it depends on the double analogy between, on the one hand, the logical steps of a program and mental states, and, on the other, between the physical states of a computer and brain states. Functionalists hold that the vocabulary meant to describe, explain and predict the states qualified as "brain states" are not ipso facto appropriate to describe, explain and predict states qualified as "mental states". This non-reductionist position enables a rational division between neurosciences (the hardware part) and cognitive sciences (the software part). But it does not make room for an explanation of the qualitative character of mental states, which are defined less by their computational and inferential role than by their correlated phenomenal experience. Moreover, in the embryogenesis of biological cognitive machines like the brain, one cannot distinguish between software and hardware.

Thus for the symbolic paradigm, cognitive sciences must be founded on a computational theory of formal manipulations of symbolic representations. These representations process information, particularly information from the external world, and acquire in this way a semantic content. But the natural causality of the operations in which they are implicated is a strictly formal and syntactic one. In other words—and this is a problem— they are opaque with respect to their semanticity.

5.2.2. *The physical level.* Insofar as the informational input of mental processes is concerned, computational mentalism of the classical paradigm is inseparable from a standard physicalist objectivism. According to the latter, what is objective in the environment is reduced to what standard fundamental microphysics (atoms, rays, sound-waves, etc.) teaches. Hence a *dualism* (strongly reminiscent of traditional philosophical dualisms) between the symbolic and the physical levels. In a seminal book, *Computation and Cognition* [318], Zenon Pylyshyn gave an excellent exposition of this. Conceived in a physicalist manner, external information is a priori without relevant meaning for the cognitive system. As we have already seen in Sections 2.1.3 and 3.2, it is converted into computationally relevant internal information by transducers (e.g., neurons' firing frequencies). There exists of course a nomologically describable causal correlation between external physical information and internal computational information produced by transduction, but this does not entail, however, the possibility of a nomological science of the meaningful relations that the subject holds with his environment. Indeed, on the one hand, the physically and causally described transduction is cognitively opaque. Its function is non-symbolic and is part of the functional architecture that constrains the structure of mental algorithms. On the other hand, linguistic "meaning" results from the operations executed by the symbolic representations, and these are not causally determined by the objective physical content of the external states of affairs. Hence, according to Pylyshyn, an irreducible gap between the internal cognitive representations and the external physical world. There is a universal physical language composed of physical terms. But there are not, *in this language*, any *physical* descriptions of what is relevant and meaningful in the environment for a cognitive subject. Quotations relating to this "strongest constraint" and this "extremely serious problem" could be listed together in Pylyshyn [318] (pp. 166-167):

> the relevant aspects of the environment are generally not describable in physical terms,
>
> psychological regularities are attributable to *perceived*, not physically described properties,
>
> [there is a] general failure of perceptual psychology to adequately describe stimuli in physical terms.

Therefore, we must use *functional* perceptual and cognitive concepts lacking physical content. Physical lexicon and cognitive lexicon do not match. They are compatible only by way of transductions.

We shall observe that such assertions are acceptable only on the basis of certain hypotheses:

(i) What exists as objective in the environment is reducible to what standard fundamental micro-physics describes; all the results of macro-qualitative physics are completely ignored.
(ii) What is relevant must be first represented symbolically in order to be meaningful.
(iii) Representations are equated with computation: the mind is computational.

5.2.3. *Difficulties of the classical symbolic paradigm.* As we have already noted, according to several specialists (Putnam, Searle, Dreyfus, etc.) at least two great problems remain unresolved in the classical paradigm.

(i) On the side of the cognitive subject: the problem of meaning and intentionality. How can symbolic mental representations acquire a meaning, a denotation, an intentional orientation towards the external world? How can a cognitive system operate in accordance with the meaning of its symbols and symbolic expressions if it is causally related only to their syntactic form? It is not enough to say that meaning results from a subject-world "interaction", since this interaction is not nomologically describable and explainable.
(ii) On the side of the external world: the problem of its qualitative and morphological manifestation. As we have seen in Section 3.1, we cannot assume with Ray Jackendoff that the phenomenal world is a simple

construction of the computational mind. Even if the phenomenal consciousness constitutive of the qualitative structuring of the projected world into things, states of affairs, events, processes, etc., perceptually apprehensible and linguistically describable is only a tiny part of the computational mind (what Jackendoff calls the "Mind-Mind problem"), it must be supplemented with morphological and qualitative objective structures that emerge from the external physical substrata by means of a self-regulating dynamical process. Without such a supplement, phenomenality would remain incomprehensible. This properly "morphogenetic" level is distinct from the "projective" one. It rests on the demonstrated existence of a morphodynamical level of reality that may well be called, as suggested by Per Aage Brandt, "pheno-physical".

5.3. Mathematization vs. Formalization

If we accept the emphasis on the "naturalness" of natural languages, we can state that, until now, the different types of formalization of linguistic structures have not at all modeled their naturalness. It is then necessary to rework formalization from the outset, without any sort of formalist prejudice, as a problem of mathematizing a specific natural kind of phenomena. It is not because mathematics is also a logical language that mathematical linguistics should be conceived as a game of more or less adequate translation between mathematical logic and natural languages.

Thus if we adopt an anti-logistic stance, we can conclude that *there exists a conflict between formalization and mathematization* in linguistic matters. Instead of having to develop the possibilities of translating natural syntaxes into formal ones, mathematical linguistics, on the contrary, must seek out the specific mathematical theories that are conform to the eidetic characteristics of the cognitive linguistic phenomena.

5.4. Modeling and schematization

In the formalist perspective inspired by Hilbertian axiomatics, one starts with primitive concepts of the descriptive-conceptual theory of a given empirical domain and applies the axiomatic method of "implicit definitions". This method consists of substituting these primitive terms with syntactic rules that regulate and normalize their use. The descriptive-conceptual theory is then translated into a formal language, whose logical syntax can be analyzed, its coherence tested and inferences controlled. This point of view construes the relation between pure mathematics and empirical reality as analogous to the relation between syntax and semantics in model-theoretic logic.

But in truly mathematized sciences like physics, the situation is completely different: primitive concepts are not axiomatized but *interpreted* by specific (possibly very sophisticated) mathematical structures. Of course, these structures themselves belong to axiomatized mathematical theories, but the mathematical axioms involved here have in general nothing to do with the primitive concepts of the descriptive-conceptual theory under consideration. To do homage to Kant, we have called this peculiarity a mathematical *schematization* of concepts.

By schematization, the content of fundamental theoretical concepts is converted into a universe of *specific* mathematical objects that can serve as models for the empirical phenomena at stake. Thus, theoretical concepts are converted into sources of mathematical models that can themselves achieve a *computational synthesis* of the phenomena. Thanks to an appropriate mathematical interpretation, concepts are translated into *algorithms* able to generate a wide diversity of *constructed* models, which can be compared to the empirically *given* phenomena.

We will show at length in the following chapters that morphodynamical models provide a good schematization of the primitive concepts of cognitive linguistics.

5.5. Morphodynamical models and connectionist models

We have already seen in Section 2.2 that morphodynamical models generalize subsymbolic connectionist models. As Smolensky rightly insists:

> subsymbolic systems are dynamical systems with certain kinds of differential equations governing their dynamics. (Smolensky [355])

There is a large number of neuro-mimetic morphodynamical models. We shall return to some of them in the course of this work.

1. The phenomena of categorization and (proto)typicality. The dynamics defines attractors, which can be assimilated to prototypes, as well as basins of attraction, which can be assimilated to categories. The models can be even further refined by viewing categorization as a process that results in a bifurcation of attractors. A good example is categorical perception in phonetics (see Petitot [269]).
2. The complementarity between the syntagmatic and paradigmatic axes of language. If we teach (by means of supervised learning) a connectionist network the statistical syntactic regularities of a corpus of sentences and if we then observe how, in its hidden layers of internal states, it has organized the lexicon so as to be able to respond correctly, we observe that it has constructed *semantic paradigms*. This stunning result is due to Jeffrey Elman [97]. In fact, many linguistic problems (grammatical inference, anaphora, ambiguity, polysemy, etc.) can be treated in this manner (see, e.g., Fuchs-Victorri [115]).

3. The relationships between semantics of natural languages and perceptual scenes. For example, Terry Regier [320] has constructed connectionist networks capable of learning the prepositional system of different languages and applying them correctly to static or dynamic configurations of objects.
4. The phenomena of learning. Learning resolves an *inverse* problem. The direct problem is to deduce the dynamical behavior of the network (for example, its attractors in a categorization task), given the synaptic weights of a neural network. Learning consists, on the contrary, in modifying the weights in such a way that the network can carry out certain a priori fixed tasks.
5. Induction and generalization, i.e., the discovery of general rules from a finite series of examples. It is rather surprising to see a neural network learn rules. The step where inductive generalization happens has the status of a phase-transition.[20]
6. The material bases of the constituency of mental representations. This constituency, evident in the case of language, is very easy to describe and very difficult to explain. It raises a hard problem for mental representations implemented in a distributed manner on a large number of microscopic units. A hypothesis currently debated is that the fine temporal structure of interactions enables binding and constituency: when neural oscillators are *synchronized*, their common phase can act like a label for a constituent.
7. The manner in which attention-focusing and pattern-recognition make the initially chaotic dynamics of the system bifurcate towards a simpler dynamics. It is this simplification of dynamics that would correspond to the recognition processes.

6. Semantic realism and pheno-physics

We have seen the deep relations that exist between cognitive grammars and morphodynamical models. But morphodynamics is more realistic, more "ecological" in Gibson's sense. It is not limited to elaborating cognitive models. It correlates cognition, through perception and action, to a phenomenology of the morphological organization of the phenomenal world, the latter being thought of as objective. We would like to clarify this point in this section.

6.1. Thom's squish and pure "etic" linguistics

Until now, we have distinguished between the "formal" and the "dynamical" paradigms. But this opposition really shows *complementarity* between two

20 See, e.g., the two *Interdisciplinary Workshops on Compositionality in Cognition and Neural Networks* COMPCOG I, II [64], [65], and the conference *From Statistical Physics to Statistical Inference and Back* [130].

of the constitutive dimensions of language. In two important papers (Thom [380] and [381]), inspired by the continuous dimensions introduced by Hansjacob Seiler [339], [340] and by the concept of "squish" introduced by John Robert Ross [328], René Thom proposed to distribute grammatical categories in a *bidimensional squish.* The X-axis of the squish arranges the categories in the following order: Nouns – Verbs – Adjectives – Numerals – Possessives – Deictics – Logical functors and Quantifiers. If we then put the semantic variability of the categories (i.e., the interval between the maximal concretion and maximal abstraction of their representatives) along the Y-axis, we can make the following observations:

(i) Semantic variability decreases along the squish and breaks down when crossing the numeral zone. It is considerable for nouns and verbs and null for logical functors.
(ii) The squish extends from a "categorematic" to a "syncategorematic" pole. Thom discovered therefore for his own purpose the fundamental distinction between open classes (with semantic variability) and closed classes (without semantic variability). Taking Pike's etic/emic opposition, he hypothesized furthermore that open classes are an "objective" etic pole concerned with the simulation of phenomenal reality, while closed classes are a "subjective" emic pole concerned with the automatisms of competence. Linguistic entities can therefore be of very different kinds: "With the noun, we are dealing with an entity endowed with a certain autonomy: the referent occupies a portion of space that it defends against disturbances from the environment; (...) the grammatical auxiliaries, on the contrary, owe their meaning only to an all but ritualized activity of the speaker, totally immersed in the automatisms of language (Thom [381], p. 79)".
(iii) The central zone of the squish where semantic variability collapses represents a sort of threshold—a critical value, a bifurcation—between the etic-objective pole and the emic-subjective one.

These remarks enable us to better understand why formal linguistics is only, so to say, *half* a linguistics. It is derived from a "pure" emic linguistics which cannot account for the etic dimension of language. According to Thom, the principles of "pure" *etic* linguistics were still to be worked out and developed mathematically. Their complementarity with those of pure emic linguistics were still to be established, and from there an attempt was still to be made in catching up with empirical linguistics. We can now posit the contribution of early morphodynamics to linguistics in saying that just as Chomsky laid the foundations of a pure emic linguistics, so Thom has laid the foundations of a pure etic linguistics.

What is first brought out in pure etic linguistics is the *regulation* of the three basic grammatical categories: nouns, verbs, and adjectives.

For the regulation of nouns (and more precisely of concrete nouns), a realist hypothesis is assumed:

> there exists a certain isomorphism between the psychological mechanisms that ensure the stability of a concept C and the physical and material mechanisms that ensure the stability of the real object R represented by C. (Thom [380], p. 247)

Therefore, contrary to the common wisdom dominating formal linguistics at that time, Thom conceived the neuropsychological mechanisms of performance not in a reductionist but rather in a gestaltic way, and the referential function not as a denotative language→reality correspondence, but as a constraint imposed on language by the reality that it can simulate (semantic realism). The content of a concrete noun is a complex, regulated, dynamical mental entity whose regulation is partially isomorphic (Thom said "isologous") with the regulation of its referent. Such an affirmation refers to a phylogenetic hypothesis concerning the origin of language. Primitive concrete concepts must have been entities whose recognition was fundamental to survival (prey, predators, sexual partners) and, in this sense,

> the logos of living beings has served as a universal model for the formation of concepts. (Thom [382], p. 131)

Thom therefore correlates semantics with a general theory of regulation (especially biological regulation). His thesis is very akin to the theses of Charles Osgood and Alexander Luria, which were adopted by Jackendoff (see Section 3.1).

The figure of regulation of a concrete noun C is intimately linked to its *verbal spectrum*. The catastrophes of regulation bounding its domain of viability can be identified with the verbal interactions in which C can play the role of an actant. We recognize here a neo-Tesnierian conception of the verb as a node distributing actantial places (semantic roles) (see Section 4). Insofar as they describe interactive processes, verbs possess in themselves the reason for their stability. They take their source in the simulation of elementary actantial interactions realizable in space-time.

Finally, the regulation of adjectives that localize substantives in qualitative spaces reduces to the *categorization* of semantic spaces, as for example the field of colors.

The development of a mathematical etic linguistics thus relies upon the mathematical theories of

(i) regulation of concepts,
(ii) verbal valence, and
(iii) categorization of semantic spaces.

It must, moreover, respect the cognitive organization of perception and action.

As Wolfgang Wildgen emphasized, this requirement is the core of "catastrophe theoretic" linguistics:

> We assume that the dynamic principles governing the semantics of words are intricately connected with basic propositional structures. This is especially true for

> verbs. Our dynamic treatment of verbs starts with a consideration of the dynamic principles underlying the perception of space and time and of changes, motion, locomotion and action in space and time. (...) In a general semantic theory our archetypal and dynamic component would be a basic stratum whose influence becomes weaker as we progress to the levels of syntax and text (conversation). (Wildgen [405], p. 235) [21]

We see how close these early catastrophe-theoretic steps were akin to the later trends of cognitive linguistics. For the etic perspective, as for the cognitive one, linguistic objects are no longer autonomous. Regulation of concepts is linked to biological regulation, verbal valence to a "physics" of actantial interactions, and categorization of semantic spaces to the existence of critical phenomena of phase-transitions.

6.2. The phenomenological question

Let us further investigate the semantic realism mentioned above. We shall call "phenomenological question" the question of the relationships between the mathematically reconstructed "deep" structures of physical objectivity and the phenomenologically manifested "surface" structure of the phenomenal world. A fundamental belief of modern rationality is that physical objectivity does not possess *in itself* the resources for being phenomenalized and qualitatively structured.

This belief implies that, in one way or the other, the qualitative morphological level must result from a "projection" of the subject onto the physical external world.

Jackendoff's conception (Section 3.1) is a good example. The projected world PrW appears as a purely cognitive construction and is separated from the real physical world RW by an ontological gap. As we have seen, at no point Jackendoff hypothesizes that there could exist a natural non-cognitive process of phenomenalization of RW into a phenomenal world PhW. At no point does he consider the possibility that the qualitative structuration of the world into things, forms, states of affairs, places, paths, states, events, processes, etc., can partially emerge from an objective macroscopic morphological organization to which living organisms should, each in its own specific way, be phylogenetically adapted. In short, $PhW = PrW$.

This type of perspective involves a prejudice about physics, the prejudice that "we know well" that physics "cannot explain" the qualitative organization of the world. Now—and this is the key point here—such an evidence inherited from the history of classical physics has proven completely wrong today. In fact, in the last thirty years, considerable progress has been made, both in physics and mathematics, in the understanding of the phenomena of macroscopic self-organization of material substrata.

21 See also Wildgen [408].

(i) In mathematics: algebraic, analytic, differential theory of singularities and their universal unfoldings (with or without symmetries); structural stability, genericity, transversality and stratifications; qualitative theory of non-linear dynamical systems (Hamiltonian or non-Hamiltonian), their (possibly "strange") attractors and bifurcations, and their ergodic properties; turbulence, routes towards chaos, etc.

(ii) In physics and in non-equilibrium thermodynamics: theories of critical phenomena, phase-transitions, spontaneous symmetry breakings (Landau's mean field theory, renormalization group, etc.); diffraction catastrophes and dislocations of wave-fronts in wave-optics with their consequences for the semi-classical approximation of quantum mechanics (caustics and singularities of Lagrangian manifolds, asymptotic solutions to the wave equation and approximation of geometrical optics, oscillating integrals and stationary phase method, etc.); theories of elasticity, buckling phenomena, shockwaves, singularities of many systems governed by variational principles (from Hamiltonian mechanics up to equilibrium theories in economics); dissipative structures in chemical kinetics and in non-linear and non-equilibrium thermodynamics, etc.; defect theories in ordered media and mesomorphic phases (liquid crystals), etc.[22]

All these convergent technical works (involving a constellation of eminent scientists, Fields medals, Nobel prizes, etc.) have radically modified the image of physics. They have developed in a flourishing manner three leading ideas:

(i) Macroscopic natural systems are systems with (at least) three levels of objective reality: a fine-grained and very complex "micro" level, which corresponds to the fundamental physics of the system; an intermediate "meso" level where global interactions of local "micro" units are generated; and finally a coarse-grained "macro" level, in general easily describable, which is more morphological than physical. The "macro" morphological level emerges from the underlying "micro" level and one can mathematically control the change of levels in the models (see, e.g., statistical mechanics in thermodynamics, aggregation theory in economics or connectionism in cognitive sciences).

(ii) The "macro" level is essentially articulated around the *singularities* (caustics, phase-transitions, shock waves, defects, symmetry breakings, etc.) of the underlying physical processes. These singularities bear *relevant* information and are phenomenologically dominant (salient, as Thom says, see Petitot [279]).

(iii) There are mathematically expressible abstract constraints governing critical and morphological phenomena. Analysis reveals strong properties of *universality* of critical behaviors, that is, a relative independence

22 For clarifications on these questions, see *Physique du sens* [279] and its bibliography.

of the organization of the "macro" level with respect to the underlying fine-grained physics.

It is very important to stress this last point. One of the main reasons that symbolic cognitivism called for a rejection of dynamical conceptions stems from a misunderstanding of the epistemology of emergence. When a system possesses many levels of organization, the higher "macro" level is *causally* reducible to the lower ones, yet at the same time its structures can be largely independent from the underlying "micro" dynamics. It benefits from a certain objective autonomy. As John Searle emphasized, only if we identify a phenomenon with its causal genesis—in other words, if we surreptitiously move from a justified causal reductionism to a dogmatic and unjustified ontological reductionism—that we are led to deny the autonomy and objective reality of the higher levels of organization.

So, the phenomenological world PhW is the result of several independent processes:

(i) on the side of the external world, the emergence of a macro pheno-physical morphological level outside of the underlying micro-physical level, which can be adequately called "geno-physical";
(ii) on the side of the cognitive subject, the processing of this morphological information by low-level, modular, bottom-up and data-driven routines;
(iii) again on the side of the cognitive subject, the emergence of a symbolic conceptual structure outside of an underlying subsymbolic dynamical level;
(iv) the projection of this "computational mind" onto the pheno-physical world, which produces the projected world PrW.

The relation between the symbolic and physical levels avoids the ontological gap encountered by almost all cognitive theories. Indeed, the morphological level constitutes *a mean term* between the physical and the symbolic ones: it is of physical origin (emergent) but without being material; it is formal but without being symbolic; it is topologically and geometrically formal and not logically formal. As David Marr [224] has noted, the morphological information passes through transduction. It is encoded and carried by light and sound signals and decoded-recoded by transducers. But during the transduction it remains in large part isomorphic to itself. Qualitative discontinuities are in some sense "contagious" and can be transferred from substrata to substrata. Hence what we call the *double* organization (cognitive-projective and pheno-physical) of the phenomenal world.

The physicalist prejudice can now be formulated as an eliminativist thesis: there does not exist any pheno-physical level. For classical cognitivism the consequences are considerable. Most of the theoretical difficulties it encounters come from the fact that it seeks to derive the morphological level from a symbolic conception of syntax and semantic, while this is clearly impossible. But consequences are also rather important for subsymbolic cognitivism. The most

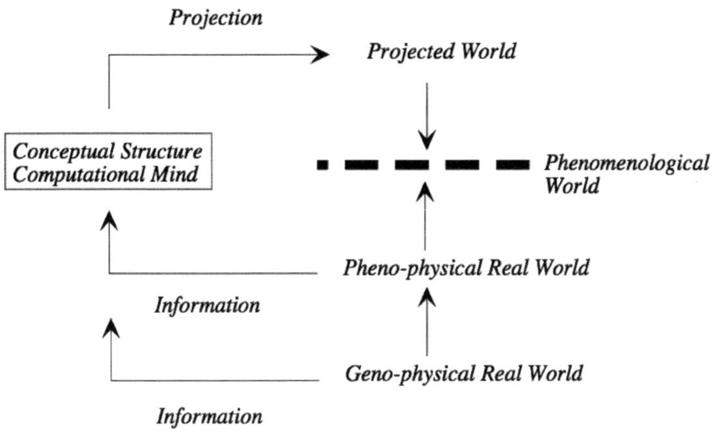

FIGURE 5. Contrary to figure 1, the phenomenal world does not reduce to a projected world separated by an ontological gap from the real world. It also emerges out of a self-phenomenalizing real world.

immediate obviously concerns the projectivist thesis $PhW \equiv PrW$. As we have just seen, the phenomenological world PhW is *both* projected and emergent. Thus there is not any real ontological gap between PrW and RW. The diagram proposed above in Section 3.1 must be revised in the following way (see Figure 5).

6.3. Pheno-physics and ecological information

The semantic realism justified by pheno-physics consists in enlarging the very concept of nature in such a way as to include processes involving a phenomenalization of physical objectivity. As such, it is very close to James Gibson's perceptual ecologism (see Gibson [122] and [123]).

Barry Smith has already shown how Gibsonian ecologism allows a realist interpretation of phenomenology.[23] As a result of evolutionary adaptation, perceptions and actions of a living organism are pre-tuned to the qualitative structure of its *Umwelt*: forms, qualities (colors, textures), etc. These have become intrinsically meaningful for it. Of course, the world is a product of cognitive processes, but it is so only "ontogenetically". "Phylogenetically", on the contrary, it is cognition which is adapted to the world. By virtue of the general principle that what is ontogenetically *a priori* is phylogenetically *a posteriori*, a neo-Gibsonian realist conception can perfectly well be developed from an evolutionary standpoint.

23 Smith [351]. See also, Petitot-Smith [307].

As we have seen, the morphodynamical approach emphasizes the fact that such a perspective presupposes the possibility of physically describing the morphological and qualitative structures of the *Umwelten* as emergent macroscopic organizations.

In the absence of such descriptions, ecological approaches are open to the well known criticism brought against James Gibson by Jerry Fodor and Zenon Pylyshyn in their celebrated paper [112] "How direct is visual perception? Some reflections on Gibson's 'ecological approach' ". Gibson's basic thesis is that perception is able to pick up from the environment invariants possessing *both* an objective content and a perceptual meaning. J. Fodor and Z. Pylyshyn [112] have tried to refute this theory using the classical hypothesis that, being cognitive, perception must necessarily be an inferential and symbolic computational process. According to them, the main inconsistency of ecological theory is to postulate the existence of an objective information which, though non-physical, would nevertheless be present in the light medium (discontinuities, forms, deformations, reflectance of visible surfaces, etc.). But what can this enigmatic "information in the light" be, they ask. (p. 143)? For them, by virtue of the dualism between the physical and the symbolic levels, information has to be either physical or symbolic. Consequently, if it is not really physical but "ecological", it must necessarily be symbolic. In short, Gibson appeals to an ecological information that can be found nowhere. He criticizes, on the one hand, physical reductionism and, on the other, psychological mentalism. But he provides no alternative, and this leads him to a vicious circle. And the authors conclude:

> What we need, of course, is some criterion for being ecological *other than perceptibility*. This, however, Gibson fails to provide. (p. 146)

What would be needed is an ecological objectivity different from physical objectivity and able to characterize what is phenomenologically relevant. But, in the authors' view, this is precisely what cannot be provided.

The whole argumentation boils down to the following. The only possible direct extraction of invariants is that processed by sensorial transduction. Transducers, even compiled transducers that operate up to post-retinal levels, can only be sensitive to the physical properties of the light signal, for their functioning is ruled by laws (it is nomological) and the only available laws are physical laws. Transducers extracting from the signal ecological properties that are not physical cannot therefore exist. Thus Fodor and Pylyshyn condemn what they consider to be a category mistake on the part of Gibson. For Gibson, information is contained *in* light. But, according to the authors, the concept of information is *relational*. Light contains information about the environment and "containing information about" means "being correlated with". Properties of the environment are then *inferred* from the structure of the light signal on the basis of prior knowledge about these correlations. By replacing "containing information about" by "information contained in", Gibson would have surreptitiously reified the relational concept of information. He would have treated it "as a thing rather than a relation" (p. 167). A perceptual system cannot be affected by a piece of information but only by the physical properties of things. The latter can certainly provide "information about something" but only by means of inferences. Information cannot be the state of a receptor. So the problem is to determine:

> how (by what mental process) does the organism get from the detection of an informative property of the medium to the perception of a correlated property of the environment?

For Fodor and Pylyshyn, there can be only one answer: by inference. "X contains information about Y" is a *semantic* relation and so depends upon the way in which X is mentally represented as a premise in inferences from X to Y.

However, this critique (carried much further by the authors) can itself be criticized because it rests on the double prejudice that physical reality possesses no macroscopic emergent properties and that what is significant for a cognitive system must necessarily be symbolic. But from the morphodynamical standpoint, the existence of objective morphological information invalidates this prejudice and justifies ecologism. Gibson was on the right track with his notion of extraction of invariants. But Fodor and Pylyshyn are equally right in identifying a vicious circle in Gibson's thinking. In other words, it is indeed true that "what we need is some criterion of being ecological other than perceptibility". But, precisely, this criterion is afforded by the morphological level of pheno-physics.

6.4. Realist phenomenology

Philosophically, the semantic realism of morphodynamics is not without precursors and is akin to naturalist versions of Husserlian phenomenology. The problem of naturalizing phenomenology has been well formulated by Roger Chambon in his seminal work *Le Monde comme Perception et Réalité* [58].[24]

The central question is:

> What does the world have to be if it is to bear within itself the potentiality for its own appearing? (p. 17)
>
> What does the unperceived universe have to be if perception (...) is to be "possible" to the point of becoming effectively real? (p. 45)

The qualitative structuring of the phenomenal world presupposes a major ontological commitment and yields a "crucial ontological clue" (p. 28) as to the nature of objectivity:

> the appearing, the phenomenon, is anchored in being. (p. 21)

24 For an introduction to Chambon's realist phenomenology, see Petitot [296].

Naturalizing Husserlian phenomenology is therefore justified just as long as it succeeds in integrating

> the event of the appearing of the world into a deep and detailed recasting of our scientific conception of nature. (p. 23)

The same ideas can be found in Merleau-Ponty's late works concerning *Naturphilosophie* reframed in phenomenological terms. In some of his last lectures at the Collège de France (1952-53, 1959-60, see [231]), Merleau-Ponty sought to understand the relationship between natural organizations, phenomenological appearing, lived experience and meaning. In order to do this, he showed that, in addition to an eidetic cognitive description, one needs *a dynamical theory of forms and structures* to explain the "morphogenetic gradients" of natural morphologies and to account, on a physical, biochemical, thermodynamical and even "cybernetic" basis, for the manner in which "organization invests physical space". One needs a "phenomenal topology" and a "phenomenological phusis" (admirable expressions) to understand

> the emergence of original macro-phenomena, which are singular spatial loci, from among micro-phenomena.

According to Merleau-Ponty, it is moreover in a phenomenology which leads up to a *Naturphilosophie* and a qualitative physics of ecologically and biologically relevant emergent structures that *meaning* has to be founded. Natural forms and their correlative perceptual *Gestalts* are intrinsically meaningful. They display "figuratively" "a *force* which can be read into a *form*". Semiotic ideality is built up on the basis of a morphological ideality which is not merely realized in subjectivity but also in the qualitative macro objectivity of the world.

7. Morphodynamics and complex systems

Morphodynamical cognitive models belong to the larger class of complex systems. Complex systems possess emergent global properties stemming from underlying collective interactions—whether of a cooperative or a competitive kind. They are singular, largely contingent (not concretely deterministic), historical, and result from processes of evolution and adaptation. They also possess an internal regulation allowing them to remain within their domain of viability. Sophisticated mathematical tools such as those of non-linear dynamical systems (attractors, structural stability, bifurcations), self-organizing criticality, algorithmic complexity, genetic algorithms, or cellular automata, have become central to any understanding of the statistical and computational properties of complex systems. Thanks to the engineering of distributed, non-hierarchical and self-organizing artificial systems, it becomes possible to model and accurately simulate biological (immunological, neural, etc.), ecological, societal or economic systems.

This scientific revolution—which is at the same time theoretical, mathematical and technological—provides for the first time *an experimental method* to disciplines which cannot concretely reconstruct their objects in the laboratory. This concerns not only physical sciences like astrophysics or geophysics, but also biological and human sciences. The only way for these sciences to have access to an experimental method is to construct virtual systems allowing to test the models. This is what the founders of the Santa Fe Institute called a "computational synthesis" of complex phenomena, as opposed to their "conceptual analysis". They claim that

> this approach represents a shift from the deductive reasoning of analysis to the inductive reasoning of synthesis.

8. The problem of Universals

Before concluding this introductory chapter, we would like to add a few remarks on the problem of linguistic universals. Indeed, the semantic realism that we have sketched led us to postulate that the universals of language are rooted in cognitive universals which are themselves dependent on the qualitative morphological structure of the natural phenomenal world.

The problem of universals in linguistics is particularly delicate, not only for empirical reasons, but also and above all for epistemological reasons. If one wishes to reduce this problem to a mere inductive generalization of data coming from comparative cross-linguistic studies, one would encounter many difficulties, whether methodological (i.e., concerning their search), theoretical (i.e., concerning their explanation), or mathematical (i.e., concerning their modeling).

When René Thom was working out his morphodynamical linguistics, universals were the subject of several important research programs. We cite here four of them.

1. The program directed by Joseph Greenberg at Stanford University. It was initiated in April 1961 during the *Conference on Language Universals* held at Dobbs Ferry (N.Y.). Its two reference publications are *Universals of Language* [133] and the four volumes of *Universals of Human Language* [134], edited by J. Greenberg.
2. The Generative Grammar program represented among others by the *Symposium of Universals in Linguistic Theory* held in April 1967 at the University of Austin. One main publication is *Universals in Linguistic Theory* edited by E. Bach and R. J. Harms [25].
3. The Relational Grammar program (Comrie, Keenan, Perlmutter, Postal, Johnson). A good introduction is the volume *Syntax and Semantics: Grammatical Relations* edited by P. Cole and J.M. Sadock [63].[25]
4. The Universalien Projekt directed by Hansjakob Seiler at the University of Cologne. It was initiated in the summer of 1977 at the *Linguistic*

25 For an introduction to relational grammars, see Petitot [261].

Institute of Buffalo and has been prepared by the work *Language Universals* edited by H. J. Seiler [339].[26] Two reference events are the Gummersbach Conferences of October 1976 and September 1983.[27]

Three major dimensions unfold in these programs:

(i) the *empirical* dimension: looking for empirical law-like generalizations;

(ii) the *transcendental* dimension: universals as definitional, essential, analytic or synthetic a priori, i.e., as conditions of possibility;

(iii) the *innatist* dimension: genetic, mentalist, cognitivist, or ethological explanations of the linguistic universals.

These three dimensions correspond respectively to the three basic epistemological principles identified by Paul Garvin: behaviorist, rationalist and innatist.

It is clear that every search for universals presupposes to go beyond the Sapir-Whorf relativist point of view according to which languages are irreducibly diverse and do not manifest any universal constraint. But, though non-relativistic, it can nonetheless be empiricist and agree with Bloomfield's precept that

> the only valid generalizations about languages are inductive generalizations.

The search is then concerned with the surface structures, and its methodological tools are comparative analysis and the study of learning. Such is the case of deep researches by Greenberg and Osgood, Keenan and Comrie or Seiler. They obtain non-definitional, statistical and implicational universals. Let us recall that the so-called *definitional* universals (Fergusson, Hockett) or *conceptual* universals (Coseriu), also called universally operative (Saporta), are logical consequences of a definitional model characterizing human language. When they are analytically included in the general concept of language, they are called *essential* or *analytic* (Moravcsik). On the other hand, *statistical* universals are, according to Charles Osgood, dynamical universals acting as principles and norms for linguistic activity and represented in the existing languages as tendencies (for example the tendency of phonological systems toward symmetry). As for the *implicational* universals, they are laws of the type "if ... then". We find several cases of them in relational grammars. Let us give a few examples.

Let us recall that the basic hypothesis of relational grammars is that the grammatical relations of subject, direct object, indirect object (O, DO, IO), are primitives of the syntax of natural languages.[28] They stand for the most basic case universals: Agent, Accusative, Dative, and the actants supporting other cases such as Instrumental, Locative or Benefactive are called oblique objects

26 See also the last book of Hanjakob Seiler on Universals [340].

27 We could also consider formalist conceptions of universals such as Montague's or Shaumjan's. But we restrict here ourselves to reflections on universals coming from empirical linguistics.

28 See [63], [256], and [261].

(OO). The main technical hypothesis is the *relational hierarchy* RH: $S > DO > IO > OO > Possessives >$, etc.[29] A transformation like passivization is then understood as a $DO \to S$ promotion of the DO to position S with a correlative "demotion" of the initial S; dative transformation in English is an $IO \to S$ promotion with a correlative "demotion" of the initial S; the Subject-to-Object raisings $S \to DO$ extract the subject from a subordinate clause supporting the grammatical relation DO relatively to the verb of the main clause and transform it into a DO. These transformations satisfy a number of implicational universals expressible in terms of RH and of the law of increasing rank: every rule that transforms the grammatical relation of a term makes it move up in RH or raises it to a hierarchically superior constituent. For example:

(i) if a language can relativize (in the sense of embedding in a relative clause) a *NP* into a position X in *RH*, then it can also relativize all the positions of the interval $[S, X]$;

(ii) if a language can promote a *NP* occupying a position X in RH to the position S, then it can also promote to S every position of the interval $[S, X]$;

(iii) if a language (like Japanese) can subjectivize indirect objects ($IO \to S$), then it cannot objectivize oblique objects (no $OO \to OD$);

(iv) if a language (like English or Indonesian) has the promotion $IO \to DO$, then it cannot have promotions $OO \to S$; etc.

But the search for universals can be also conducted on rationalist bases and be concerned with deep structures. Such is the case with Chomsky. Since *Aspects of the theory of syntax* [59], Chomsky has insisted that

> the main task of linguistic theory must be to develop an account of linguistic universals,
>
> the study of linguistic universals is the study of the properties of any generative grammar for a natural language.

In this perspective, universals allow us to define the "humanly accessible" languages within the set of all formal languages. In Chomsky's terminology they are *substantial* when they are concepts necessary for the construction of a grammar, and *formal* when they impose constraints on the general form of a grammar.[30] They are intrinsic properties of the language acquisition system, and constrain the form and the substantial features of the grammars that can be learnt from a finite set of empirical data. They are definitional, non-statistical and non-implicational and, according to Chomsky, can be explained

29 *RH* is an example of continuum (i.e., of relation of total order) in linguistics. Tesnière's translations or Ross' squishes are other examples.

30 For Eugenio Coseriu, substantial universals are *specific* (i.e., concern specific facts such as: all languages have the category of the noun), while formal universals are *generic* and concerned with general principles and norms of language without specifying the facts in which they are manifested (Coseriu [66], p. 55).

on the basis of an innatist hypothesis. And in effect, the universalist perspective is inseparable from the attempt of justifying theoretically empirical laws. It is therefore very tempting to adopt an innateness hypothesis making universals the expression of genetic constraints. In this context, we can first identify *intrinsic* explanations in which universals are specifically linguistic (for Chomsky, humans speak like spiders build their web or beavers their house). The environment is only an "ignition key" and universal grammar should describe, in the form of a multipurpose master program including many sub-routines, the steady state of an adult speaker starting from a biologically constrained initial state. We can also identify *extrinsic* explanations in which universals are not specifically linguistic and manifest constraints (innate or not) of a cognitive and perceptual nature. For instance, for Edward Keenan, the explication should follow from the functions that the language is required to fulfill as well as from the structure of the cognitive and perceptual apparatus of the *homo sapiens*.

As for McNeill, he proposed to call *strong* the linguistic universals reflecting a specifically (intrinsic) linguistic ability, and *weak* the linguistic universals reflecting a universal (extrinsic) cognitive ability.

Hermann Parret [254] has criticized empiricism and innatism from a rationalist point of view. For

> a strong linguistic theory elaborated in rationalistic terms is obviously the best heuristic framework for universals research.

His idea is that universals are *normative a priori* conditions of possibility, and that they are empirically valid due to this transcendental status. Their approach should be therefore epistemological. Parret also observes that there exist definitional universals only for *formal* languages. Between, on the one hand, the deductive principles derived from a definitional model of language and, on the other hand, the inductive principles (particularly the implicational universals) derived from comparative studies, there must exist a third term, in the form of *a priori synthetic* principles (see Figure 6).

This last point is critical. Indeed, as noted, among others, by Osgood and Moravcsik, the essential analytic a priori universals are essentially trivial. Besides, universals obtained inductively are *synthetic a posteriori*. As Eugenio Coseriu had insisted at the *International Congress of Linguists* in Bologna in 1972, their universality is not a logical necessity and they should be motivated, either in an innatist or in a rational manner. But in the latter case, their motivation would be synthetic a priori. Now, according to Coseriu, no one would accept a synthetic a priori in matters of language. As he insists in [66]:

> an a priori synthesis is not conceivable in this case.

In the same congress in Bologna, Hans-Heinrich Lieb proposed a classification of universals into types:

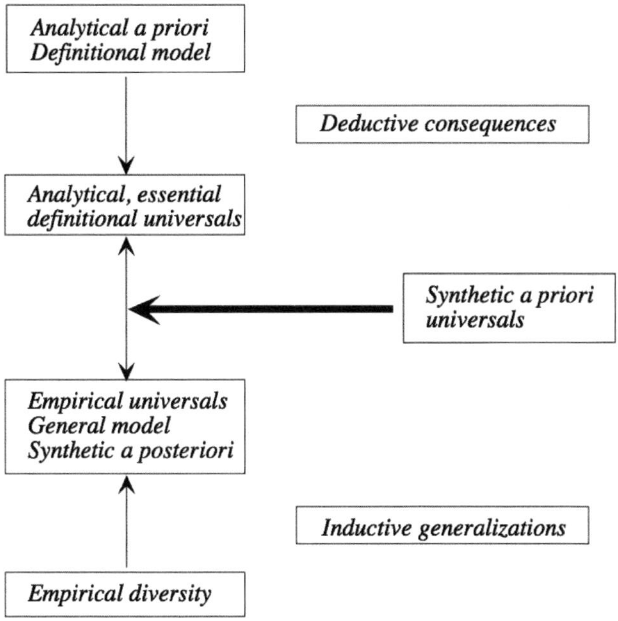

FIGURE 6. The relations between analytic-deductive, empirical-inductive, and synthetic a priori universals.

1. properties of the system that constitutes language;
2. conditions of possibility of communication in natural language (deictics, tense, aspect, etc.);
3. developmental properties of language;
4. properties of the relation between language and thought;
5. properties of the relation between language and society;
6. properties of the ontogenetic development of language (psychological and physiological basis of language learning);
7. properties of the phylogenetic development of language;
8. properties of the relation to the semiotic field in general.

(1) corresponds to the static universals, (3), (4), (6) and (7) to the dynamical universals as "principles and norms of activity which produce languages".

Note that in this whole debate, no one ever makes the hypothesis that universals might in fact reflect the structure of the phenomenal world and its synthetic a priori forms. Yet, it is precisely such a point of view that we will develop in this work. For us, it is no longer an issue of formal, definitional, intrinsic analytic (strong) universals, but rather of weak extrinsic universals, grounded by way of perception and action both in the structures of cognition and in the phenomenology of the natural world. As such, they are synthetic a priori universals constraining the semantics of natural languages.

9. Morphological schemata and *proto-linguistics*

In the following chapters, we present in technical detail several morphodynamical schemata associated with the three main classes of entities represented in natural languages (see Section 3.2 on Langacker): *objects*, *relations* and *processes*, focusing above all in each case on the perceptual basis. That is, the schemata that we present—even if they must be conceived, according to what we have seen, as phenomenological and cognitive schemata universally constraining the semantics of language— should not in any event be considered linguistic models in the proper sense of the term. They are, we may say, only *proto*-linguistic schemata concerning deep structures and not surface structures.

It is essential to keep this point in mind. Indeed, our proposals pertain to a constellation of ideas where, as was strongly emphasized by Claude Vandeloise in his last works (see, e.g., [390] and [391]), grammaticized structures rooted in perception (such as prepositions) are not only of a localist nature (be they topological, geometrical, and/or dynamical), but also "functional" and "utilitarian", depending on the use of the items in particular contexts or circumstances.

In his beautiful tribute [207] to Claude Vandeloise, Ron Langacker clarified with great acuteness "the relative weight of spatial and functional factors" in cognitive grammar. It is a general feature of *conceptual archetypes* to be highly polysemic.

> Conceptual archetypes are experientially grounded concepts so frequent and fundamental in our everyday life that we tend to invoke them as anchors in constructing our mental world with all its richness and levels of abstraction. (...) Archetypes are basic conceptual units readily grasped in gestalt-like fashion. (...) Conceptual archetypes represent salient, essentially universal aspects of every day experience, as determined by the interplay of biological and environmental factors. Their emergence is a natural consequence of how we interact with the physical and social world, having evolved to cope with it successfully.

Even if conceptual archetypes are difficult to describe due to their evolutionary depth, they share a fundamental linguistic role: they are lexicalized, widely used metaphorically in their abstract sense, and constitute one of the main sources of grammaticizing.

> A key to understanding grammar lies in the recognition that particular conceptual archetypes—especially salient due to their prevalence in moment-to-moment experience—provide the prototypical values of basic categories and canonical constructions.

So we can make with Bernard Victorri [394] the hypothesis that, at the evolutionary level, *proto*-linguistic structures of a localist nature enriched by

> dynamical, functional and intentional properties coming from our everyday life experience,

have been complexified not only by syntactic recursivity but also by

> many grammatical specificities, such as the systems of modal or aspectual markers, or lexical properties such as polysemy.

In this book, we will consider only basic proto-linguistic structures anchored in perception, independently of their other complex, abstract, metaphoric and polysemic uses. This restriction will be compensated for by elaborating rigorous and operational mathematical models.

CHAPTER 2

Things

1. Introduction

In this first technical chapter we address the first class of entities from the Things/Relations/Processes trilogy of cognitive grammars. As we have seen in the previous chapter, every form (and, in particular, every shape) results from the emergence of qualitative discontinuities on a substrate space. Rather evident for physical and visual forms, this rule is also a universal semiolinguistic principle. As it was strongly emphasized by Louis Hjelmslev, and restated later by Algirdas J. Greimas, every form results from the articulation of an amorphous "materia", that is, from the discretization of an underlying continuum by means of discontinuities.

We will show that even at the most primitive and elementary levels of the phenomenology of visual perception, the constitution of boundaries that delimit homogeneous regions—what is called the *segmentation problem*—already raises considerable difficulties when attempting to elaborate segmentation models conform to the advances of visual neuroscience and computational vision. Ronald Langacker's concept of "scanning" (Chapter 1, Section 3.2.2) pointed towards a difficult technical problem requiring sophisticated mathematical tools. We want to introduce these tools in the present chapter, as we will use them again in subsequent parts of this book.

- In Section 2 we begin with a pure phenomenological description of 2D image segmentation. We think that the best description was given by Edmund Husserl in the early twentieth century, following ideas from his master Carl Stumpf, one of the founders of the Gestalt theory.
- We then present several models of segmentation: René Thom's topological model is briefly recalled in Section 3 and an example of a physicalist implementation on the side of the natural world is given. Here, segmentation is essentially viewed as a type of information processing that is also objectively grounded in pheno-physics. In Section 4 we turn to the subjective side of the subject-world complex and concern ourselves with retino-cortical neural processing and computational models of vision. The main problem is to obtain a *geometrical* formatting of signals picking up qualitative discontinuities. The signal-to-geometry transformation process undeniably constitutes one of the deepest enigmas in

vision. At the retinal level, elements of an answer can be provided by the use of *wavelet analysis*.

- Section 6 takes a closer look at a major computational model of segmentation, the *variational model* due to David Mumford. Related or alternative works based on non-linear partial differential equations and multiscale space analysis (especially those by Jan Koenderink, Luc Florack and Jean-Michel Morel) are further described in Section 7.
- After drawing a brief link between these results and possible applications to the famously difficult and old gestaltic puzzles (Section 8), we set about to deal with the problem of *mereological constituency*, that is, the canonical decomposition of a form into its constituent parts. While several techniques can be used to achieve a mereological analysis, we focus in Section 9 on one of them, *skeletonization*, which is in our opinion particularly far-reaching and efficient, setting aside a few technical difficulties.
- Returning to natural vision, we then shift from the peripheral (retinal) level to the central (cortical) level and reformulate the mereological constituency problem in terms of the so-called "binding problem" (Section 10). How can representations that are neurally implemented in a distributed fashion nevertheless exhibit constituent structures? The most convincing solution is found in phenomena of *temporal synchronization*, and their main class of associated models are fields of coupled oscillators.
- Finally, after commenting on "filling-in" models in Section 11, we conclude this chapter with Section 12, stressing the fact that there is still obviously a wide gap between elementary segmentation processes in early vision and the perception of 3D objects in the 3D objective space. We propose a few ideas to address these higher levels of processing.

2. The eidetic kernel of the concept of form

What is a form?

2.1. *Verschmelzung* and *Sonderung*

Let us start with the phenomenologically eidetic description of what a "form" is, given by Edmund Husserl in his third *Logical Investigation* [160]. The central problem analyzed by Husserl is the dependence relation between qualitative moments (e.g., color) and spatial extension (*Ausdehnung*). At a basic underlying level there exists a functional dependence connecting the immediate moments of quality and extension. Every point x of the substrate's extension W is associated with a value $q(x)$ of the quality q. According to Husserl, qualities are abstract essences (species) and constitute categorized manifolds. He thought of the *quality* $\rightarrow$ *extension* dependence as an eidetic law binding generic *abstracta* or types:

> The dependence [*Abhängigkeit*] of the immediate moments concerns a certain relation conform to a law existing between them, a relation which is determined only by the immediately super-ordered abstracta of these moments. (p. 233)

The dependence is objectively legalized by a pure law that acts only at the level of these essences. This "ideal a priori necessity grounded in the material essences" is, according to Husserl, a typical exemple of synthetic a priori.

Now, in the two chapters of this third *Logical Investigation* Husserl presents in reality two different types of eidetic analysis: a morphological one and a formal ontological one. The formal ontological level became the basis of formal mereology, which was further developed by the Polish school of logic (Ajdukiewicz, Lesniewski, Twardowski, etc.).[1] At the morphological level, Husserl introduces a key idea: in §§8-9, he analyzes the difference between the contents that profile themselves intuitively against the background and the contents that are intuitively merged and fused together. Perception presupposes a global unity of the intuitive moments and a *phänomenale Abhebung*, that is, a "saliency" in René Thom's sense. It is such a saliency that is expressed by the difference between, on the one hand, contents intuitively separated from neighboring contents and, on the other hand, contents merged with neighboring contents.

The concept of "fusion" or "merging" (*Verschmelzung*) is crucial. It signifies an active process of spreading of qualities, that is, a topological transition from the local level to a global level. Its complementary concept is the "separation" or "disjonction" (*Sonderung*). Therefore, *Sonderung* is an obstacle to *Verschmelzung*: it generates boundaries that delimit parts. At the intuitive synthetic a priori level, the mereological "whole vs. part" opposition, which belongs to formal ontology, is grounded in the topological "*Verschmelzung* vs. *Sonderung*" opposition.

2.2. The fit with some current ideas

It must be pointed out that Husserl's pure phenomenological description fits very well with contemporary cognitive research. For instance, through his numerous works in computational vision, Stephen Grossberg [136] concludes that there exists two fundamental systems in visual perception:

1. The *Boundary Contour System* (BCS), which controls the emergence of the segmentation of the visual scene. It detects, sharpens, enhances and completes edges, especially boundaries, by means of a "spatially long-range cooperative process". It groups textures and generates a boundary web of form-sensitive compartments that simultaneously encode smooth

1 For a discussion of formal ontology and mereology, see Smith [348] and Poli, Simons [311].

shading, discrete boundary and textural elements (p. 40). The boundaries organize the image geometrically (morphologically). They are virtual, since "contrast-sensitivity does not imply visibility" (p. 32).

2. The *Featural Contour System* (FCS), which performs featural filling-in (lateral spreading), i.e., diffusion. It stabilizes qualities such as color or brightness. The diffusion processes are triggered and limited by the virtual boundaries provided by the BCS. "These filling-in processes lead to visible percepts of color-and-form-in-depth at the final stage of the FCS (p. 5)".

Therefore, according to Grossberg,

> Boundary Contours activate a boundary completion process that synthesizes the boundaries that define perceptual domains. Feature Contours activate a diffusion filling-in process that spreads featural qualities, such as brightness or color, across these perceptual domains. (p. 35).

These ideas are conceptually equivalent to Edmund Husserl's descriptions but complete them with neurophysiological expertise.

2.3. The origin of the *Verschmelzung* concept

In reality, the concept of *Verschmelzung* does not come directly from Stumpf but from the German psychologist Johann Friedrich Herbart (1776-1841), who has developed a continuous theory of mental representations. Essentially in the same vein as Peirce after him, Herbart was convinced that mental contents are vague and can vary continuously. For him, a "serial form" (*Reihenform*) was a class of mental representations that undergo a graded fusion (*abgestufte Verschmelzung*) gluing them together via continuous transitions. He coined the neologism of *synechology* for his metaphysics (Peirce's neologism of *synechism* is clearly parallel).

It is not sufficiently known that Herbart's point of view was one of the main interests of Bernhard Riemann when he was elaborating his key concept of a Riemannian manifold. Even if Riemann did not agree with Herbart's metaphysics, he strongly claimed that he was a "Herbartian in psychology and epistemology". Erhard Scholtz [337] has shown that in Riemann's celebrated work *Über die Hypothesen, welche der Geometrie zu Grunde liegen* [325] the role of the differentiable manifold underlying a Riemannian manifold

> is taken in a vague sense by a Herbartian-type of "serial form", backed by mathematical intuition.

2.4. Qualitative discontinuities and segmentation

Still in §8 of the third *Logical Investigation*, Husserl claims that

> *Sonderung* is based on discontinuity (Sonderung *beruht auf Diskontinuität*).

Whereas *Verschmelzung* corresponds to a continuous (*stetig*) spreading of qualities in an undifferentiated unity (p. 244), *Sonderung* thus corresponds to qualitative discontinuities in the way extension is covered (*Deckungszusammenhang*) by qualities.

These qualitative discontinuities are salient only if

1. they are contiguously unfolded against the background of moments that vary continuously, namely the spatial and temporal moments;
2. they present a sufficient gap (threshold of discrimation).

Husserl's morphological description is precise and remarkable:

> It is from a spatial or temporal limit (*Raum- oder Zeitgrenze*) that one springs from a visual quality to another. In the continuous transition from a spatial part to another, one does not progress also continuously in the covering quality (*in der überdeckenden Qualität*): in some place of the space, the adjacent neighboring qualities present a finite (and not too small) gap. (p. 246)

This pure eidetic Husserlian description of the dependence relation *quality* → *extension* therefore yields the following homologations:

Totality (Whole)	**Parts**
Verschmelzung	*Sonderung*
Spreading activation (featural filling-in)	Boundaries
Continuity	Discontinuity

3. Objective correlates of phenomenological descriptions

We now want to show that Husserl's pure phenomenological eidetic description can be adequately modeled within the framework of remarkably convergent scientific explanations. We begin with the "realist" pheno-physical models explaining the emergence of qualitative discontinuities in the natural world.

3.1. Topological-geometrical explanation

Let us first recall the basic model of René Thom already sketched in Chapter 1, Section 3.2.2. Throughout this section "internal" will refer to "internal to the material system under consideration".

Phenomenologically, a material system S occupying a spatial domain W manifests its form through a set of observable and measurable qualities $\{q_1(w), ..., q_n(w)\}$ that are characteristic of its actual internal state A_w at every point $w \in W$, and take their values in typical quality spaces $Q_1, \ldots, Q_n$ (color, texture, etc.). When the spatial control w varies smoothly in W, A_w varies smoothly, too. If A_w subsists as the actual state, then the q_i functions also vary smoothly. But if the actual state A_w bifurcates towards another actual state B_w when w crosses some *critical value*, then some of the q_i functions must present a discontinuity. Thom called *regular* the points $w \in W$ where locally all the qualities q_i vary smoothly and *singular* the points $w \in W$ where locally some of the q_i present a qualitative discontinuity. The set R of regular points is by definition an open set of W and its complementary set $K = W - R$, the set of singular points, is therefore a closed set. By definition, K is the *morphology* produced by the internal dynamical behavior of the system S. It decomposes W into homogeneous regions.

The singular points $w \in K$ are critical values of the control parameters and, in the physical cases, the system S presents in them a critical behavior. Thom was one of the first scientists to stress the fact that qualitative discontinuities are phenomenologically dominant, that every qualitative discontinuity is a type of critical phenomenon, and that a general mathematical theory of morphologies manifested by general systems had to be an enlarged general theory of critical phenomena and symmetry breaking.

Now, it is clear that Thom's description is in fact an exact topological equivalent of Husserl's description: regular points literally correspond to *Verschmelzung*, and singular ones to *Sonderung*. In short, the geometrical structure (W, K) geometrizes a phenomenological segmentation:

- Regular points $\equiv$ *Verschmelzung*,
- Singular points $\equiv$ *Sonderung*,
- Saliency $\equiv$ *phänomenale Abhebung*.

3.2. Morphodynamical explanation and pheno-physics

Let S be a material substrate. The problem is to explain its obervable morphology, that is, to generate a segmentation (W, K) from its underlying physics. First, we choose some level, or *scale*, of observation. The main morphodynamical model is then the following: in S, an internal dynamical mechanism X defines the internal states. There exists a phase space M of S, which is a differentiable manifold and whose points x represent the instantaneous transient states of S. M is called the "internal space" of S, and X is a flow on M, i.e., a system of ordinary differential equations $\dot{x} = X(x)$ with good regularity properties. The smooth vector field X is called the "internal dynamics" of S. The internal states of S are then modeled by the (asymptotically stable) *attractors* of X.

The intuitive (informal) definition of an attractor is the following: a subset $A \subset M$ of the internal space M is an attractor of the flow X if it is topologically closed, X-invariant (i.e., the trajectories of the points of A stay in A), minimal for these properties, and if it attracts asymptotically every point x belonging to one of its neighborhoods U. Additionally, A is asymptotically stable if it also confines the trajectories of the points belonging to a sufficiently close neighborhood. If A is an attractor of X, its basin of attraction $B(A)$ is the set of points $x \in M$ attracted by A.

Now, of course, as only one internal state A can be the actual state of S, there exists necessarily some criterion I (e.g., a physical principle of minimization of energy) that selects A from among the possible internal states of S.[2]

The system S is also controlled by control parameters that vary in the extension W of S. W is called the "external space" of S. The internal dynamics X is therefore a dynamics X_w that is parametrized by the external points $w \in W$ and varies smoothly relative to them.

With such a morphodynamical model we can easily explain the topological description *physically*. The segmentation K is generated by the critical values of the control W, i.e., by the values for which the internal dynamics X_w displays a catastrophe or a bifurcation, one internal state being replaced by another internal state.

We see that we can go from deep physics to surface phenomenology through the following steps:

1. the choice of a scale that allows us to use a smooth approximation for formatting the physical phenomena;
2. the use of a qualitative approach to the dynamical models of the internal physics;
3. the bracketing of the fine-grained structure of internal physics and the coding of the internal attractors by observable external qualities.

These three steps lead to the morphological (topological-geometrical) description. They are followed by a fourth step:

4. the phenomenological eidetic description of the morphological description.

3.3. Example: fields of oscillators

Many examples of such models involving the emergence of spatial and temporal patterns out of material substrates are now widely known. Think for instance about chemical waves or dissipative structures, such as the Rayleigh-Bénard convection cells. Pierre Coullet [67] produced beautiful collections of such examples in his study of morphological properties emerging from continuous fields of coupled oscillators. At every point of W there exists an oscillating internal dynamics coupled with the dynamics of its immediate neighbors.

In a first approximation (mean-field approximation) the system can be locally described using an "order parameter", the average phase Z, which depends on the spatial position. The typical differential equations are

$$(1) \qquad \frac{\partial Z}{\partial t} = \lambda Z - \mu |Z|^2 Z + \gamma_n \overline{Z}^{n-1} + \nu \Delta Z$$

2 For details, see Petitot-Smith [307].

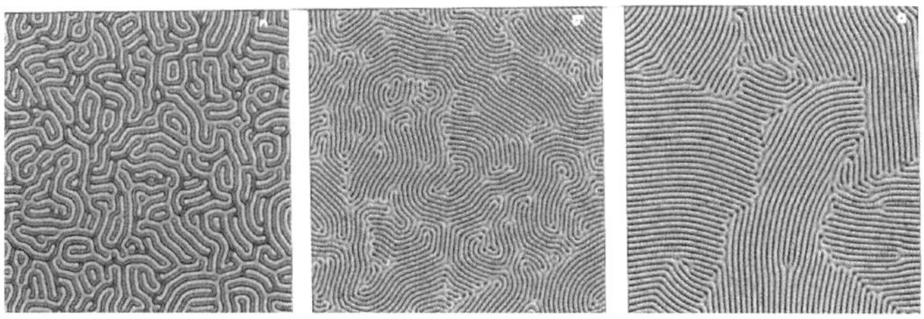

FIGURE 1. Three examples of patterns emerging in a continuous network of oscillators: stripes with defects and domains of stripes (from Coullet, Emilsson [67]).

FIGURE 2. Two more examples of emerging patterns in continuous fields of oscillators: rotating waves and turbulence (from Coullet, Emilsson [67]).

where λ, μ and ν are complex parameters, γ_n is real, $|Z|$ and $\overline{Z}$ are respectively the module and the complex conjugate of Z, and Δ is the Laplacian operator of spatial dispersion.

Pierre Coullet [67] showed that such partial differential equations can model a rich variety of spatio-temporal patterns similar to those observed in various macroscopic domains (see Figures 1 and 2): stripes with defects, domains of stripes, rotating waves, turbulence, etc.

We will see in Section 10.3 that dynamical physical models of this kind have recently been applied, for fundamental theoretical reasons, to cognitive neuroscience.

4. A first cognitive explanation: Marr and wavelet analysis

From an external physical explanation, we now shift to an internal cognitive explanation—i.e., "internal to the subject". At its most peripheral level (the retina), visual processing is a kind of signal analysis much more complex than the well-known Fourier transform.

David Marr [224] introduced the hypothesis that the main function of the retina's ganglion cells is to extract qualitative discontinuities encoded in the signal. He called these discontinuities "zero-crossings" and claimed that all higher levels of visual processing are grounded in this early stage of morphological organization of the image, the 2D *primal sketch*. In fact, it has been shown that the convolution of the signal by the receptive profiles of the ganglion cells is truly a *wavelet analysis*, that is, a spatially localized and multiscale Fourier analysis. Now, wavelet analysis is also the best known method for extracting discontinuities. Let us briefly discuss this point.

The retina performs an important compression of the visual information delivered by the photoreceptors. It essentially consists of three layers:

(i) the photoreceptors,
(ii) the bipolar cells, and
(iii) the ganglion cells.

Their are also two plexiform layers: the horizontal cells and the amacrine cells. The photoreceptors carry out the *transduction*, or neural coding, of the physical properties of the optic signal. Bipolar cells, together with the horizontal cells (on the receptors' side) and the amacrine cells (on the ganglion cells' side), are intermediate structures. Ganglion cells (GCs), whose axons constitute the fibers of the optic nerve, are characterized by the structure of their receptive field (RF), defined as the set of receptors to which a cell is connected.

In first approximation, GCs can be considered as linear *filters* which convolute the original signal with their RF profiles. Now, it is a fundamental fact of neurophysiology that RFs have a center/periphery antagonist structure. If a ray of light hitting the center of a cell's RF excites (inhibits) it, then a ray of light hitting the periphery will have the opposite effect of inhibiting (exciting) the cell. Cells of this type are called ON-center (and OFF-surround); cells of the inverse type are called OFF-center (and ON-surround). After being processed by the RFs of the ganglion cells, the signal is retinotopically transmitted, via the thalamic relay of the lateral geniculate body, to the hypercolumns of the primary visual cortex (striate area).[3]

What type of information processing is performed by the layers of retinal ganglion cells? Marr introduced the key idea of a "zero-crossing criterion".[4] Mathematically, the center/periphery profiles of the RFs approximate *Laplacians of Gaussians*, ΔG (Figure 3).

Let $f(x)$ be a smooth function on $\mathbb{R}$ presenting a "discontinuity" at x_0, i.e., a sharp variation. At x_0, the first derivative $f'(x)$ shows a peak (it is a Dirac distribution δ, if x_0 is a true discontinuity) while the second derivative $f''(x)$ has two peaks (one positive, the other negative) surrounding a zero-crossing (Figure 4).

Let $I(x, y)$ be the input pattern (the pixelized retinal image). The convolution $G * I$ of I by a Gaussian G

$$G(x, y) = \frac{1}{2\pi\sigma^2} e^{-\frac{r^2}{2\sigma^2}} \tag{2}$$

3 For details see, e.g., Buser-Imbert [48].

4 For a technical discussion of Marr's theory, see, e.g., Haralick [143] and Richter-Ullman [324]. For an epistemological discussion, see, e.g., Kitcher [180].

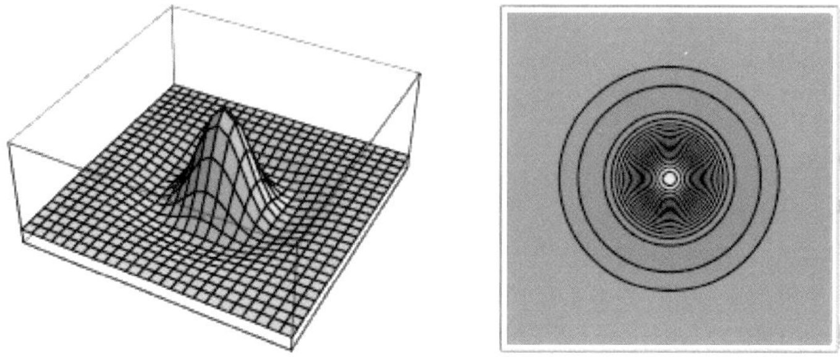

FIGURE 3. A Laplacian of Gaussian. Its level curves are represented on the right.

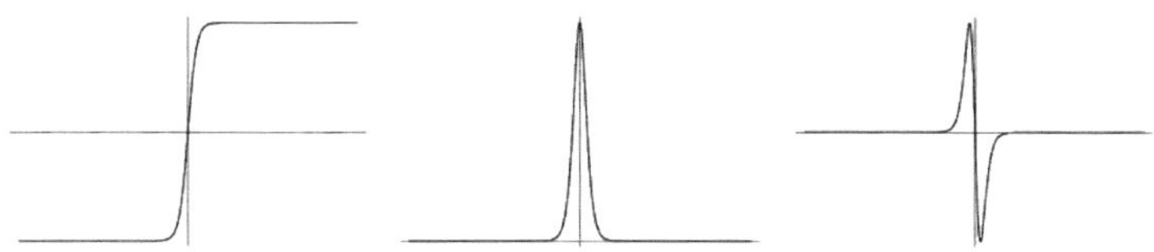

FIGURE 4. If a function presents a sharp variation at 0, its first derivative presents a peak at 0 and its second derivative two peaks of opposite signs surrounding a zero-crossing.

(where $r^2 = x^2 + y^2$ is the distance to the center of G, and σ its width) corresponds to the smoothing (fusion, merging, *Verschmelzung*) of I at a certain scale. Taking the Laplacian of this convolution, $\Delta(G * I)$, corresponds to the second derivative. It locally extracts the zero-crossings of the G-smoothed signal $G*I$. Now, it is a well-known mathematical result that $\Delta(G*I) = \Delta G*I$, where

$$\Delta G(x,y) = -\frac{1}{\pi\sigma^4}\left(1-\frac{r^2}{2\sigma^2}\right)e^{-\frac{r^2}{2\sigma^2}}\,. \tag{3}$$

Therefore, the two operations of smoothing and zero-crossing extraction at the corresponding scale, that is edge detection, can be performed in one shot: the convolution by a RF profile of the form ΔG (Figure 5).

With its many layers of ganglion cells operating at different scales, the retina can thus perform a local and multiscale extraction of qualitative discontinuities. In fact, Marr's algorithm was one of the first known examples of what is now called *wavelet analysis*.[5] As we already mentioned, wavelet analysis is a type of Fourier analysis that is spatially localized, multiscale, and capable of extracting the singularities encoded in a signal. Let us very briefly and roughly explain the main idea in the 1D case.

In general one considers that the space of (1D) signals can be modeled by the Hilbert space $L^2(\mathbb{R})$ of square-integrable functions on $\mathbb{R}$ (this condition

5 For an introduction to wavelet analysis, see Meyer [232] and Mallat [219].

FIGURE 5. Edge detection. Left: the initial image. Right: the extraction of contours using zero-crossings.

expresses the fact that the energy of the signal is finite). Fourier analysis is a frequency analysis of f which provides an orthogonal decomposition of every function $f \in L^2(\mathbb{R})$ along the orthonormal basis of trigonometric functions $e^{-i\omega x}$ and the Fourier transform (FT) of $f(x)$, denoted by $\widehat{f}(\omega)$, is given by the formula:

$$\widehat{f}(\omega) = \frac{1}{\sqrt{2\pi}} \int_{\mathbb{R}} f(x)\, e^{-i\omega x}\, dx\ . \tag{4}$$

It can be shown that $\widehat{\widehat{f}} = f$ and that the norms $\|f\|$ and $\|\widehat{f}\|$ are equal (i.e., FT is an isometry).

The problem is that the information provided by $\widehat{f}$ is spread out and delocalized because of the infinite range of the plane waves $e^{-i\omega x}$. In order to obtain an information processing of a more local type, the British-Hungarian physicist Dennis Gabor (the inventor of holography) introduced in the late 1940's the idea of the *window Fourier transform* (WFT):

$$Gf(\omega, u) = \frac{1}{\sqrt{2\pi}} \int_{\mathbb{R}} f(x)\, g(x-u)\, e^{-i\omega x}\, dx\ . \tag{5}$$

$Gf(\omega, u)$ is localized through a spatial "window" $g(x)$ translated along the x axis. The WFT depends not only on the frequency ω but also on the position shift u. It generalizes the classical FT. Similarly, the inverse WFT transform is given by:

$$f(x) = \frac{1}{2\pi} \int_{\mathbb{R}^2} Gf(\omega, u)\, g(u-x)\, e^{i\omega u}\, du\, d\omega\ . \tag{6}$$

It can be shown that the WFT transform G is an isometry between $L^2(\mathbb{R})$ (with coordinate x) and $L^2(\mathbb{R}^2)$ (with coordinates (ω, u)), that is, $\|f\| = \|Gf\|$.

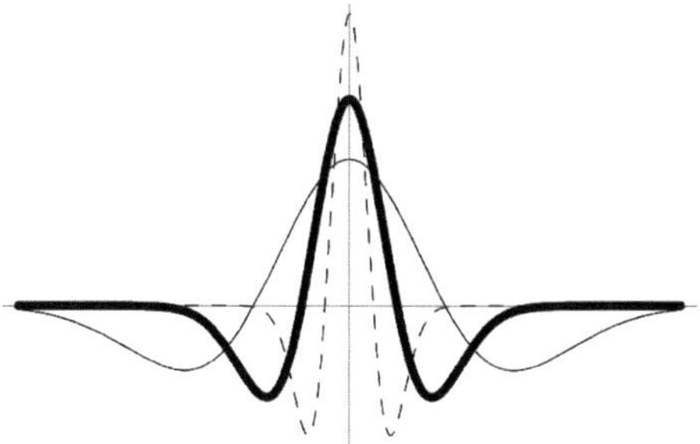

FIGURE 6. A wavelet (thick curve) and two of its rescaled transforms (thin and dashed curves).

The WFT is in general highly redundant and it is therefore possible to restrict it to a discrete sample of u and ω.

The problem with Gabor's WFT is that it operates at only one level of resolution. Let σ_u be the standard deviation of the window $g(x)$:

$$\sigma_u^2 = \int_{\mathbb{R}} x^2 \, |g(x)|^2 \, dx \tag{7}$$

and σ_ω the standard deviation of $\widehat{g}(\omega)$:

$$\sigma_\omega^2 = \int_{\mathbb{R}} \omega^2 \, |\widehat{g}(\omega)|^2 \, d\omega \; . \tag{8}$$

One gets a description of f and $\widehat{f}$ in resolution cells

$$[u_0 - \sigma_u, u_0 + \sigma_u] \times [\omega_0 - \sigma_\omega, \omega_0 + \sigma_\omega] \tag{9}$$

but inside these cells, the information is still delocalized and spread out. Consequently, it is impossible to localize boundaries at a scale smaller than σ_u. If the signal is multiscaled (e.g., fractal) this is a drastic limit.

The idea of the *wavelet transform* (WT) is to find decompositions of $L^2(\mathbb{R})$ using a single function $\psi(x)$ (the "mother" of the wavelets), its translated transforms $\psi(x-u)$, and its rescaled transforms $\psi_s(x) = \sqrt{s}\psi(sx)$ (or, alternatively, $\psi_s(x) = \frac{1}{s}\psi(\frac{x}{s})$) (Figure 6).

Its formula is

$$\begin{aligned} Wf(s,u) &= \int_{\mathbb{R}} f(x)\, \psi_s(x-u)\, dx \\ &= f * \tilde{\psi}_s(u) \end{aligned} \tag{10}$$

with $\tilde{\psi}(x) = \psi(-x)$. It is well defined if the following admissibility condition (C) on the FT $\widehat{\psi}$ of the "mother" wavelet ψ is satisfied:

$$(C) : \widehat{\psi}(0) = 0 \quad \text{and} \quad C_\psi = \int_0^\infty \frac{|\widehat{\psi}(\omega)|^2}{\omega} \, d\omega < \infty \; . \tag{11}$$

(C) says that $\widehat{\psi}$ must be sufficiently flat near 0.

For WT the resolution cells are then:

$$[u_0 - \frac{\sigma_u}{s}, u_0 + \frac{\sigma_u}{s}] \times [\omega_0 - s\sigma_\omega, \omega_0 + s\sigma_\omega] \,. \tag{12}$$

The main result of the theory is that a convenient ψ *does* exist and a typical example of this is precisely Marr's wavelet ΔG. A theorem by Jean Morlet and Alex Grossmann [137] states that, up to some coefficient, the WT transform W is an isometry from $L^2(\mathbb{R})_x$ to $L^2(\mathbb{R}^* \times \mathbb{R})_{s,u}$. The formula for the inverse transform is

$$f(x) = \frac{1}{C_\psi} \int_{\mathbb{R}} \int_{\mathbb{R}^*} Wf(s,u)\, \psi_s(x-u)\, ds\, du \,. \tag{13}$$

The amplitude $|Wf(s,u)|$ of the WT is an indicator of the *singularities* encoded in the signal. More precisely, the Lipschitz order of f at x can be deduced from the asymptotic decreasing of $|Wf(s,u)|$ in the neighborhood of x when the scale tends towards 0. As was emphasized by Stéphane Mallat [218]:

> The ability of the WT to characterize the type of local singularities is a major motivation for its application to detect the signal's sharper variations.

The WT is generally highly redundant (at zero redundancy, wavelets are termed "orthogonal"). Using the following reproducing kernel:

$$K(s,s',u,u') = \frac{1}{C_\psi} \int_{\mathbb{R}} \psi_s(x-u)\, \psi_{s'}(x-u')\, dx \tag{14}$$

the redundancy can be expressed as:

$$Wf(s',u') = \int_{\mathbb{R}} \int_{\mathbb{R}^*} Wf(s,u)\, K(s,s',u,u')\, ds\, du \,. \tag{15}$$

As for Marr's wavelet, $\psi = \Delta G$, redundancy corresponds to the heat equation:

$$\frac{\partial W_s f}{\partial t} = \Delta W_s f \ \ (\text{with } t = s^2) \,. \tag{16}$$

Due to the redundancy of the WT, it is possible to discretize it by sampling variables u and ω. For instance, discretizing the scale s yields the following discrete WT:

$$\begin{aligned} W_{2^j} f(x) &= f * \tilde{\psi}_{2^j}(x) \\ &= \langle f(u) \,|\, \tilde{\psi}_{2^j}(u-x) \rangle \,. \end{aligned} \tag{17}$$

With the condition:

$$\sum_{j \in \mathbb{Z}} |\widehat{\psi}(2^j \omega)|^2 = 1 \tag{18}$$

one obtains:

$$f = \sum_{j \in \mathbb{Z}} W_{2^j} f * \tilde{\psi}_{2^j}(x) \tag{19}$$

$$\|f\|^2 = \sum_{j \in \mathbb{Z}} \|W_{2^j} f\|^2 . \tag{20}$$

In this case, one speaks of "dyadic" wavelets.

Thanks to transformation tools of this kind, it becomes possible to compress an image in an intrinsic way, that is, according to its specific geometrical structure. Most importantly, it is also possible to reconstruct the whole image from the detection of its edges (Marr's conjecture). The fidelity of the reconstruction can be very high because it relies upon the morphological structure of the image. Only the finest details, such as textures, are smoothed in the process.

More precisely, if we use as wavelets the Laplacian ΔG of a Gaussian G (Marr) or the Laplacian $\Delta\vartheta$ of a compact-support smooth function ϑ (Mallat), we can extract the qualitative discontinuities using the zero-crossing criterion. If we use as wavelet the gradients of these functions, ∇G or $\nabla\vartheta$, we can extract the qualitative discontinuities using the maxima of the wavelet transforms.

The discontinuities that are stable under large scale variations can be interpreted as *objective* and, in particular, as boundaries of external objects.

The main result we want to stress here regarding Marr's conception and wavelet analysis is that the compression of information, which is an information processing constraint, is identical to a morphological analysis, which is a geometrical objective fact. The morphological representation of the images, obtained in a bottom-up and data-driven manner by extracting qualitative discontinuities through wavelet analysis, provides the basis for more symbolic and higher-level representations. As was pointed out by Marr [224]:

> Zero-crossing provides a natural way of moving from an analogue or continuous representation like the two-dimensional image intensity values $I(x, y)$ to a discrete, symbolic representation. (p. 67)

Wavelet analysis can be refined—in particular for the application to data compression problems—by means of wavelet packet algorithms. Many wavelets are used in parallel in order to adapt in the best way their choice to the particular structure of the signal. See, e.g., Wickerhauser [403].

In more recent works, Stéphane Mallat introduced maximal redundancy. He defines a three-parameter family of "time-frequency atoms":

$$g_\gamma(t) = g_{s,u,\xi}(t) = \frac{1}{\sqrt{s}} g\left(\frac{t-u}{s}\right) e^{i\xi t} \tag{21}$$

where u is the position, s the scale, and ξ the frequency. An adaptive algorithm is then iterated to search for the function g_γ that approximates f the best.

Starting with:

$$f = \langle f \mid g_{\gamma_0} \rangle \, g_{\gamma_0} + Rf \tag{22}$$

where the rest Rf is orthogonal to g_{γ_0}, one then iterates the formula:

$$R^n f = \langle R^n f \mid g_{\gamma_n} \rangle \, g_{\gamma_n} + R^{n+1} f \, . \tag{23}$$

5. Global contour extraction

Wavelet analysis is an excellent tool for modeling the *local* detection of pieces of edges, and understanding how signal filtering can also amount to local geometric analysis. But in order to construct global contours, the visual system has to *integrate* local contrasts into global edges and this processing is implemented in the *cortical* areas. To date, there is a lot of experimental data on the *functional architecture* of primary visual areas—i.e., on the specific organization of their neural connections—and in particular of the first area V1 (where the optic nerve projects through the thalamic lateral geniculate body). The so-called "simple" neurons of V1 selectively detect not only positions in the visual field, but also *orientations*, and are organized in "pinwheels". Moreover, there exist long range cortico-cortical connections linking neurons in different pinwheels that are selective to the *same* orientation. Due to this very special "neurogeometry" immanent to the functional architecture, local edges can be integrated into global contours. As we have extensively shown in [295], [298], [302], and [304], in order to correctly model this "neurogeometry" we are naturally led to use sophisticated mathematical tools such as *contact geometry* and *sub-Riemannian geometry*. We do not elaborate any further on this topic here, however, and instead shift to the interpretation of edge detection as a segmentation problem.

6. Variational segmentation models in low-level vision

Following our previous discussion of edge detection models, which described how qualitative 2D discontinuities can be detected at the low levels of natural vision, we move towards the *segmentation models* of computational vision.

6.1. Transforming the signal into a geometric observable

The question of "how signal processing can be at the same time a geometrical and a morphological analysis" is a central theoretical problem in low-level vision and occupies an increasingly significant position in both natural and computational vision problems. Why is that? Because, starting as early as the low processing levels (bottom-up and data-driven), the visual system imposes a *geometrical format* on the signal. This geometrical format is necessary for further processing by higher-level (cognitive, symbolic, inferential, top-down) routines. There exists a "geometrical syntax" of images that is put in place from the very beginning and extracts geometrical invariants from physical measurements of the signal, so that these invariants can be passed on to higher-level representations.

The main difficulty is that the signal is not naturally, as such, a geometrical object and therefore must be appropriately transformed to become a geometrical observable. Conversely, one can also say that geometrical differential operators are not well-suited for direct application to a physical signal and must be made more "physical".

Multiscale and multichannel (color, texture, etc.) edge detection can serve as an input for multiscale image segmentation and space-invariant algorithms ("space-invariant" meaning here that all points of the image are processed in the same way). The problem is then to segment an image $I(x_1, x_2)$ (or $I(x)$ if we interpret x as a multi-variable), defined on a domain W of $\mathbb{R}^2$, in an optimal way, that is, to partition W in maximally homogeneous domains limited by boundaries K.

Let $I(x)$ be a rough image signal. It is an unstructured hyletic datum without any geometrical structure. How can we go from this rough signal to a morphologically organized perceptual image? What is the "geometry machine" providing the morphological formatting?

A great number and variety (more than a thousand) of segmentation algorithms have been worked out by mathematicians and engineers. All these algorithms share the goal of merging local data into homogeneous regular regions and separate the regions through regular, crisp edges. The main problem is that 2D regions and 1D edges are geometrical entities of different dimensions that are in competition and interact in a subtle way. Fortunately, behind this proliferation of models there seems to exist a methodological unity, as pointed out by Jean-Michel Morel [239]:

> most segmentation algorithms try to minimize (...) one and the same *segmentation energy*.

This concept of "segmentation energy" allows to compare one segmentation with another and measure how well they approximate the original rough signal I. The most popular segmentation energy is given by the so-called Mumford-Shah model.

6.2. The Mumford-Shah model

In a seminal 1994 paper, *Bayesian rationale for the variational formulation*, David Mumford [242] writes:[6]

> one of the primary goals of low-level vision is to segment the domain W of an image I into parts W_i on which distinct surface patches, belonging to distinct objects in the scene, are visible.

6 David Mumford is a Fields Medal in Algebraic Geometry who has become a specialist of vision.

The mathematical problem is to use low-level cues

> for *splitting* and *merging* different parts of the domain W.

Bayesian models contain two parts: a *prior* model and a *data* model. Although references to phenomenology are generally not found in computational vision, we can say that the prior model takes as an a priori the Husserlian *Verschmelzung*/*Sonderung* complementarity discussed previously (Section 2.1). It imposes the constraint of approximating the signal I by piecewise smooth functions u on $W-K$ which are discontinuous along a set of piecewise regular edges K.

Now, we rise from an eidetic description to true mathematical modeling by introducing a method for selecting the best possible approximation of I among all the possible (u, K) models. To this aim, Mumford and Shah use a functional $E(u, K)$ that provides a comparison measure between two segmentations $E(u_1, K_1)$ and $E(u_2, K_2)$ and contains three terms:

1. one term measuring the variation of u on the connected components of $W - K$ and controlling their smoothness;
2. one term controlling the quality of the approximation of I by u;
3. one term controlling the length, smoothness, parsimony and location of the boundaries, and penalizing spurious situations of "oversegmentation".

The Mumford-Shah energy functional writes (see [243]):

$$E(u, K) = \int_{W-K} \|\nabla u\|^2 \, dx + \int_W (u - I)^2 \, dx + \int_K d\sigma \, . \tag{24}$$

Minimizing E consists in finding a compromize between three properties embodied by these three terms:

1. the *homogeneity* of the connected components of $W - K$: if u was constant, then ∇u would vanish, therefore, minimizing the first term tends to bring u as close as possible to a constant function;
2. the *approximation* of I by u: if u was exactly I, the second term would vanish, therefore, minimizing this term tends to bring u as close as possible to I;
3. the *regularity* and parsimony of the boundaries: the third term measures the total length of K, therefore, minimizing it tends to avoid oversegmentation.

The model can be made *multi-scale* by introducing coefficients λ and μ :

$$E(u, K) = \int_{W-K} |\nabla u|^2 \, dx + \lambda \int_W (u - I)^2 + \mu \int_K d\sigma \, . \tag{25}$$

If μ is small, one gets a "fine-grained" segmentation, if μ is large, one gets a "coarse-grained" segmentation. The sensibility to contrast is $\left(4\lambda^2\mu\right)^{\frac{1}{4}}$, the scale

$\lambda^{-\frac{1}{2}}$, the thresholds for ramp effects (the segmentation of a regular increase of I) $\left(\frac{\lambda^2\mu}{4}\right)^{\frac{1}{4}}$, and the resistance to noise $\lambda\mu$.

Note that variational models of this kind based on an energy functional can also be interpreted in a probalistic framework. Using the equivalence $E(u,K) = -\log(p(u,K))$, where p is a probability function defined on the space of all possible segmentations (u,K), the goal is then to maximize the likelihood $p(u,K)$ of a configuration.

The above algorithm proposes an optimal way to merge neighboring pixels into homogeneous domains separated by qualitative discontinuities, thereby providing a variational approach to the *Verschmelzung/ Sonderung* complementarity. It transforms the segmentation problem into a particular case of physical "free boundary problem". This problem is extremely difficult to solve in its generality and no complete solution was found yet. A partial solution exists, however, under the simplifying assumption that the approximant u is locally constant, that is, $\nabla u = 0$. In this case, the first term vanishes on $W - K$ and the solution is given by a theorem of Mumford.

> **Mumford's theorem**: *The minimum of E is attained for some boundary set K containing piecewise C^1 curves whose curvature is bounded by $8\,\mathrm{osc}(I)^2$ and singular points are 120°-angle triple points or 90°-angle boundary points on ∂W.*

Here $\mathrm{osc}(I)$ is the amplitude oscillation of I, defined as $\mathrm{osc}(I) = \max(I) - \min(I)$.

It should be noted that one drawback of the Mumford-Shah model is precisely to possess only 120°-angle triple points, while it is a well-known gestaltic fact that triple points relevant to perception are generally T-shaped junctions that provide fundamental cues for the detection of occluded contours and depth information. We return to this point later (Section 7.4).

Many beautiful works have been dedicated to the Mumford-Shah model by the Italian school (Ennio De Giorgi, Luigi Ambrosio, Gianni Dal Maso, Sergio Solimini, Antonio Leaci, Massimo Gobbino, Franco Tomarelli, Alessandro Sarti, Giovanna Citti, etc.), and in France by Jean-Michel Morel, Alexis Bonnet and Guy David.[7] They try to prove *Mumford's conjecture*, which says that in the general case where u is no longer locally constant, Mumford's theorem remains valid but with a supplementary type of "end point" singularities (end points of branches of K also called "cracktips") whose normal form is given in polar coordinates by the formula :

$$u(r,\theta) = \left(\frac{2r}{\pi}\right)^{\frac{1}{2}} \sin\frac{\theta}{2} + C \tag{26}$$

for $-\pi < \theta < \pi$ (see Figure 7).[8]

7 See David, Semmes [80] and Bonnet, David [40]. For a synthesis, see [239].

8 C is a constant.

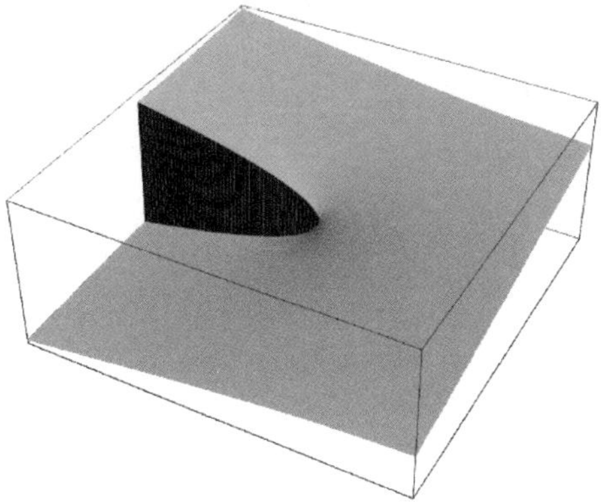

FIGURE 7. The structure of the function u near a crackpit point.

6.3. The link with diffusion equations

There is a fundamental link between variational models and *diffusion equations*. This is due to the key fact that the natural gradient-descent method associated with the minimal variation of the energy function:

$$E(u) = \frac{1}{2} \int_W \|\nabla u\|^2 \, dx \tag{27}$$

(the first term of the Mumford-Shah energy) is actually given by the heat equation:

$$\frac{\partial u}{\partial t} = -\frac{\delta E}{\delta u} \tag{28}$$

$$= \Delta u \,. \tag{29}$$

Indeed, let us compute the functional derivative $\nabla E = \frac{\delta E}{\delta u}$ defined by

$$E(u+g) = E(u) + \int_W \frac{\delta E}{\delta u}(u(x))\, g(x) dx \,. \tag{30}$$

For a smooth test function φ with compact support $C \subset W$, we get

$$\frac{E(u+t\varphi) - E(u)}{t} = \frac{1}{2} \int_W \frac{\|\nabla(u+t\varphi)\|^2 - \|\nabla u\|^2}{t} \, dx \,. \tag{31}$$

But at first order we have

$$\|\nabla(u+t\varphi)\|^2 - \|\nabla u\|^2 \approx 2t \nabla u \bullet \nabla \varphi \tag{32}$$

(where $X \bullet Y$ is the scalar product of vectors) and therefore

$$\frac{E(u+t\varphi) - E(u)}{t} = \int_W \nabla u \bullet \nabla \varphi \, dx \,. \tag{33}$$

Let us consider the vector field $\varphi \nabla u$. Applying Stokes' theorem we get:

$$\int_{\partial W} \varphi \nabla u \, ds = \int_W \operatorname{div}(\varphi \nabla u) \; dx \; . \tag{34}$$

But $\operatorname{div}(\varphi \nabla u) = \nabla u \bullet \nabla \varphi + \varphi \operatorname{div}(\nabla u) = \nabla u \bullet \nabla \varphi + \varphi \Delta u$. As $\varphi = 0$ on the boundary ∂W (since the support C of φ is included in W),

$$0 = \int_W \nabla u \bullet \nabla \varphi \; dx + \int_W \varphi \Delta u \; dx \tag{35}$$

and therefore

$$E(u + t\varphi) - E(u) = -\int_W t\varphi \Delta u \; dx = \int_W \frac{\delta E}{\delta u}(u(x)) \, t \, \varphi(x) \, dx \; . \tag{36}$$

As φ can be any test function, we must have $\frac{\delta E}{\delta u} = -\Delta u$.

Note that, when applied to vision, the variable t is not a time parameter but a *scale* parameter. Diffusion is a multi-scale smoothing process that takes place in a *scale-space* and is the physical process associated with the phenomenological concept of *Verschmelzung.*

7. Scale-space analysis

7.1. Multiscale differential geometry

We saw earlier that differential geometry is not well suited for signal analysis since the application of differential operators to signals is an *ill-posed* problem. One of the great ideas that emerged during the 1980's was that making differential geometry operational for physical signals could be possible at the cost of introducing a new degree of freedom, namely, the *scale* parameter.

To better grasp the meaning and impact of this notion, let us go again over the results of David Marr (previously discussed in Section 4). One way to look back at these results is to reinterpret them by saying that Marr actually introduced a multiscale version of a differential operator, the Laplacian operator, and doing so, opened the way to a *multiscale* differential geometry, which is well adapted to the "physicality" of the signal.

From this perspective—especially well developed by Jean-Luc Florack [109], after Jan Koenderink—the signal s is identified with a *distribution* T_s. We recall that distributions are generalized functions introduced by Laurent Schwartz. The most famous one is the Dirac peak δ, a "function" on $\mathbb{R}$ which is everywhere null except at $x = 0$ and whose integral is nevertheless 1. δ can be described as the limit of a family of Gaussian functions, whose width σ tends towards zero while their height grows to $+\infty$ under a constant integral surface 1. A much better way to conceptually characterize δ is to see how it operates on functions. Under good conditions of regularity, any function φ on $\mathbb{R}$ verifies:

$$\int \delta(x)\, \varphi(x)\, dx = \varphi(0)\,. \tag{37}$$

Thus, δ associates to functions their value in 0. As a functional defined on functions, the δ operator is obviously linear and it is this simple fact that Schwartz so powerfully exploited to create the generic concept of distributions: distributions are continuous linear functionals T on a space of test functions φ (C^∞ functions with compact support, or rapidly decreasing for "temperate" distributions). We denote by $T(\varphi)$—or, equivalently, $\langle T \mid \varphi \rangle$ to make a link with Dirac's brackets—the action of T on φ. It generalizes the integral:

$$T_f(\varphi) = \langle T_f \mid \varphi \rangle = \int f(x)\, \varphi(x)\, dx \tag{38}$$

corresponding to the case of a function f.

Generalizing to $\mathbb{R}^n$, with $x = (x_1, ..., x_n)$, it is possible to show that the natural formulas for derivatives of distributions are:

$$\langle D^p T \mid \varphi \rangle = (-1)^{|p|} \langle T \mid D^p \varphi \rangle \tag{39}$$

where p represents a multi-index $(p_1, ..., p_n)$ and $|p|$ its length $p_1 + ... + p_n$. The *convolution* of a distribution, for its part, is defined as:

$$T * \varphi(x) = \langle T_t \mid \varphi(x - t) \rangle$$

where T_t means that T operates on functions of t.

Now, the central idea of multiscale analysis is to consider that the processing system contains *wired-in* test functions that act like regularizing filters on the signal s represented by T. Knowing, on the other hand, that the derivatives $D\delta$ of the Dirac distribution δ (where the generic notation D represents any differential operator with constant coefficients) act by convolution like the differential operators D:

$$\delta * T = T \tag{40}$$

$$\delta' * T = T' \tag{41}$$

$$\delta^{(m)} * T = T^{(m)} \tag{42}$$

$$D\delta * T = DT \tag{43}$$

we only have to know what δ becomes from a multiscale perspective to finally give a meaning to the concept of *scale-space*. The most natural choice is given by the family of Gaussians:

$$G_\sigma(x) = \frac{1}{\sigma\sqrt{2\pi}}\, e^{-\frac{x^2}{2\sigma^2}} \tag{44}$$

which verifies

$$G_\sigma * G_\tau = G_{\sqrt{\sigma^2+\tau^2}}\,. \tag{45}$$

These Gaussian kernels constitute the multiscale version of identity. They express what a point becomes. But Gaussians are also the kernel of heat diffusion; therefore we can say that the multiscale approach actually takes the signal T as initial condition for a solution to the heat equation:

$$\left(\frac{\partial}{\partial s} - \Delta\right) I = 0 \tag{46}$$

(where $2s = \sigma^2$). This *diffusion equation* achieves the link between a "pure geometry" and its "physical" counterpart (the multiscale aspect). It expresses the operational constraint of transforming the signal into a geometrical observable.

In short, we have replaced the concept of *infinitesimal* neighborhood by the concept of *local multiscale* neighborhood. In this framework, classical differential geometry corresponds to the limit case of scale 0.

7.2. Scale-space morphogenesis of images

This idea of multiscale analysis, or "scale-space" filtering, is now commonplace in theories of geometrical image processing. It dates back to Witkin [413] and Koenderink [181], [184], and consists in embedding the rough image I in a smooth family of images, $F(x, y; s) = I_s(x, y)$, where (x, y) are the coordinates of the image plane and $s \in [0, 1]$ is a scale parameter such that

(i) $I_0 = I$ (initial condition);
(ii) I_1 is an undifferentiated "blob", an image without any internal structure;
(iii) the smooth process of simplification from $s = 0$ to $s = 1$ (or "undifferentiation") is as canonical and straightforward as possible, i.e., in some way optimal among all possible paths; in other terms, when scale s increases from fine to coarse-graining, the image I_s must be consistently less and less differentiated; this constraint is called the *causality principle*.

Assuming that we have found such a deformation path $I_0 \to I_1$, the inverse deformation path $I_1 \to I_0$ can then be identified with a *morphogenesis* of I, that is, a process of progressive and successive differentiations leading from an initial blob-like shape to the full and detailed morphology of I.

According to Koenderink, constraint (iii) can be interpreted as a causality condition in the following sense. Let us consider the graph G of F defined by the equation $z = F(x, y; s) = I_s(x, y)$ in the 4D space $\mathbb{R}^4 = \mathbb{R}^2 \times \mathbb{R} \times \mathbb{R}$ of coordinates (x, y, z, s). Let $F = z_0$ be a section of G. The causality condition states that at the extrema of $I_s(x, y)$, the section $F = z_0$ turns its convex side in the direction of decreasing s. It can be shown that this implies $\operatorname{sign}(\partial_s F) = \operatorname{sign}(\Delta F)$, where $\Delta F = \partial^2 F/\partial x^2 + \partial^2 F/\partial y^2$ is the spatial Laplacian. Hence, to satisfy the causality condition, F must verify $\Delta F = \alpha\, \partial_s F$ at the extrema. The simplest way to do this is of course to let the deformation $F = I_s$ satisfy the diffusion equation (heat equation) $\partial_s F = \Delta F$ that we discussed in previous

FIGURE 8. Gaussian blurring as an example of an undifferentiation process.

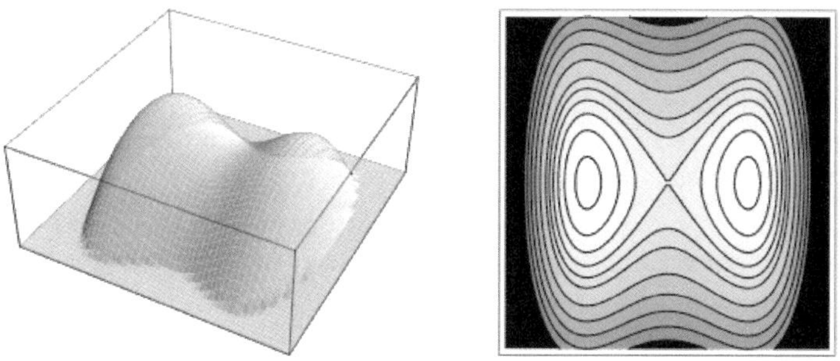

FIGURE 9. Left: a surface with two maxima and a saddle. Right: some of its level curves. When the height of the levels crosses the saddle, the level curve splits into two curves.

sections (see 6.3 and 7.1). The path $I_0 \to I_1$ is then nothing else than the Gaussian blurring of the image (see Figure 8).

Yuille and Poggio [415] have shown that a diffusion process driven by the heat equation is the easiest way to globally simplify an image *without introducing new zero-crossings*. Indeed, as was also pointed out by Robert Hummel and Robert Moniot [159], the causality constraint on the evolution of zero-crossings expresses the maximum principle for the heat equation.

Now, again, the kernel of the heat equation is a Gaussian, thus the family of deformations I_s is obtained through successive convolutions of the initial image I_0 by a family of Gaussians G_s, which have the effect of progressively blurring the image. This Gaussian blurring is a type of multi-resolution (multiscale) blurring. It leads from a fine-grained initial image to a coarse-grained final blob. As Jan Koenderink [181] claims :

> Gaussian blurring is the only sensible way to embed a primal image into a one-parameter family. (p. 365)

Consider the decomposition of the surface $z = I(x, y)$ (the graph of function I) into level curves L_z (Figure 9).

If $I(x, y)$ is a smooth function, these level curves are constituted by nested and juxtaposed topological circles and, as proven by Morse theory, the crossings of the critical points of I as z increases correspond to transformations of the topological type of the level curves L_z: when z crosses a minimum (resp. a maximum) of I, a new component appears (resp. vanishes); when z crosses a saddle point, two components merge into one or vice-versa. When we blur $I(x, y)$ in a Gaussian way, the components of the level curves L_z progressively fuse until one reaches a unique blob whose L_z are concentric topological circles. The fusion process corresponds to a well-defined sequence of *bifurcation* events: successive vanishings of components through collapse of minima and maxima with saddles. As Koenderink [181] explains:

> the image can be described unambiguously as a set of nested and juxtaposed light and dark blobs that vanish in a well defined sequence on progressive blurring. (p. 369)

Such a dynamic and morphogenetic analysis of the image yields a constituent-structure analysis. As is explained by Koenderink in another article *Dynamic Shape* (Koenderink-van Doorn [184]),

> The perceptual approach is dynamic (...) in the sense that a partial order is apparent that relies on a hypothetical evolution or morphogenesis that is an integral part of the shape description: the shape is thought of as having been formed from a primeval, shapeless, ovoid blob that was articulated in first rough then finer steps, finally leading to the present object. (p. 384)
>
> When you blur an object the passes and pits or summits meet in pairs and annihilate, thus simplifying the system of subobjects and ridges. Where the subobject boundaries meet the object boundary transversally you have concave vertices which are exactly the singular boundary points proposed as object transitions. There exists psychophysical evidence that human observers segment the object boundary at exactly these points. (p. 390)

In fact, such a morphogenetic analysis can easily be implemented. Singularity theory provides normal algebraic forms for the possible bifurcations occurring in the geometric analysis of I and of the I_s by means of their level-curves $L_{s,z}$. These normal forms make use of partial derivatives of the convolutions $I_s = I * G_s$ up to degree 4. We only have to generalize Marr's use of the second derivatives of $I * G_s$, and consider layers of cells whose receptive profiles are higher order partial derivatives of Gaussians (see an example Figure 10).

In one dimension, we consider for instance profiles of the form:

$$G_n(x, s) = \partial^n_{x^n} \frac{e^{-\frac{x^2}{4s}}}{\sqrt{4\pi s}} = \partial^n_{x^n} G_0(x, s) \tag{47}$$

and convolution products:[9]

$$\partial^n_{x^n} (I * G_0) = (\partial^n_{x^n} I) * G_0 = I * G_n \ . \tag{48}$$

9 Such layers of RFs implement a low-order jet calculus.

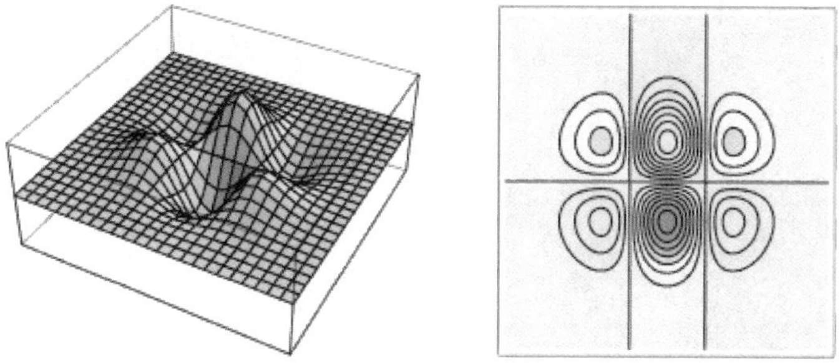

FIGURE 10. A partial derivative of third order (second order w.r.t. x and first order w.r.t. y) of a Gaussian. Left: its graph. Right: its level curves.

With such profiles it becomes possible to implement the *geometrical* features that occur in the morphogenetic dynamical analysis of an image. According to Jan Koenderink (Koenderink, van Doorn [185]), there are experimental evidences that certain cortical (hyper)columns do perform such a task:

> The modules (like "cortical columns" in the physiological domain (...)) of the sensorium are local approximations (nth order jets) of the retinal illuminance that can be addressed as a single datum by the point processor. (p. 374)

7.3. Anisotropic diffusion and morphological analysis

The problem is that the heat equation is an isotropic and homogeneous diffusion equation, therefore cannot be *morphological.* It is well adapted to the regularization of the image (*Verschmelzung*) but not to its segmentation (*Sonderung*). It also blurs boundaries. To be morphological, a diffusion equation must preserve boundaries from blurring and, even better, enhance them. It may seem rather impossible because merging and splitting are opposite processes. But it is nevertheless possible if the diffusion equation is *anisotropic*, inhibited in the direction of large gradients.

Several solutions have been considered (this is a technical mathematical debate). The first idea (Perona, Malik [257]) was to modify the possibility of diffusion near the boundaries. This led to equations of the type

$$\partial_s I_s = \operatorname{div}\left(g\left(\nabla I_s\right)\nabla I_s\right) \tag{49}$$

where g is a positive decreasing function such that $g(x) \to 0$ when $x \to \infty$. Figures 11 and 12 show two examples of this method.[10]

But the most radical solution is to use a diffusion equation that inhibits the diffusion transversely to the level lines of the image. We obtain (L. Alvarez,

10 Images are made using NIH-Image routines developped by our student Stéphane Brault.

FIGURE 11. A first example of Malik and Perona's anisotropic diffusion.

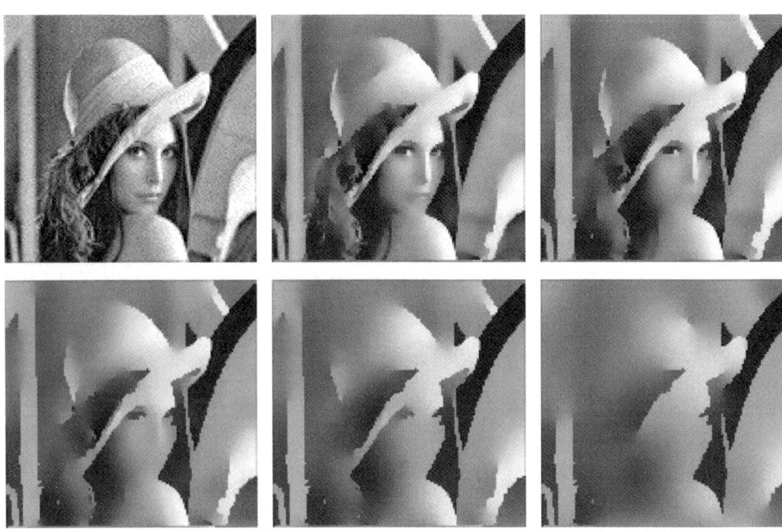

FIGURE 12. A second example of Malik and Perona's anisotropic diffusion. It shows that even when the diffusion becomes very strong the main boudaries are preserved.

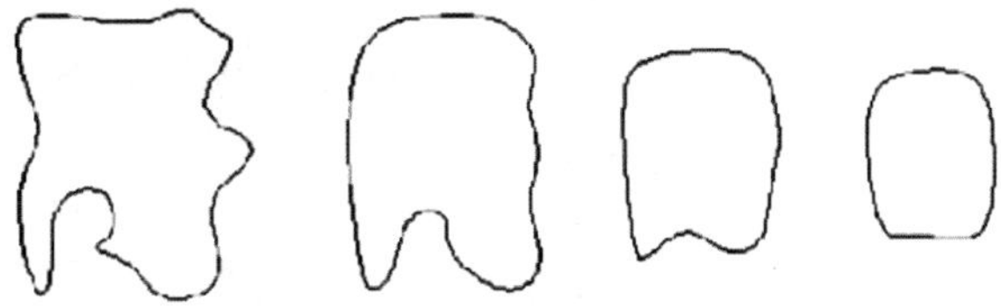

FIGURE 13. An example of Grayson's theorem. The initial contour is progressively convexified and converge metrically to a circle.

P.-L. Lions, J-M. Morel [8]) the equation:

$$\partial_s I_s = \partial^2_{\xi^2} I_s \tag{50}$$

where ξ is a coordinate normal to the gradient, i.e., tangential to the level line at any point. This equation can be written in the form:

$$\partial_s I_s = \|\nabla I_s\| \operatorname{div}\left(\frac{\nabla I_s}{\|\nabla I_s\|}\right) = \Delta I_s - \frac{H(\nabla I_s, \nabla I_s)}{\|\nabla I_s\|^2} \tag{51}$$

where H is the Hessian of I_s (the quadratic form given by its second partial derivatives). It is uniformly parabolic along the level curves of I but totally degenerate in the gradient direction. It makes the level curves evolve as fronts with a normal velocity equal to their curvature.

This non-linear diffusion process is known as "curve shortening", "flow by curvature", "heat flow on isometric immersions"; see the works of differential geometers such as M. Gage, R. Hamilton [116], M. Grayson [132], and also S. Osher [249] or J. Sethian [342]. It can be applied to isolated contours in a very interesting fashion.

The following result can be shown (see Figure 13).

Grayson's Theorem. *A contour (a non-self-intersecting smooth closed curve), as winding it may be, is strictly simplified during the process (no singularities appear), is progressively convexified and converges metrically to a circular point.*

During the process, inflexion points successively disappear by bifurcation and this sequence of bifurcations provides a morphological analysis of the contour. Jean-Michel Morel and colleagues (Alvarez et al. [8]) introduced additional Gaussian smoothing in these partial differential equations (PDEs), with the goal of controlling the speed of diffusion by coupling it with the image geometry. The basic equation thus becomes:

$$\partial_s I_s = g\left(\|G * \nabla I_s\|\right) \|\nabla I_s\| \operatorname{div}\left(\frac{\nabla I_s}{\|\nabla I_s\|}\right) \tag{52}$$

FIGURE 14. Denoising an image. From left to right: 1. the noisy image, 2. Gaussian blurring, 3. Denoising through anisotropic diffusion.

where G is a Gaussian kernel and $g(x)$ is a decreasing function such that $g(x) \to 0$ when $x \to \infty$. They also showed that, if an affine invariance constraint is imposed (instead of the Euclidian invariance), the typical PDE becomes:

$$\partial_s I_s = \|\nabla I_s\| \operatorname{div}\left(\frac{\nabla I_s}{\|\nabla I_s\|}\right)^{\frac{1}{3}} . \tag{53}$$

These algorithms of anisotropic diffusion are now widely used for *denoising* images. Indeed, as is shown in Figure 14, they allow to smooth an image while preserving its edges.

7.4. Generalizations

The previous ideas can be generalized in order to noticeably improve the morphological analysis of the image. We saw that, to reconcile *Verschmelzung* and *Sonderung* and to format the image geometrically in a multiscale fashion, one can use diffusion to smooth the signal while at the same time *preventing* the smoothing of critical geometrical features, such as gradients. But other geometrical features can be protected against diffusion. For example, we saw that one of the main drawbacks of the Mumford-Shah model was that triple points are forced to be different from T-junctions. It is possible to adapt anisotropic diffusion to this bias in such a way that T-junctions are preserved. The results are spectacular (see Morel's works).

7.5. A few mathematical remarks about contour diffusion

In fact, the problem of contour diffusion in scale-space analysis is a problem of propagating fronts. Let C_0 be a closed curve in $\mathbb{R}^2$. The general problem of propagating fronts is to analyze the evolution C_t of C_0 when each point P of C_t is moved at a speed depending on the curvature K of C_t at P. Let $P(r,t) = (x(r,t), y(r,t))$ be the parametrization of C_t. The equations of motion are: $\partial_t P = F(K)N$ (where N is the outward pointing unit normal vector of C_t

at P).[11] We have:

$$\left\{\begin{array}{l}\partial_t x = F(K)\frac{\partial_r y}{\left((\partial_r x)^2+(\partial_r y)^2\right)^{\frac{1}{2}}} \\ \partial_t y = -F(K)\frac{\partial_r x}{\left((\partial_r x)^2+(\partial_r y)^2\right)^{\frac{1}{2}}}\end{array}\right. \tag{54}$$

with $K = \frac{\partial_{r^2}^2 y \partial_r x - \partial_{r^2}^2 x \partial_r y}{\left((\partial_r x)^2+(\partial_r y)^2\right)^{\frac{3}{2}}}$.

Two opposite main cases have been studied the most. When $F(K) = V =$ constant, e.g., $V = 1$, we get the class of "grassfire" models $\partial_t P = N$: the velocity of the propagating fronts is constant as in optics (wave equation). It can be shown that $\partial_t K = -K^2$, that is $K(r,t) = K(r,0) \;/\; (1 + tK(r,0))$. The curvature becomes therefore singular when $t = -1/K(r,0) > 0$, which will always be the case if $K(r,0) < 0$. We come back to this point in the next section.

The other fundamental case corresponds to $F(K) = -K$, i.e., to a normal velocity equal to the curvature. It has drawn much attention and received the names, already mentioned above, of "curve shortening", "flow by curvature", "heat flow on isometric immersions". In terms of the immersions $j_t : S^1 \to \mathbb{R}^2$ defining the curves C_t, it corresponds to the heat equation. Indeed $\Delta j_t = -KN$ and, therefore, the heat equation $\partial_t j_t = \Delta j_t$ is nothing else than the equation $\partial_t j_t = -KN$.[12] If s is the arclength along $C = C_t$, the equation for the curvature is:

$$\partial_t K = \partial_{s^2}^2 K + K^3 \;. \tag{55}$$

A first fundamental "shortening curve" theorem was proved by M. Gage and R.S. Hamilton in 1986 [116]. It says that if C_0 is a convex embedded curve, then the (inward oriented) heat equation shrinks C_0 to a point, C_t becoming asymptotically a metrical circle as it shrinks. It is this result that was generalized by Matthew Grayson in 1987 [132]. He showed (see above) that the same result holds if C_0 is only an embedded (not necessarily convex, and eventually very winding) smooth curve. Stanley Osher [249] and James Sethian [342] have studied the mixed cases where $F(K)$ is neither constant (wave equation) nor equal to $-K$ (heat equation). One can consider for instance the cases where $F(K) = 1 - \varepsilon K$, $\varepsilon \geq 0$. The equation for K is then:

$$\partial_t K = \varepsilon \partial_{s^2}^2 K + \varepsilon K^3 - K^2 \tag{56}$$

which is a reaction-diffusion equation where the reaction term $\varepsilon K^3 - K^2$ (driving C_t towards singularities) is balanced by the smoothing effect of the diffusion term $\varepsilon \partial_{s^2}^2 K$.

To take into account possible changes of the topological type of the C_t curves, i.e., the occurrence of singularities, Osher and Sethian considered C_0

11 The notation is $\partial_t f = \partial f / \partial t$, etc.

12 The heat equation is applied here to the map $j_t : S^1 \to \mathbb{R}^2$ and not to a map $f : \mathbb{R}^2 \to \mathbb{R}$.

as the 0-level curve of a function u_0 defined on $\mathbb{R}^2$:

$$C_0 = \{P = (x, y) \in \mathbb{R}^2 \mid u(P) = 0\} \tag{57}$$

and searched for a partial differential equation (PDE) whose solutions u_t possessed the property that their level curves (and in particular their 0-level curves C_t) evolve according to the equation of propagating fronts $\partial_t P = -KN$. This equation can be rewriten $\partial_t P = -(\operatorname{div} N)N$. As

$$P(t) \in C_t = \{P \in \mathbb{R}^2 \mid u_t(P) = 0\} \ , \tag{58}$$

we have:

$$0 = \frac{d}{dt} u_t\left(P(t)\right) = -\left(\nabla u \bullet N\right) \operatorname{div} N + \partial_t u \ . \tag{59}$$

But as $N = \nabla u / |\nabla u|$, the equation writes (Evans, Spruck [103]):

$$\partial_t u = \|\nabla u\| \operatorname{div}\left(\frac{\nabla u}{\|\nabla u\|}\right) = K \|\nabla u\| = \Delta u - \frac{H(\nabla u, \nabla u)}{\|\nabla u\|^2} \tag{60}$$

which is nothing else than Morel's equation (51).

8. Gestaltic applications

From the previous sections, we can see that the segmentation problem in early perception is open to many different physicalist approaches. These frameworks and methods confer to the descriptive morphological concepts (from Husserl to Thom and Langacker) a technical scientific content and offer a remarkable example of what it means to convert theoretical concepts into algorithms for the computational synthesis of phenomena.

8.1. Examples

Let us evoke briefly a few applications of such techniques.

1. Convexification of shapes. Grayson's theorem yields of course a very powerful tool for convexifying a shape, that is, to reduce it to a trivial domain. Convexification provides a morphogenesis of the shape.
2. Hierarchical constituent analysis. Using non-linear diffusion equations, one can solve the mereological constituency problem. The segmentation and simplification of a complex shape individuates its constituents progressively, and moreover these are hierarchically organized.
3. Grouping. One of the oldest gestaltic puzzles is grouping scattered objects into super-ordered objects. For individuating the latter, grouping has to construct virtual boundaries, and one can legitimately wonder if this is possible. The Perona-Malik algorithm is able to overcome this difficulty.

8.2. Crest, ridges and cut locus

We are particularly interested in another example, namely skeletonization. If S is a 2D shape (an outline figure), one can make the boundary of S diffuse, which defines an intensity function $u(x, y)$. Then, the crests and ridges of u can be extracted by using an anisotropic diffusion equation that preserves and enhances the sharp variations in the direction of the gradient ∇u.

Skeletonization is fundamental for constituent analysis. As we will see below, it can be obtained more naturally via other PDEs that are *hyperbolic* equations of propagation of wave fronts. In this theory, the *cut locus* is the singular locus of the propagation. It is a 1-manifold whose generic singularities are triple points and end points. These singularities can be detected *locally* and automatically yield an immediate mereological decomposition of the shape into parts.

9. Skeletonization

We sketch some aspects of skeletonization.

9.1. Harry Blum's contour diffusion and grassfire models

The idea of analyzing a shape by looking at its skeleton is an old and powerful one. Mathematically, it can be implemented using not a parabolic diffusion equation (such as the heat equation) but a propagation equation such as the wave equation

$$\frac{\partial^2 I_s}{\partial s^2} = \Delta I_s \; . \tag{61}$$

As we have seen in Section 7.5, the characteristic rays of this hyperbolic PDE propagate orthogonally to the initial contour B and wave fronts propagate parallel to it.

More precisely, for the wave equation $\partial^2\varphi(x,t)/\partial t^2 = \Delta\varphi(x,t)$ in $\mathbb{R}^n$ one considers solutions of the form $a_\tau(x)e^{i\tau(\psi(x)+t)}$ where τ is a frequency, $a_\tau(x)$ an amplitude and $\psi(x)$ a *spatial* phase. The manifolds $\psi(x) = \text{constant}$ are the wave fronts. What is called the "eikonal" equation $1 - \|\nabla\psi\| = 0$ is an Hamilton-Jacobi equation expressing the fact that the norm of the gradient of the spatial phase ψ is constant $= 1$.[13] The rays are its integral lines.[14]

In such an "optical" propagation or "grassfire" model, two sorts of singularities may occur:

(i) caustics, that is envelopes of rays;
(ii) cut loci, that is, points reached at the same time by two rays coming from two different points, or equivalently the centers of the maximal

13 See the equation $\partial_t P = N$ in Section 7.5.

14 See Petitot [279] for an introduction to this basic formalism.

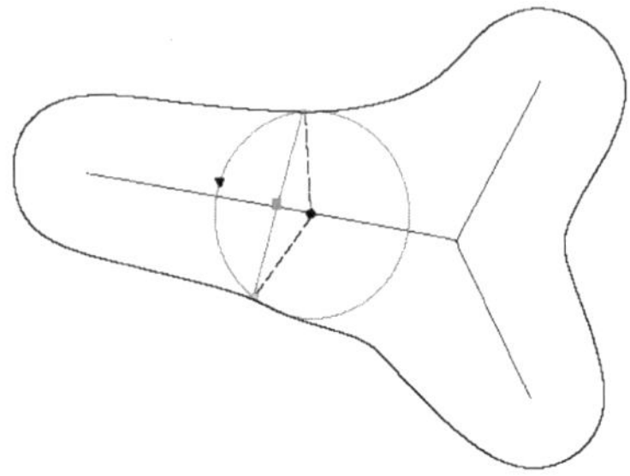

FIGURE 15. The cut locus of a contour as the singular locus of an "optical" propagation: it is the locus of the centers of the bitangent circles (from Kimia [179]).

discs included in the surface S bounded by the contour B (see Figure 15).

We shall consider here only cut loci. They satisfy a natural "entropy condition", which is equivalent to the well known Huyghens principle.

Historically, the contour propagation routine was introduced in vision theory by Harry Blum in 1973 [38] as a fundamental and ubiquitous visual routine. In his pioneering work *Biological Shape in Visual Science*, he addressed the question:

> How do organisms describe and characterize other organisms' shapes?

and tried to find the implicit biological geometry underlying this very common type of performance. For

> biology has no set of statements for its everday spatial relations

and

> without a proper shape mathematics for biology, we are in the position that physics would have been in trying to develop mechanics without Euclidean geometry.

He started from the basic fact that "shapes are normally described by their boundaries". He introduced then the key idea of contour propagation—the "grassfire" model: every point of the boundary B of the shape S becomes the center of diffusion of spherical waves, and so the contour propagates as a wave front. The wave fronts can be identified with level-curves of a surface. The fundamental geometrical object characterizing a shape S is, according to this perspective, the "symmetry axis" of S, that is, precisely its cut locus CL.

9.2. Neural implementation of cut loci

In 1991, our student Hugh Bellemare implemented the contour diffusion and propagation algorithms in neural networks. The difficulty of implementing such PDEs in neural nets is due to the fact that the units are non-linear and the symmetry is broken by the network geometry. For propagation, the network is composed of five layers:

1. The first layer is a "retina" that enters the input.
2.-3. The second and third layers compute the X and Y components of the field of rays.
4. The fourth layer computes the singular points of the propagation.
5. The fifth layer computes the cut locus using as geometrical criterion the discontinuities of the divergence of the field.

Figure 16 shows the progressive construction of the CL of a rectangle. Three steps are represented: the CL starts at the vertices and propagates until the central component is constructed. Figure 17 represents the activity of the fifth layer of the network for these three stages. When a shape's contour is irregular, the CL gives a double information: the internal components of the CL correspond to convex parts and the external components to concave parts (see Figures 18 and 19).

9.3. Properties and structure of cut loci

The cut locus CL of a shape S is a very interesting structure:

(i) It is a *singular* locus and allows to reconstruct the global shape S from the radius function (i.e., the radius $r(x)$ of the maximal disc centered at $x \in CL$).
(ii) It is a *dynamical* object built from the propagation of the wave fronts, i.e., following the direction of increasing radius $r(x)$.
(iii) Its *topological* properties (and in particular its singularities: triple points, end points) are fundamental indicators for the geometrical properties of the shape S, e.g., the convexity.

Blum's theoretical propositions have been partially confirmed experimentally by J. Psotka in 1978 [317]. Psotka presented several outline figures to subjects who were unaware of the purpose of the experiment and told them to place an interior dot in the first place that comes to mind. The dots cluster spectacularly on the cut locus.

9.4. Leyton's works

There are many variants of the concept of cut locus: Blum's (centers of maximal discs bitangent to the contour of the shape), Brady's (middle points of the cords joining points of bitangency), and Leyton's (middle points of the arcs joining points of bitangency).

Michael Leyton [212] did extended research on the concept of axis of symmetry. He has shown that the cut locus of a shape constitutes a fundamental information for reconstructing a virtual causal temporal and dynamical process giving rise to the shape, that is, a morphogenesis. Indeed, the cut locus of a

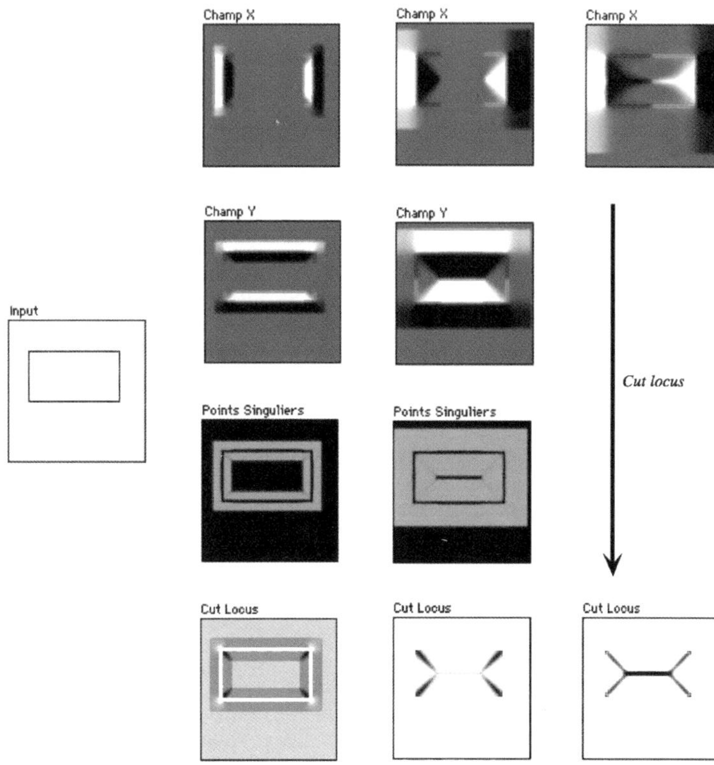

FIGURE 16. The progressive construction of the CL of a rectangle. From top to bottom: the X and Y components of the field of rays, the singular points of the propagation, the cut locus. Three steps are represented from left to right: the CL starts at the vertices and ends at the central component.

trivial domain, a metrical circle, is a point, and in some sense the CL measures how non-trivial a shape is.

9.5. The neurophysiological relevance of skeletonization

It must be emphasized that there is neurophysiological evidence showing that the cut locus is actually processed by the visual areas. Figure 20 (from [209]) shows the response to a bar up to 400ms. We see how the cut locus, i.e., the symmetry axis of the bar, is rapidly constructed after 150-200ms.

9.6. Skeletonization and mereological constituency

Skeletonization provides a very natural way of decomposing hierarchically a global shape into parts. It is one of the most powerful mereological algorithms.

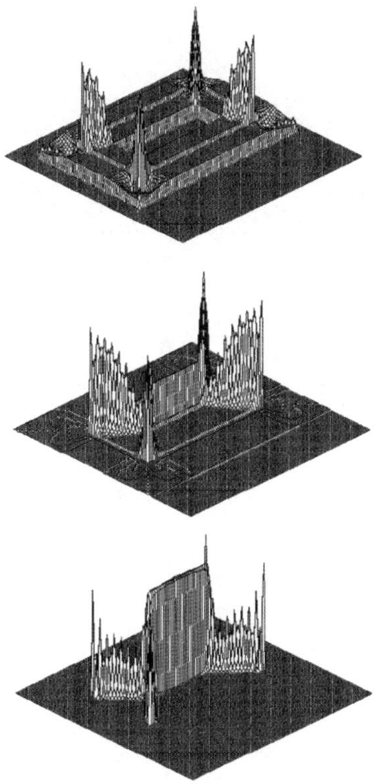

FIGURE 17. The activity of the fifth layer of the network for the three stages of the construction of the CL in Figure 16.

In fact, the skeleton of a shape is a one-dimensional manifold that is very easy to analyze. Generically, it can present as only singularities triple points and end points, and it is composed of pieces of curves joining such singular points. It provides a decomposition of a shape, as irregular as this shape may be, into a compound of generalized cylinders in the sense of Marr and Biederman.

Moreover, it is easy to associate with the skeleton a *symbolic structure*, namely an abstract graph. This is particularly relevant and important for the *external* CL_{ext} of a configuration A of domains A_i: CL_{ext} evolves and progressively partitions—categorizes, stratifies—the ambiant surrounding space into regions R_i associated with the domains A_i. In this case, the CL is a 1-dimensional singular structure that is locally computable (as are its singularities, essentially triple points) and whose geometry characterizes the configuration A.

We return later to this fondamental point in Chapter 3.

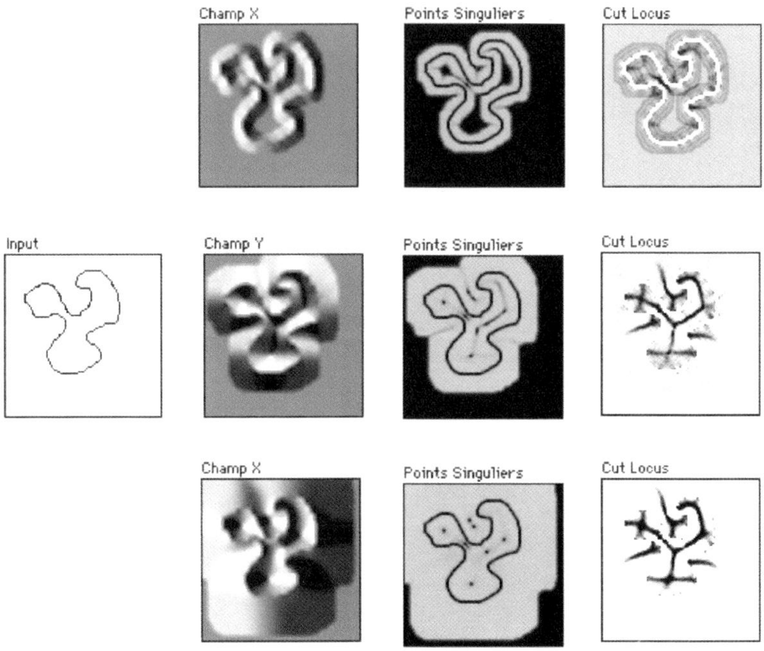

FIGURE 18. The internal and external CL of an irregular shape.

9.7. Multi-scale cut locus

One of the main problems raised by the use of the cut locus structure is that it is very sensitive to noise: each bump of the boundary generates a new branch of the CL. But if we regularize the shape using some of the previously presented routines we can eliminate this drawback by pruning.

It is especially interesting to use a *multiscale* version of the cut locus, for instance using curve shortening, i.e., flow by curvature. According to Grayson's theorem (Section 7.3), the shape is convexified and converges metrically to a circle. But the cut locus of a circle is trivial: it is a point. It means that the cut locus of the shape contracts into a point via a sequence of bifurcations (successive fusions of end points and triple points eliminating the successive branches). If we reverse this process, we get an unfolding of a point into a cut locus, and this unfolding explains how the shape can be constructed from a trivial circle (Figure 21).

We tackle now another class of neural models.

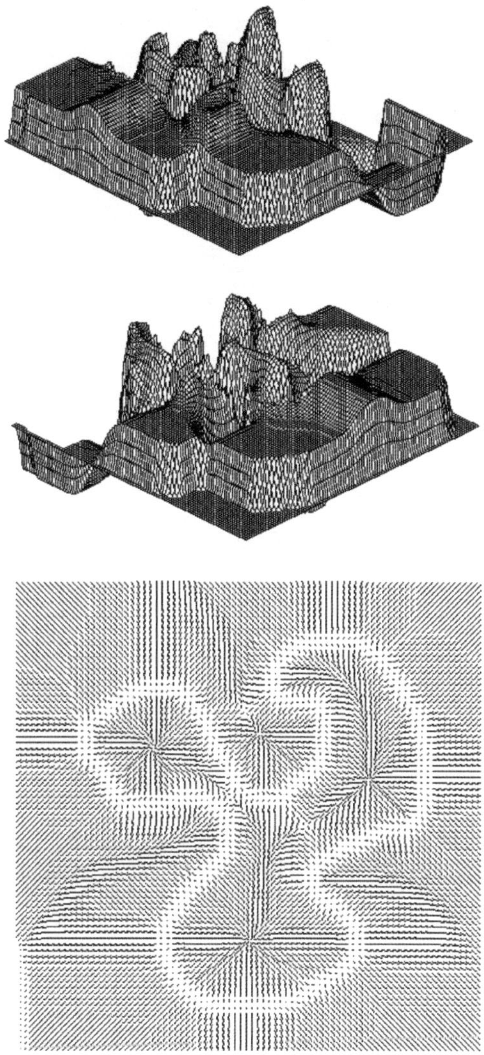

FIGURE 19. From top to bottom. The activity of the X and Y layers of the network for the CL in Figure 18 and the vector field (X, Y).

10. The binding problem and oscillator networks

Thus far we have examined segmentation and constituency models pertaining to either low-level natural vision (retina and wavelets) or computational vision (variational model and diffusion/propagation equations). In this section, we approach the same problem through higher-level natural vision (cortical levels).

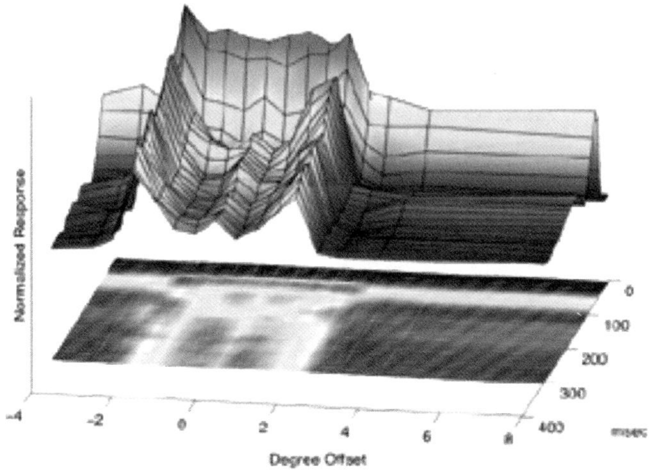

FIGURE 20. The neurophysiological reality of the cut locus. The stimulus is a bar. Initially the response is sensitive only to the boundaries of the bar. But rapidly (after 150ms) the cut locus (the symmetry axis) is detected in spite of its absence from the stimulus. From Lee [209].

10.1. Cortical fibrations

Let us begin by recalling some well known neurophysiological facts concerning the "columnar" structure of the primary visual cortex (area 17 or V1) extensively investigated since the pioneering works of Hubel, Wiesel and Mountcastle.

The basic functional module is a "cube" organized in a retinotopic, columnar and layered fashion:

(i) The *retinotopic* structure preserves the topographic connections of the retinian ganglion cells through the lateral geniculate body. We obtain that way a retinotopic *fibration* (in the geometrical sense, see below), whose base space is constituted by glued local receptive fields.

(ii) The *columnar* structure (which is orthogonal to the cortex surface) is essentially constituted by orientation columns, columns of ocular dominance, and color blobs. Orientation columns (of diameter about 50μ) are arranged somewhat orthogonally to the columns of ocular dominance. Their preferential orientation varies from $0°$ to $180°$ in $10°$ steps. They constitute hypercolumns of size 0.8-1mm.

(iii) The *layered* structure is composed of six layers. Its depth is about 1.8mm. The geniculate fibers project onto layer IV. Layer VI is a feedback layer. Layer V projects onto the colliculus. Layers II and III receive the axons of layer IV and project their efferent fibers onto other cortical

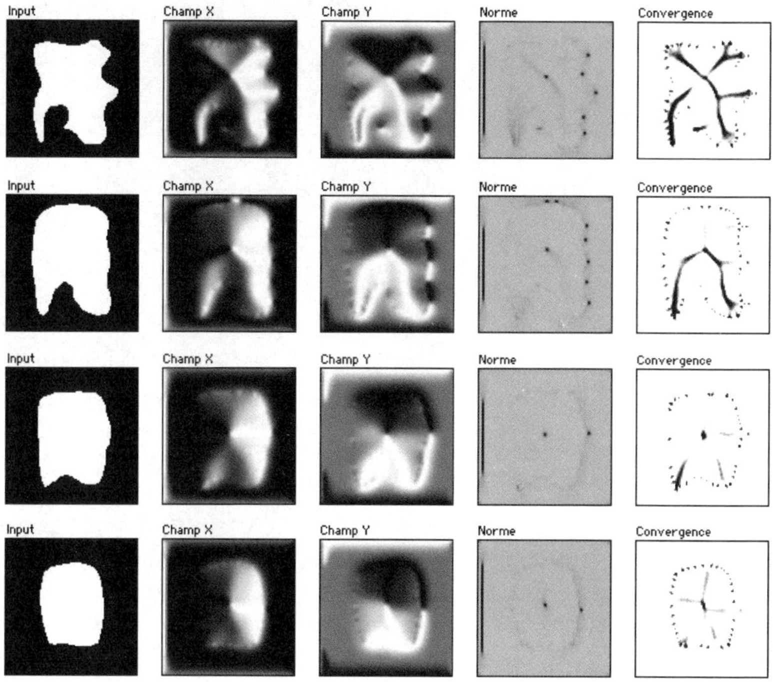

FIGURE 21. A multiscale cut locus. Four steps are represented. The convexification of the shape makes the branches of the cut locus progressively disappear in a sequential order reflecting the hierarchical mereology of the shape. The five layers of the neural implementation are explained above.

regions where different attributes (shape, color, movement, stereopsis, etc.) are further processed, eventually in a non-retinotopic way.

The orientation columns provide a beautiful example of a neurally implemented fibration, that is a geometrical structure where "above" each point of a base space W there exists a copy of another space F called a fiber. Actually, they implement the fibration $\pi : E \to W$ whose base space is the retina W and whose fiber is the projective line $F = \mathbb{P}^1$ of directions in the plane (Figure 22).

As was emphasized by William Hoffman [156]:

> fibrations (...) are certainly present and operative in the posterior perceptual system if one takes account of the presence of "orientation" micro-response fields and the columnar arrangement of cortex. (p. 645)

For colors, the situation is a little more complex. There are parvocellular layers in the lateral geniculate body coming from ganglion cells sharing a spectral antagonism $+/-$ between red, green, yellow, and blue: (R^+/G^-), (G^+/R^-),

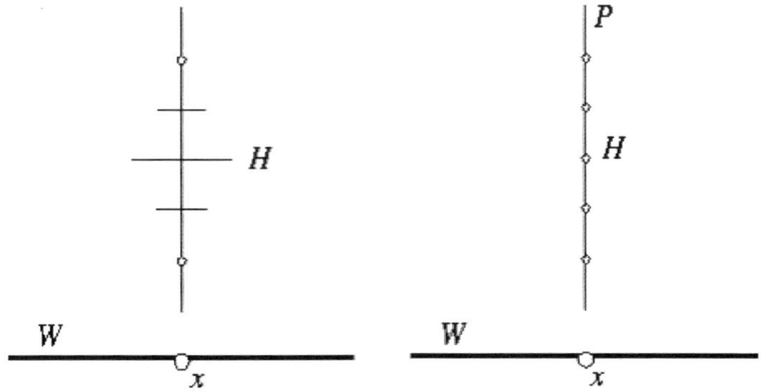

FIGURE 22. A schematized fibration. Above each point x of the base space W there exists a copy of the fiber F. Here the fiber is the projective line P of the orientations p in the plane. Left: orientations p shown as lines in perspective. Right: orientations shown as points in P.

(Y^+/B^-), (B^+/Y^-); there are cortical blobs in V1 initiating color processing; and, according to Semir Zeki, the area V4 is specialized in color processing. But it seems nevertheless plausible to hypothesize that color is also processed through fibrations (with a dynamical antagonist spectral structure in the fibers).

10.2. The binding problem and the labeling hypothesis

At the cortical level of processing, the problem of parts and wholes, that is the mereological problem of constituency, is also called the "binding" problem. At the early stages of perception, features of objects are extracted in a local, distributed and parallel way. How can such localized features be integrated into higher level units (constituents) in spite of their distributed encoding? One must avoid the "superposition catastrophe": if two constituents A and B are coded by activity patterns P_A and P_B of some neural network, and if the composition of A and B is coded by the superposition $P_A + P_B$, then A and B lose their individuation.

The most promising solution is now considered to be found in the fine temporal coding by means of coherent neural oscillations. The coherence of features and constituents would be encoded in the *synchronization* (phase locking) of oscillatory neuronal responses to stimuli. And therefore different phases can code for different constituents. This hypothesis is also called the labeling hypothesis.

There is a large amount of experimental evidence concerning synchronized oscillations in the cortical (hyper)columns (in the frequency range of 40 – 70Hz) which are sensitive to the coherence of the stimulus (see e.g., Andreas Engel, Peter König, Charles Gray and Wolf Singer [99]).

As was emphasized by the authors:

> According to this model, coherence in and between feature domains may be encoded by transient synchronization of oscillatory responses and thus permit a binding of distributed features of an object.

When one uses the new parameter of synchronization, the binding resulting from functional coupling becomes dynamic and purely transient. It is no longer the consequence of a fixed anatomical wiring and the existence of higher level integrating "grandmother" neurons.

To corroborate this hypothesis

(i) one must have enough experimental evidence at disposal, and
(ii) one must show that the functional coupling of oscillators whose frequency is stimulus-dependent can effectively reflect the coherence of the stimuli patterns.

With such a local mechanism one can explain in a new way the global phenomena of fusion and separation:

$$\text{Fusion} \equiv \text{Synchronization}, \text{ Segmentation} \equiv \text{De-synchronization}.$$

10.3. Networks of oscillators

10.3.1. *Hopfield equations and Hopf bifurcations.* The problem is now to model the fundamental fact of synchronization. For this, we need the theory of networks of weakly coupled oscillators whose frequency depends on the intensity of the stimulus.

Even if it is a very complex system, a cortical (hyper)column can be approximated by a single oscillator via some mean field approximation. Let S be a system of formal neurons u_i satisfying the Hopfield equations:

$$\dot{x}_i = -x_i + \sigma\left(\sum_j w_{ij}x_j + \tau_i\right) \tag{62}$$

where x_i is the state of activity of the i-th neuron u_i, w_{ij} the synaptic weight between u_i and u_j, τ_i the threshold of unit u_i, and σ a sigmoidal gain function. Averaging over the excitatory neurons E and the inhibitory neurons I, one gets a system of two equations (Wilson-Cowan equations) for the mean activities X_E and X_I. Under retinal stimulus, the equilibrium state of this system can bifurcate spontaneously, via what is called a Hopf bifurcation, towards a cyclic attractor (attracting limit cycle). Moreover, the frequency of this limit cycle depends on the intensity of the stimulus. One then observes a synchronization of the columns coding for the *homogeneous* parts of the stimulus.

The mathematical explanation of this phenomenon is difficult. Let us start with a network of N oscillators F_i $(i = 1, \ldots, N)$ of respective frequencies ω_i (periods $T_i = 2\pi/\omega_i$). If θ_i are their phases and φ_{ij} are their phase differences

$\varphi_{ij} = \theta_i - \theta_j$, the differential equations of the network are of the form:

$$\dot{\theta}_i = \omega_i - H\left(\varphi_{ij}\right) \tag{63}$$

where frequency ω_i depends on the intensity of the stimulus at position i.

Many works have been devoted to the analysis of such systems using qualitative dynamics (e.g., George Bard Ermentrout and Nancy Kopell) or statistical physics (e.g., Yoshiki Kuramoto). We mention only one simple example concerning the former. Take a linear chain of oscillators with linearly decreasing frequencies. Under the weak coupling hypothesis, one can show that *plateaus* are formed. This means that the oscillators effectively try to synchronize (phase locking), but as the total difference of phases is too large, they can only partially synchronize. Homogeneous synchronized zones are formed (plateaus), which are delimited by sharp discontinuities (jumps between plateaus). Now if there are discontinuities in the stimulus, they naturally constitute preferential boundaries.

In a nutshell, the theory of weakly coupled oscillators:

(i) shows that such systems enhance and complete existing boundaries;
(ii) can generate new virtual boundaries (which are not in the inputs);
(iii) confirms the labeling hypothesis.

10.3.2. *Kuramoto model and phase transitions.* In statistical physics, the most common systems are of the type

$$\dot{\theta}_i = \omega_i - \sum_{j=1}^{j=N} K_{ij} \sin\left(\theta_i - \theta_j\right) , \tag{64}$$

where K_{ij} are coupling constants. In the case of uniform coupling and full connectivity, Y. Kuramoto and I. Nishikawa [195] gave a detailed analysis of the system

$$\dot{\theta}_i = \omega_i - \frac{K}{N} \sum_{j=1}^{j=N} \sin\left(\theta_i - \theta_j\right) . \tag{65}$$

To this aim, they introduced the mean phase *order parameter*, defined as follows:

$$Z(t) = |Z(t)|\, e^{i\theta_0(t)} = \frac{1}{N} \sum_{j=1}^{j=N} e^{i\theta_j(t)} \tag{66}$$

and studied the equivalent system

$$\dot{\theta}_i = \omega_i - K\,|Z| \sin\left(\theta_i - \theta_0\right). \tag{67}$$

If frequencies ω_i are selected randomly according to a distribution law $g(\omega)$ representing the statistical regularities of the environment (we can assume that g is centered on 0 by regularly rotating the phase circle), global synchronization is a phase transition that happens at critical value $K_c = 2/\pi g(0)$ of the coupling constant K.

First, Kuramoto searched solutions for Z = constant. He classified oscillators into two groups, the S-group of oscillators that could synchronize, i.e., satisfy

$$\dot{\theta}_i = 0 \text{ hence } \left|\frac{\omega_i}{KZ}\right| \leq 1 \tag{68}$$

and the D-group of oscillators that could not, because in their case

$$\left|\frac{\omega_i}{KZ}\right| > 1 , \tag{69}$$

and he showed that only the S-group contributes to the synchronization. Writing

$$Z = \int_0^{2\pi} n_0 (\theta, t) \, e^{i\theta} d\theta , \tag{70}$$

where $n_0(\theta, t)$ is the phase distribution at equilibrium at time t and

$$n_0 (\theta, t) \, d\theta = g(\omega) d\omega , \tag{71}$$

where $\omega = K |Z| \sin (\theta - \theta_0)$, he obtained a self-consistency equation $Z = S(Z)$ that he developed in the neighborhood of $Z = 0$. He thus produced the following cubic equation:[15]

$$\begin{cases} \varepsilon Z - \beta |Z|^2 Z = 0 \\ \varepsilon = \frac{K - K_c}{K_c} \\ \beta = -\frac{\pi}{16} K_c^3 g''(0) \end{cases} \tag{72}$$

A stability analysis of the solutions shows that the trivial solution $Z = 0$, which is stable for $K \approx 0$ (uncoupled oscillators), becomes unstable when crossing $Z = Z_c$.

Kuramoto then established the evolution of the order parameter under quasi-adiabatic conditions. He obtained an equation such as:

$$\xi \frac{dZ}{dt} |KZ|^{-1} = \varepsilon Z - \beta |Z|^2 Z . \tag{73}$$

He then studied fluctuation terms, in particular near the critical point, when they become huge and pull the system into phase transition.

10.3.3. *Daido model and renormalization.* For his part, Hiroaki Daido [75] studied the following systems:

$$\dot{\theta}_i = \omega_i - K \sum_{j \in V_i} \sin (\theta_i - \theta_j) , \tag{74}$$

15 We have met a more complicated equation with Pierre Coullet in Section 3.3.

where V_i is the set of i's nearest neighbors on a cubic lattice of dimension d. Using renormalization group methods, he showed that if the frequency distribution $g(\omega)$ behaves asymptotically like a power law

$$g(\omega) \approx |\omega|^{-\alpha-1}, \alpha \in]0,2[\ , \tag{75}$$

then the system is equivalent to an uncoupled system (i.e., not synchronizable) for

$$\beta = 1 - \frac{1}{\alpha} - \frac{1}{d} < 0 \, . \tag{76}$$

More precisely, if the lattice is partitioned into $M = L^d$ blocks B_k of size L (the unit length is the elementary edge of the initial lattice), and if frequencies ω_i and phases θ_i are averaged over blocks B_k (giving frequencies Ω_k and phases Θ_k), then the renormalization operation reads

$$\begin{cases} \tau = tM^{\frac{1-\alpha}{\alpha}} \\ \Theta_k^* = \Theta_k - \frac{\gamma_M}{M} t \end{cases} \tag{77}$$

where γ_M is defined by the fact that the frequency

$$\omega_n^* = \frac{\sum_{i=1}^{i=n} \omega_i - \gamma_n}{n^{\frac{1}{\alpha}}} \tag{78}$$

obeys a stable distribution law of characteristic function

$$\left\langle e^{iz\omega^*} \right\rangle = e^{-|z|^\alpha} . \tag{79}$$

Recall that if X is a random variable with distribution function

$$F(x) = p\,(X < x) \ , \tag{80}$$

its characteristic function is the Fourier transform

$$G(z) = \left\langle e^{izx} \right\rangle = \int e^{izx} dF(x) \, . \tag{81}$$

Stable laws are indefinitely divisible laws (i.e., which can be considered as sums of independent, infinitely small variables: all $(G(z))^\alpha$ for $\alpha > 0$ are characteristic functions), whose class is stable by linear combination. Then it can be shown that their characteristic function is of the type

$$G(z) = \exp\left[\left(-c_0 + \frac{iz}{|z|} c_1\right) |z|^\alpha\right] \tag{82}$$

with $\alpha \in (0,2)$, $c_0 \geq 0$ and $\left|c_1 \cos\left(\frac{\pi}{2}\alpha\right)\right| < \left|c_0 \sin\left(\frac{\pi}{2}\alpha\right)\right|$. We have here $c_0 = 1$ and $c_1 = 0$.

Daido thus obtained the renormalized equations

$$\frac{d\Theta_k^*}{d\tau} = \omega_{M,k}^* - KM^\beta \sum_{l \in J_k} \sin_{lk}^* \left(\Theta_l^* - \Theta_k^*\right) \tag{83}$$

with $\beta = 1 - \frac{1}{\alpha} - \frac{1}{d}$, J_k ={blocks B_l neighbors of block B_k}, where the effective coupling writes:

$$\sin^*_{lk}(\Theta) = M^{\frac{1-d}{d}} \sum_{\substack{(i,j) \\ i \in B_l, j \in B_k}} \sin\left(\Theta + \psi_{l,i} - \psi_{k,j}\right) \tag{84}$$

where, for $j \in B_k$, $\psi_{m,j}$ is the deviation of phase θ_j from mean phase Θ_m over B_m: $\theta_j = \Theta_m + \psi_{m,j}$.

The basic fact is that for $\beta < 0$, the system is attracted toward the trivial fixed point of the renormalization group:

$$\frac{d\Theta^*_k}{d\tau} = \omega^*_k \tag{85}$$

$$\omega^*_k = \lim_{M \to \infty} \omega^*_{M,k} \,. \tag{86}$$

The interactions converge to 0 and synchronization cannot happen. This does not prevent, of course, *clustering*. As the coupling K increases, a greater number of oscillators synchronize; however this is not a phase transition anymore.

Once the synchronization properties of such oscillator systems are better understood, models of higher-level cognitive functions based on the "labeling hypothesis" can be elaborated. For example, Erik Lumer [215] proposed a theory of *attention*, where one focuses on one constituent of a perceptual scene. It consists of extracting the phase of one synchronized group with a "phase tracker" and use it as a tag.

10.4. Synchronized oscillations and segmentation

More recently, Alessandro Sarti and Giovanna Citti [330] have shown that such oscillator networks can converge towards the Mumford-Shah variational segmentation model presented in Section 6.2. The idea is to generalize the Kuramoto model. The phase $\theta(x,t)$ is now also a function of the spatial position x. Let ξ be the distance between oscillators. We look at a PDE of the form:

$$\begin{aligned}\frac{\partial\theta(x,t)}{\partial t} &= \omega(x) + \\ &\frac{1}{|\xi|^2}\left\{K(x+\xi)\left[\varphi\left(\theta(x+\xi,t)-\theta(x,t)\right)\right] - \right. \\ &\left. K(x)\left[\varphi\left(\theta(x,t)-\theta(x-\xi,t)\right)\right]\right\}\end{aligned} \tag{87}$$

where the function φ generalizes the function *sin* and where the sum Σ is taken over the $x + \xi$ and $x - \xi$ neighbors of x. To take into account the different ξ, we introduce a probability law on the connections. The simplest case is a Gaussian isotropic law. If we introduce the mesh ε of the lattice and if we encode in the coupling function $K(x)$ the *anisotropy* induced by the neural functional architecture, we get a model that converges (in a technical sense

	Totality (Wholes)	**Parts**
Phenomenological description	*Verschmelzung*	*Sonderung*
Topological-morphological description	Continuity	Discontinuity
Morphodynamical-physical explanation	Stability of internal attractors under spatial control	Bifurcation of internal attractors
Neuro-cognitive explanation I: wavelet analysis	One behavior of the amplitude of the wavelet transform	Another behavior of the amplitude of the wavelet transform
Computational vision I: Variational models	*Merging*: homogeneous 2D regions	*Splitting*: regular 1D boundaries
Computational vision II: Non-linear PDE	Diffusion	Inhibition of diffusion
Neuro-cognitive explanation II: oscillator networks	Synchronized oscillations (phase-locking)	Unsynchronized oscillations

adapted to varitional models) towards the gradient flow associated with the Mumford-Shah model endowed with the metric defined by $K(x)$.

10.5. Returning to the morphological nucleus

There is therefore a remarkable *convergence* of several different models of the morphological nucleus: physical, morphodynamical, geometro-topological, sensorial (wavelet analysis), cortical (networks of oscillators). All these models confirm and naturalize the eidetic phenomenological description of forms.

The table above sums up the convergence.

We see that, when approached from a physicalist viewpoint, a phenomenon in appearance as simple as 2D signal segmentation requires sophisticated tools to be modeled properly. In the next chapters, starting from perception and "climbing up" the levels toward semantics, we will need to simplify somewhat these models to preserve their practical aspect, while remaining as close to them as possible. It is the fundamental thesis of this work that the schematization of semantic contents is rooted in perceptual Gestalts.

11. Models for the Clark/Pylyshyn debate

Models for the filling-in of spatial extensions with qualities shed some light on a problem that is still at the core of cognitive studies. We give an example here.

In a 2004 article *Feature-Placing and Proto-objects* [62], Austen Clark discusses with Zenon Pylyshyn, who, in his own 2001 *Cognition* article [319] entitled *Visual indexes, preconceptual objects, and situated vision*, discussed another article by Clark from 2000. We begin with Clark's work.

11.1. A. Clark: Feature-placing and proto-objects

The issue at stake is low-level, pre-attentive vision, from retinal transduction to early processing in the primary visual areas. Stimuli coming from various sensory modalities lead to sensible qualities that constitute feature spaces (such as color space) possessing a certain number of dimensions.[16] It is an empirical fact of neuroscience that these features organize in "cortical feature maps", which are topographical. This means that the features are spatially localized. In other words, there exists a *spatial format* and, moreover, this format is common to different sensory modalities. This format organizes the objects of visual perception by means of segmentation, figure-background segregation, contour extraction, etc. (p. 13).

The theoretical problem raised by these facts is formulated by Clark as follows:

> This old picture [that we can describe perception only with feature terms] contains a big mistake, which is still doing damage in neurophysiology. Not only do we need more terms than just feature terms; we need a totally different *kind* of terms—one which has logical and semantic properties distinct from that of any possible feature term. (p. 5)

This "big mistake" is that philosophy of mind studies high-level cognitive faculties without worrying about their grounding in finely structured low-level structures.

> Any solution to this problem [binding, feature integration], I argue, requires a distinction in kind between features and their apparent locations. (p. 6)
>
> Terms for features and terms for places must play fundamentally distinct and non-interchangeable roles, because otherwise one could not solve the binding problem. (p. 6)

Those interested in the history of philosophy will recognize an aspect of the Leibniz/Kant debate. Clark rediscovers the Kantian problem of the principled irreducibility of the spatial to the conceptual.

16 The construction of these feature spaces through similarities and differences is already, as such, a fascinating and difficult problem.

He also rediscovers that, on the basis of functional arguments, what matters in feature localization are three fundamental operations:

(i) the restriction $S \subset R$ of spatial regions,
(ii) the identity of qualities on the intersections $R \cap S$ of spatial regions,
(iii) the cross-modal identity of spatial regions.

In short, in the logic of perceptual judgments, there is need for "a new kind of term, with a distinct function" (p. 8), "terms like names" (p. 8)—singular terms—to identify localizations.

This is necessary in order for judgments to apply to individuals, since *only localization individuates.* On the one hand, qualities are conceptual, generic, abstract and non-individuating. On the other hand, feature localization is "proto-predicative" and "proto-referential". The whole problem is to put the two together.

This problem is difficult because at the linguistic level (the predicative format of judgments), localization is *deictic.* It uses *indexicals* such as "here" and "there", which means that any perceptual judgment, even the most primitive, is *pragmatic* in the sense that its reference can be determined only in an indexical fashion.

Clark then explains that perceptual judgments must be grounded in a non-conceptual form of representation corresponding to the schema "appearance of quality Q at region R" (p. 8), in which the roles of Q and R cannot be exchanged. R indexically identifies a localization and Q attributes a quality to R.

> I claim that this "appearance of quality Q at region R" is the form of non-conceptual representation employed in any vertebrate sensory modality that can solve the many properties problem. (p. 8)

Thus, one must correctly describe the "ways of filling space" (p. 9). Clark understands that there is a logical problem with elaborating "feature-placing languages" comprising indexicals and demonstrative deictics. We must understand the logic and semantics of these feature-placing languages, which also exist in animal perception (p. 11) and are therefore non-conceptual and proto-predicative.

Such languages are poor, without defined descriptions or quantification, and restricted to basic statements such as "here is red", "there is cold".

> The prototypical sentence indicates the incidence of a feature in some demonstratively identified space-time region. (p. 9)

Their elementary data are pairs (r, q), with $r \in R$, $q \in Q$, and Clark remarks that this "pairing principle" is at the origin of the subject/predicate relation in judgments:

> Specification of the content of an act of sense requires pairs of the form $([q_1, ..., q_n], [r_1, ..., r_m])$ and (I argue) the pairing principle is analogous to the tie between subjects and predicates. (p. 11)

Therefore,

> Space-time provides a simple and universally available principle of organization (p. 16)

and the feature-placing languages refer to proto-objects which are

> preconceptual, direct, and derived from the causal mechanisms of perceptual processing. (p. 21)

It is evident that the models presented above formalize "feature-placing languages".

11.2. Z. Pylyshyn: Visual indexes and preconceptual objects

In his 2001 article [319], Zenon Pylyshyn concurs with Clark that perception and action require more than just conceptual representations giving a description of proximal sensory stimulations.

In effect, such conceptual representations by definition rely on categorizations, abstractions, and discrete judgments, which cannot establish direct causal relations with individual things coming from the experience of the external world. Pylyshyn rediscovers the thesis that one cannot reach the lived experience with only categorizing concepts that are equivalence classes of an infinity of concrete cases (p. 131).

> Sooner or later the regress of specifying concepts in terms of other concepts has to bottom out. (p. 129)

Thus, there is a fundamental problem of direct reference of deictic and indexical type ("this is red"), a direct reference opposite of a causal connection (causal theory of reference).

Zenon Pylyshyn focuses on this indexical characteristic of visual perception and introduces the hypothesis that there exist in the cognitive system certain indexes that point toward "visual objects", which are individual proto-objects in a visual scene. These proto-objects are "less" than real 3D objects (p. 144) and the problem is to pick them out. This problem is linked to localization but is not reducible to it:

> The present proposal is that the grouding begins at the point where something is picked out directly by a mechanism that works like a demonstrative. (p. 129)
>
> [The problem is] to pick out an individual in the world other than by finding the tokens in a scene that fall under a particular concept, or satisfy a particular description, or that have the properties encoded in the representation. (p. 130)
>
> There would have to be a non-descriptive way of picking out the unique object in question. (p. 138)

> The visual system has a mechanism for picking out and accessing individual objects prior to encoding their properties. (p. 139)[17]

Zenon Pylyshyn conducted numerous experiments on object motion (MOT, Multiple object tracing) to show that the spatio-temporal trajectory guarantees the numerical identity of the object. He insists a lot on the central issue that there is not only localization but also translocal identity along a trajectory. One cannot consider these trajectories simply as the translation of an attentional spot within a "large panoramic display", since the latter does not exist.

His conclusion is that there must be some non-conceptual index assignations, which constitute a primitive mechanism to select and maintain the identity of visual objects (p. 141), and allow to address objects, in the computer-scientific sense. In short:

> Sooner or later concepts must be grounded in a primitive causal connection between thoughts and things. (p. 154)
>
> The principle of grounding concepts in perception remains an essential requirement if we are to avoid an infinite regress. (p. 154)
>
> Without such a preconceptual grounding, our percepts and our thoughts would be disconnected from causal links to the real-world objects of those thoughts. (p. 154)
>
> Whithout preconceptual reference we would not be able to decide that a particular description D was satisfied by a particular individual (i.e. by that individual). (p. 154)

This opposition between Clark and Pylyshyn belongs to one of the most important debates in cognitive science, which addresses the relation of non-conceptual perceptual contents to judgment.

In its current formulation, the question was launched by Gareth Evans in *The Varieties of Reference* [102], who introduced non-conceptual and proto-propositional contents in an informational relationship to the environment, whose format is much finer-grained (continuous and geometrical) than the conceptual (discrete and propositional) format. At the same time, he also posited that the subject's conscious experience required conceptual contents. Vision experts such as Peacocke are in agreement, while they also think that non-conceptual contents are not autonomous. Bermudez believes that there is autonomy, but the link with judgments then becomes a real problem. Others, like McDowell in *Mind and World* [228], think that perception is always a perceptual judgment, thus is conceptual, and try to destroy the "myth of the given."

We see therefore what is the theoretical relevance of filling-in models.

17 After having severely criticized Gibson and the thesis that perception picks out information from the environment (see Section 6.3), Z. Pylyshyn seems to have changed his mind.

12. From 2D to 3D

The models we have presented in this chapter mainly concern how a 2D image can be structured geometrically and articulated into constituents. Yet, at this point we are still far from the concept of a 3D object in the Euclidean space $\mathbb{R}^3$. To be able to access that level, we need to include many more levels of representation. We review them here, to conclude this chapter.

12.1. Looking back on David Marr's perceptual theory

David Marr [224] introduced several levels of representation that make explicit certain aspects of the information encoded in the signals $I(x, y)$. Three of them are fundamental:

(i) The first level—the 2D *primal sketch* that we mentioned previously (see Section 4)—makes explicit the *local* geometrical and morphological organization of the signal. It allows segmentation processes that will support the intermediate and final (cognitive and inferential) stages, such as interpretation, recognition, understanding, etc.

(ii) The second level—called "$2^1/_2$D" to point out that it is *intermediate* between the 2D and 3D levels—is the essential level of the theory. In fact, Marr calls it the "pivotal point" of "pure perception". It is a globally organized level, integrating several modular computations carried out on the primal sketch: visible surface contours, textures, stereopsis, movement, shading, etc. It represents the external word as composed of visible surfaces filled with sensible qualities and moving in $\mathbb{R}^3$. It is neither strictly sensory (since surfaces are external), nor strictly objective (since appearances are internal). It constitutes the *phenomenological appearing* per se and is therefore of a true *morphological* nature. In particular, this level is where certain discontinuities are interpreted as *apparent contours* of objects. The processing of apparent contours is crucial, since it allows the transition from the 2D sketch to 3D models in space-time. This is done by reconstructing forms (and spatio-temporal relations among forms) from qualitative 2D discontinuities.

(iii) The third level—the level of "3D models"—is the objective level of real things and material volumes with their real properties. It is from this level that superior cognitive tasks and constituents of conceptual structures operate, e.g., the hierarchical decomposition of shapes into parts, the constitution of prototypes, etc.

The hypothesis is that perception is a bottom-up process

$$2\text{D} \to 2^1/_2\text{D} \to 3\text{D} \to \text{conceptual structure}$$

that possesses top-down feedback mechanisms (anticipation, inferences, interpretations, etc.)

$$\text{conceptual structure} \to 3\text{D} \to 2^1/_2\text{D}.$$

The $2^1/_2$D level would thus be the last level of the bottom-up perceptual processing, hence its particular significance.

12.2. Marr's $2^1/_2$D sketch and Husserl's adumbrations

Marr's $2^1/_2$D sketch corresponds quite exactly to what Husserl called "adumbrative" perception. In many of his writings, especially *Ding und Raum* [161] and *Ideen I* [162], Husserl explains in detail how objects immersed in an external objective 3D space can be constituted from 2D adumbrations. We have studied these questions in several works [18] and will not elaborate much further here. We will only remind the reader about a few essential points.

12.2.1. *Adumbrations and perceptual intentionality.* In adumbrative perception only *a single* adumbration (*Abschattung*) of the object is given at every moment. Every adumbration is a noematic appearance (correlative of a noetic synthesis). This type of manifestation is by definition *incomplete* and *inadequate* (since a 3D object cannot be reduced to a single 2D aspect), and this very incompleteness is the origin of the external transcendence of the object. One meets here the perceptual roots of *intentionality* as a mapping from an internal representation to an external object. In his texts, Husserl treats a great number of problems, which are still remarkably current.

(i) There is an equivalence between the temporal flow of adumbrations and the object $= X$ as noema. There are "chaining rules" that are prescriptive for experience, i.e., determine "descriptive compositions" and "internal organizations" ruling the diversity of adumbrations. Those rules are submitted to the noematic unity of the object. They regulate its appearing and its mode of presentation.

(ii) The incompleteness of adumbrations implies that the complete determination of the object's givenness can only be *temporal.* Incompleteness and temporality are essentially linked and are the characteristics of the *finiteness* of consciousness. Constituted in the temporality of the pure ego, the flow of adumbrations unfolds as a dynamical order that is noematically prescribed. Hence, the crucial problem of *anticipation.* Aspect changes can be anticipated. The possibility of coherent and ruled anticipations characterizes objective transcendence. It is founded in the immanence of the acts that give access to objects.

(iii) The incompleteness of adumbrations generates a *horizon* of "co-givenness": the current givenness of an actual adumbration is inseparable from the (implicit) givenness of an infinite number of other, virtual, adumbrations.

(iv) The incompleteness generates also the gap between intuition and intention. It generates the intentional content of the noematic object and

18 See, e.g., [261] and [294].

founds is transcendence. As Ronald McIntyre [229]) insisted, it is only because the complete determination of the givenness is necessarily temporal and relying on a structure of anticipation, that the noematic rules are *semantic.* We meet here the phenomenological genesis of meaning and denotation.

Clearly, this conception is as far as it can be from the classical conceptions of reference, in which a formal symbol denotes an individual object. In fact, the traditional logical conception is valid only for mathematical idealities and is mistakenly applied to perception. The perceptual level is more primitive than the semantic level; the latter is grounded on the former, not the other way around.

12.2.2. *Apparent contours and singularities.* We already pointed out long ago the link between the Husserlian eidetic description of *Abschattungen* and the research on multiscale differential geometry and computational vision (see [260], [261], and [279]). It is especially striking in the problem of *apparent contours* (ACs) of objects. ACs are a fundamental example of what Husserl called dependent moments and condense the essentials of the problematic of adumbrations.

Suppose that ACs have already been constituted as noematic appearances (via signal analysis, stereopsis, etc.). Then remains the geometrical problem of the equivalence between an objet T (a surface) embedded in $\mathbb{R}^3$ and the functional space of its ACs. It is a highly non-trivial *inverse problem.* The direct problem gives the object T in $\mathbb{R}^3$, a projection plane P, and a direction of projection π, and requires to build the AC of T relative to the projection (π, P). We denote this specific AC of T by $\mathrm{AC}_T(\pi, P)$. This entity already involves deep mathematical concepts, as it represents the projection on P parallel to π of the *singular locus* S of this projection restricted to T. In other words, S is the set of points x of T where direction π is *tangent* to T, and $\mathrm{AC}_T(\pi, P)$ is the projection of S. Figure 23 shows the visible part of the AC of a torus T under an oblique direction, while Figure 24 represents the full AC.

Using the singularity theory developed by Hassler Whitney, René Thom, John Mather, Vladimir Arnold and others, one can classify the *generic singularities* of ACs. They contain the essential morphological information. It can be shown (with help from multiscale jet theory) that the detection of ACs can be done with cell fields whose receptive profiles have the appropriate form.

It is especially interesting to consider the links between these singularities and the Riemannian geometry of the surface: its elliptic domains, its hyperbolic domains with two asymptotic direction fields, and its parabolic curves separating the elliptic from the hyperbolic domains. Depending on how the direction of the projection is positioned with respect to these elements, one finds singularities that are more or less complex and were classified by Thom, Arnold and Kergosien in the early 1970's. For example, the swallowtail singularities of a torus correspond to the points where the direction of projection is tangent

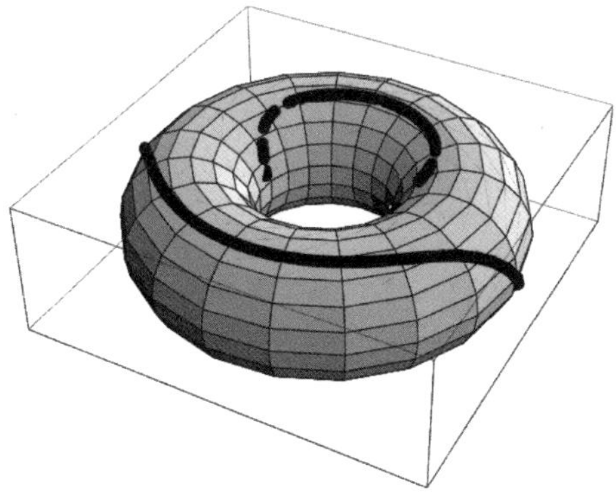

FIGURE 23. The apparent contour of a torus for a frontal direction of projection with slope $\pi/4$.

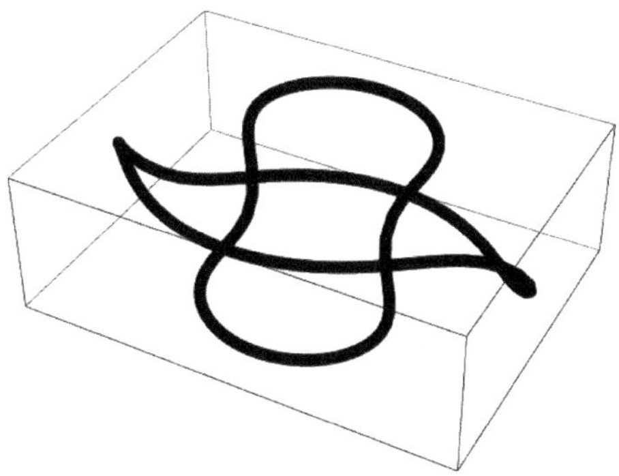

FIGURE 24. The complete apparent contour of the previous torus figure.

to an asymptotic curve in one of its inflection points ("flecnodal" point). The most complex situation happens when the direction of projection is a double asymptotic line tangent to a parabolic curve. Figure 25 shows the transformation of the aspect of a torus under rotation, while Figure 26 shows in some detail the evolution of the ACs. Notice the apparition of the two swallowtails and, later, the uncrossing of their horizontal branches.

If we now consider the space V of all projections (i.e., viewpoints), we can study the *temporal evolution* of the AC of object T as we follow a path inside V. First, V is decomposed into domains that represent ACs of the same qualitative type—called *aspects*. The classification of the different ACs in qualitative types *categorizes* (stratifies) V. Every object T categorizes V by means of a catastrophe set K_T (a system of boundaries) and defines what is called its *aspect graph* G_T. There is an *equivalence* between T and K_T. A given T generates K_T: this is the (relatively easy) direct problem. Conversely,

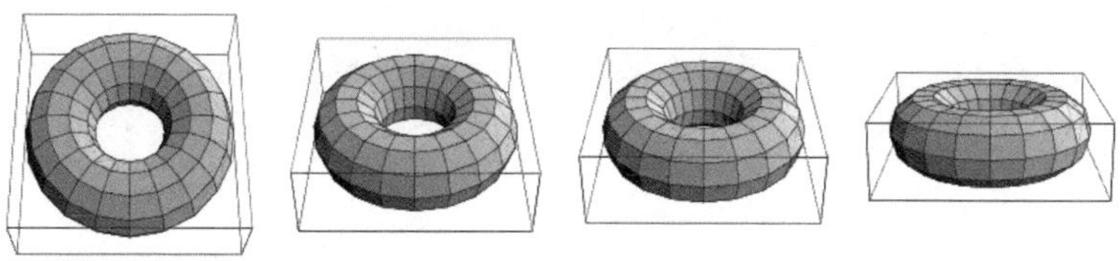

FIGURE 25. Transformation of the apparent contours of a torus under rotation around a frontal horizontal axis.

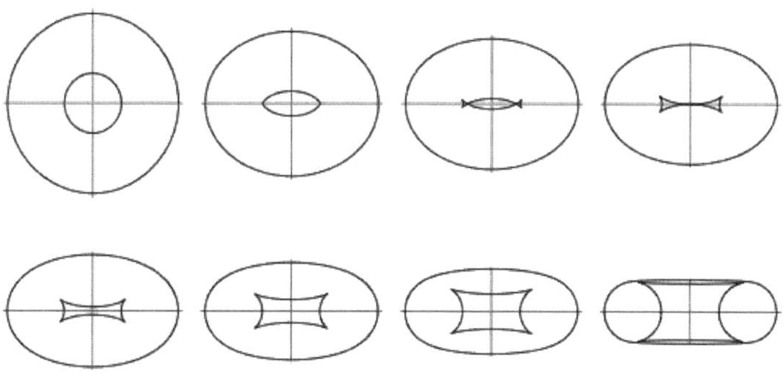

FIGURE 26. Detailed evolution of the torus's apparent contours during rotation. The two swallowtails start unfolding at step 3. Their horizontal branches become tangent and uncross at step 4.

a given K_T allows to reconstruct T: this is the (much more difficult) *inverse* problem.

The equivalence $T \Leftrightarrow G_T$ is the mathematical version of the Husserlian eidetic law of adumbrative perception. It can be expressed in a temporal way by exploring (V, K_T) and G_T through *temporal paths.* This conversion of the "synchronous" object into a sequential scanning of aspects induces the feeling of "exteriority". More precisely, it is the *paradigm* (in the structuralist sense) of the different aspects—a paradigm geometrized by the categorized space V—that *founds* intentionality. Intentionality is therefore a radically non-logical mechanism.

This example shows very well the difference between the noema as object = X and the noematic appearances (*Abschattungen*). An adumbration has a noetic side (the processing) and a noematic side (its pure geometry). As a shape (AC), it is a noematic appearance, not a noematic content. And as a content, it is an iconic, gestaltic content, not a propositional content.

As a projection of an external object (direct problem), an AC is a piece of information about the external world for the cognitive system, that is a "large content" that refers to the object according to the causal theory of reference. But as an element of the space of ACs, i.e., as associated with the categorization (V, K_T) and the aspect graph G_T (inverse problem), it is a "narrow content" functionally individuated by its relations with the other ACs of the same object. The link between these two statuses is made by the object $= X$ as a principle of coherence and unity. The object $= X$ warrants that the aspect graph G_T is effectively generated by an object. It acts as a *choice operator* that selects an AC in an intensional and indexical way.

In this conception, reference has nothing to do with the denotation of symbols. It is only when, at higher levels, the object $= X$ noema has been converted into a symbol that reference becomes denotation. But at the perceptual-morphological level, reference is a result of incompleteness and anticipation, above all of an inverse problem, and not at all of denotation.

CHAPTER 3

Relations

In collaboration with René Doursat

1. Introduction

How can the same English relationship 'in' apply to scenes as different as "the shoe in the box" (small, hollow, closed volume), "the bird in the tree" (large, dense, open volume) or "the fruit in the bowl" (curved surface)? What is the common 'across' invariant behind "he swam across the lake" (smooth trajectory, irregular surface) and "the fly zigzagged across the hall" (jagged trajectory, regular volume)? How can language, especially its spatial elements, be so insensitive to wide topological and morphological differences among visual percepts? In short, how does language drastically simplify information and *categorize*?

The previous chapter addressed the basic concept of *things* and introduced algorithms dealing with image segmentation, perceptual mereology and object constituency. We now turn to the second major component of Langacker's trilogy, the *relations* between things. Relations clearly constitute the most important problem at the core of all theories of language. Ultimately, different theoretical perspectives on syntax will be distinguished on the basis of how they construe relations. It is only after relations are expressed in a mathematical form that *processes* can be modeled as temporal evolutions of relations and *events* as changes occurring during these processes.

1.1. The gestaltic conception of relations

The various theoretical paradigms of language are broadly divided between formal or symbolic conceptions of things and relations, on the one hand, and what we here call "gestaltic" conceptions, on the other hand. In the formal framework, things are comprised of autonomous, already individuated objects, or "atoms", and relations are represented by abstract links connecting these atoms. This classical view, which has been extensively studied in the philosophical literature (for an excellent discussion, see in particular Wittgenstein [414] and Mulligan [240]), is nominalist and forms the basis of logical atomism and set theory.

In the gestaltic conception, by contrast, things and relations taken together constitute wholes, i.e., they are part of higher-order complex objects, holistic units or global configurations. First, things are separated and detached from their common background by a segmentation process (see previous chapter). Then, a mereological analysis can reveal what this "background" is made of, both in the spatial relations between the objects and in their morphological features. Here, relations between symbols are not taken for granted but *emerge together with the objects* through mereological segmentations and transformations.

One of the best presentations of this key idea, already promoted in the early beginnings of Gestalt theory and phenomenology, can be found in Husserl's 1907 *Ding und Raum* [161]. In §59 Husserl tries to understand how object configurations or "complexes" of objects are formed. Insisting that a typical feature of visual sensations is to constitute a field endowed with a spatial format, he explains that everything having a perceptual unity is rooted in the perceptual unity of the visual field (hence its temporal evolution, too). Individuated objects can only exist inasmuch as they are detached and profiled against their background. As a consequence, relations among objects are themselves profiled against the same background of spatial unity. In this sense, objects involved in relations can thus be conceived as being the parts of a *higher-order complex object*, the global configuration. The relations making a configuration of objects are therefore *mereological* and concern

> the spatial order of parts at the core of the spatial whole. (p. 256)[1]

The main challenge is then to *categorize* these relations, and this constitutes a genuinely difficult problem.

From Husserl to contemporary research on AI, human-machine interaction and robotics, the interplay between perception and language in processing spatial relations has always been a central issue. Let us cite two examples, among many others. In his 1995 paper [151] "Coping with static and dynamic spatial relations" (see also [152]), Gerd Herzog stresses that

> (the) interplay between visual perception and natural language in human-machine interaction receives growing attention since it constitutes a prominent issue in many potential application areas. The aim of language-oriented AI research in this context is to achieve an operational form of referential semantics that reaches down to the sensoric level.

Going from visually accessible information to linguistic spatial descriptions is a crucial challenge that requires modeling spatial relations

---

1 Independently of Husserl, it is well known in algebraic topology that the number of blobs X_i embedded in an ambient plane W is coded by the homotopy group of the complementary set $W - \{X_i\}$. This principle is widespread: the structure of a configuration X embedded in an ambient space E can be analyzed by looking at its complementary set $W - X$.

> as intermediate conceptual units between linguistic and sensory levels to bridge from visual accessible three-dimensional geometric data to language-oriented representations.

In much the same vein, Nikolaos Mavridis and Deb Roy from the Cognitive Machine Group at the MIT Media Lab developed "grounded situation models for robots" [226] to bridge the gap between perception, action, and language.

> Our long-term objective is to develop robots that engage in natural language-mediated cooperative tasks with humans. For example, the system can acquire parts of situations either by seeing them or by "imagining" them through descriptions given by the user: "There is a red ball at the left". These situations can later be used to create mental imagery, thus enabling bidirectional translation between perception and language.

The gestaltic conception calls into question the traditional perceptual vs. linguistic opposition. Visual perception is generally thought of as a faculty that is exclusively concerned with the shapes of objects, while language is supposed to be exclusively concerned with the abstract relations between those objects. Interestingly, this assumption, which rules over mainstream linguistics and the philosophy of language, is directly inherited from the classical gestaltic figure/background opposition. However, it starts losing its relevance when one takes into account the fact that the so-called perceptual "background" is a true *figure* in the eyes of language. As any figure, it is actively processed and organized by structuring principles—the grammatical elements (see below)—in a spatial and morphological way that is radically different from mere symbolic relationships. As we show in this chapter, the gestaltic conception of relations was remarkably revived and further developed by the recent trends in cognitive linguistics and cognitive grammars.

This conception is also very close to many psychological studies inspired by the fundamental works of Albert Michotte in the 1940's on the perception of causality. Against the traditional Humian view, Michotte [234] showed that apparently very high-level cognitive concepts such as *causality* are in fact deeply rooted in the automatic and hardwired algorithms of perception. As explained by Johan Wagemans ([399], p. 11):

> Before Michotte, nearly all writers had treated causality as a high-level cognitive concept, and tended to think of the currency of perception in terms of only lower-level properties such as color, texture, and motion. Michotte, in this context, demonstrated that even seemingly "cognitive" properties such as causality may be processed in the visual system.

Michotte was inspired by Gestalt theory and claimed for instance that the visual "inferences" explaining Kanizsa amodal completion phenomena are automatisms of low-level visual processing (see, e.g., [235]). We claim that this central thesis is valid for geometric and kinematic relations in general.

In cognitive psychology, the categorization of spatial relations in preverbal conceptual thought has also been investigated extensively (see, e.g., Mandler [222] and McDonough [227]).

In this chapter we want to apply this perspective to the analysis of the topological contents of grammatical elements. We examine the link between the spatial structure of visual scenes and their linguistic descriptions. Based on a sampler of scene/description pairs, we attempt to reconstitute the toolbox of basic dynamical routines that underlie the grammatical and spatial organization of these descriptions. We then derive effective algorithms from these routines and propose a computational model explaining how the endless diversity of schematic visual scenes can be mapped to a small set of standard semantic labels. We also show how these algorithmic principles naturally relate to new developments in the neuroscience of vision and spiking neurons dynamics. In short, our proposal is that spatial and grammatical structures are related to each other through image transformations that rely on diffusion processes and wavelike neural activity.

To this aim we focus our attention on the remarkable work of Leonard Talmy (Chapter 1, Section 3.3), unsurpassed in its precision, richness of details and systematic ordering. Our intention is to use a few of the numerous examples given by Talmy as the foundation for a new model based on gestaltic visual routines.

In the remainder of this chapter, Section 2 applies an "active semantic" approach to the analysis of spatial grammatical elements and draws a link with visual processing routines. In Section 3 we discuss spatial invariance and the notion of "linguistic topology" in the context of mathematical geometry, which leads us to reaffirm the importance of transformation routines. Section 4 illustrates the specific difficulties raised by perceptual-semantic schemata with a detailed example, the English preposition "across". Modeling principles and algorithms are then introduced in Section 5, while numerical simulations on cellular automata are described in Section 6. Finally, we suggest a natural and plausible neural implementation of expansion-based morphology, based on waveform activity dynamics in spiking neural networks, in Section 7.

1.2. Scope of this study

Before proceeding further, however, we wish to avoid any misunderstanding by pointing out the following important epistemological aspect: in this chapter, we restrict our analysis of prepositional contents to their *morphological and perceptually rooted kernel*. A full linguistic analysis of prepositional semantics would be of a much more complex nature and will not be addressed here.

Contemporary theories of semantics have shown that terms with spatio-temporal content are highly polysemous. For example, different uses of 'in' relate to qualitatively different spatial situations [150]:

(1) a. There is a cat in_1 the house.
b. There is a bird in_2 the tree.
c. There is a spoon in_3 the cup.
d. There is a crack in_4 the vase.

Spatial prepositions typically form prototype-based radial categories (see Rosch [326]). Here, 'in_1' represents the most prototypical "containment" schema, while 'in_2' departs from it slightly by the texture and openness of the container. Many top-down knowledge-based inferences are also constantly operating on the data. For instance, the analysis of the 'in_3' sentence must take into account the instrumental function of the spoon and its mereological decomposition into spoon head and spoon handle. The spoon handle is sticking out of the cup so the complete spoon is not physically inside the cup but only *metonymically*, via its head part.[2] In the subcategorical island 'in_4', the container is a surface, not a volume, etc.

Another major departure from prototypical use comes from *metaphors*, a pervasive mechanism of cognitive organization and categorization rooted in the fundamental domains of space and time (see Lakoff [197], Johnson [173], Lakoff-Johnson [201], Talmy [374], [375]). Metaphorically, the preposition 'in' grammaticizes a generalized "containment" schema that can be applied not only to a great variety of concrete objects and scenes but also to highly abstract situations:

(2) a. I am in_A a committee.
b. I am in_B doubt.

These uses of 'in' pertain to a virtual concept of space generalized from real space, respectively:

(3) a. I am in_5 a crowd.
b. I am in_6 water.

In these examples, 'in_A' metaphorically maps to an element of a discrete numerable set exemplified by 'in_5', while 'in_B' relates to an immersion into a continuous ambient substance such as 'in_6'.

With these preliminary remarks in mind, we restrict the scope of this chapter to relatively homogeneous subcategories or *protosemantic* features, i.e., low-level compared to linguistic categories but high-level compared to local visual features. Our goal is to categorize scenes into elementary semantic *subclasses*, such as 'in_1', or clusters of closely related subclasses, such as 'in_1-in_2'. We will not attempt to delineate the whole cultural complex formed by the English preposition 'in' by a single (and non-existent) universal.

---

2 Langacker also emphasized this point In [207]: "As shown by *The flower in the vase*, where most of the flower protudes", the semantic content of 'in' is not only topological but also functional.

However, even without additional semantic operators, we are still faced with the difficult problem of linking invariant semantic kernels to an infinite continuum of perceptual shape diversity. Even in their most typical, physical and concrete realizations taken independently, such as 'in_1', in_2', etc., how can these invariant semantic kernels apply to perceptual data?

This challenge will be at the core of the present chapter and the remainder of this book. As mentioned above, contributing to its solution means focusing on a collection of low-level topological and dynamical routines that structure visual scenes in a gestaltic fashion. We are not implying, however, that top-down symbolic operations are not a full part of the cognitive act of language. After all, linguistic productions are made of symbolic utterances. Moreover, there must exist a higher-level control or "intentional" level of organization deciding which low-level routines should be applied to the scene and in what sequence. Yet, before these issues can be solved, the central interface between the lower perceptual and protosemantic levels of language must first be addressed.

2. Talmy's Gestalt semantics

As we have seen in Section 3.3 of Chapter 1, a fundamental thesis of Talmy is that language involves a system of organizational elements that structure the conceptual material in much the same way as visual elements structure the perceptual material.[3] He also shows that these elements are expressed *grammatically*, as opposed to lexically. The grammatical elements (prepositions 'in', 'above', etc.; declension and conjugation suffixes, and so on) form closed classes, i.e., restricted sets that contain relatively few members and that almost never expand in the course of a language's evolution. By contrast, the lexical elements ('bird', 'box', etc.) are extremely numerous and continuously produced, therefore creating open classes. Assuming that sentences evoke cognitive representations (CRs) in the listener's mind, Talmy claims that the grammatical elements provide the conceptual structure or "scaffolding" of these CRs, while the lexical elements provide their conceptual contents.

Ronald Langacker [203] (see Section 3.2) also proposes that language essentially revolves around *categorization* at all levels of cognitive organization—as well among the things as among their relations (processes) or the temporal evolution of these relations (events). From there, the notion of *schematization* plays a central role in cognitive linguistics. Since its early Kantian origins, a "schema" is a procedure that can transform the content of a concept into a modus operandi for *constructing its referents.*

---

3 All major articles of Len Talmy published between 1972 and 1996 have been collected in the two volumes of his magnum opus *Toward a Cognitive Semantics* [374], [375].

2.1. Active semantics

Talmy's most compelling idea is that in a relational complex of objects where a figure—Langacker's trajector TR—is profiled against a ground—Langacker's landmark LM— the ground is not merely a passive frame for the figure, but *the ground's geometry is also grammatically specified.* In fact, the majority of spatial grammatical elements treat the figure TR as a mere point while making various elaborate distinctions for the reference object LM. For example, consider the following three sentences:

(4) a. The ball is in the box.
 b. The ball is on the box.
 c. The ball is 20ft away from the box.

Depending on the viewpoint, (4a) construes the box as a container volume, whereas (4b) focuses on the top surface of the box and (4c) reduces the box to a reference point. Thus, different grammatical elements scaffold reality in conceptually different ways. A classical theory of syntax, by contrast, would assimilate the box to a symbol irrespective of its actual geometrical shape or any other perceptual features, and assume that all spatial meaning is encapsulated in the abstract relationships 'in', 'on', and 'away from.'

Now, by giving a new significance to the object's geometry we are also challenged to find how this geometry varies with linguistic circumstances, i.e., how prepositions actually select and process certain morphological features from the perceptual data, while ignoring others. We propose that simple morphemes like the spatial elements 'in', 'above' or 'across', correspond in fact to visual processing algorithms that take perceptual shapes as input, transform them in a specific way, and deliver a semantic schema as output. We call these algorithms *morphodynamical routines* and the global structuring process *active semantics.*

If we take the tenets of cognitive linguistics seriously, we are therefore led to the following postulate: given an input perceptual scene that is already the outcome of low-level preprocessing tasks, such as boundary detection and object segmentation (see Chapter 2), language semantics is an active process that takes this scene to the next level of organization through a package of transformations or "morphodynamical" routines. These morphodynamical routines erase details and create new forms that evolve temporally.

- They enrich a visual scene with new *virtual* structures or *singularities* (e.g., influence zones, fictive boundaries, skeletons, intersection points, force fields) that did not originally belong to the scene but reveal the meaning it conveys.
- To this goal, they select from the scene its appropriate morphological features (surfaces, angles, segments, etc.) as the basic material to carry out the said transformations.

In summary, according to this program, semantic schemata literally trigger additional processing routines. We suggest below that these routines are akin to object-centered and diffusion-like non-linear dynamics.

2.2. Basic structuring schemata

In this section we recall and adapt a representative sample of Talmy's findings to illustrate the basic structuring operations performed by grammatical elements. As explained above, our goal is to associate grammatical configurations with data-reducing transformation algorithms essential to scene categorization.

2.2.1. *Magnitude invariance → multi-scale analysis.* We start with the linguistic properties of invariance and neutrality with respect to the detailed shape and dimensions of the perceptual data. One of Talmy's major claims ([374], p.28) is that:

> the nature of [grammatical] structure is largely relativistic, topological, qualitative or approximative rather than absolute, Euclidean, quantitative or precisional.

We will discuss the "topological" quality of grammar in more detail in Section 3. For now, we retain the fact that grammatical elements are fundamentally scale- or metric-invariant. Compare, for example (Talmy [374], p. 25):

(5) a. This speck is smaller than that speck.
b. This planet is smaller than that planet.

The above sentence pair shows that real distance and size do not play a role in the partitioning of space carried out by the deictic words 'this' and 'that'. The relative notions of proximity and remoteness conveyed by these elements are the same whether in millimeters or parsecs. This does not mean, however, that grammatical elements are insensitive to metric aspects but rather that they can uniformly handle similar metric configurations at vastly different *scales*. Multi-scale processing therefore lies at the core of Gestalt semantics and, interestingly, as we have seen in Chapter 2, has also become a main area of research in computational vision (see, e.g., Mallat [219], Perona-Malik [257], Whitaker [401], Morel-Solimini [239]).

2.2.2. *Morphology invariance.* Numerous examples collected by Talmy show that grammatical elements are largely indifferent to the morphological details of objects or trajectories of moving objects, and that these details are mostly entrusted to lexical elements. This invariance of meaning points to the existence of underlying visual-semantic routines that perform a *drastic but targeted simplification* of the geometrical data.

Data-reducing routines fall into a few different categories, partly listed below. They are crucial to the faculty of scene *categorization.*

◇ *Bulk-neutrality* → *skeleton routines*

Another form of magnitude invariance appears in the selective rescaling of specific spatial dimensions. This is the case of thickness or bulk invariance. For example (Talmy [374], p. 31):

(6) a. The caterpillar crawled up along the filament.
b. The caterpillar crawled up along the flagpole.
c. The caterpillar crawled up along the redwood tree.

These sentences show that 'along' is indifferent to the girth of LM and focuses exclusively on its main axis or *skeleton*, parallel to the direction of TR's trajectory. Like multi-scale analysis, skeleton transforms are widely used in computational vision and implemented using various algorithms (see the previous chapter). Studied by Blum [38] under the name "medial axis transform" and by others as "cut locus", "stick figures", "shock graphs" or "Voronoi diagrams" (see, e.g.,, Marr [224], Siddiqi et al. [345], Zhu-Yuille [419]), morphological skeletonization plays a crucial role in theories of perception and is considered a fundamental structuring principle of cognition by Leyton [212]. Skeletons indeed conveniently simplify and schematize shapes by getting rid of unnecessary details, while at the same time conserving their most important structural features. As we have seen in the previous chapter, there is also experimental evidence that the visual system effectively constructs the symmetry axis of shapes (see, e.g., Kimia [179] and Lee [209]).

◇ *Continuity-neutrality* → *expansion routines*

Sentences such as:

(7) a. The ball is in the box.
b. The fruit is in the bowl.
c. The bird is in the cage.

are good examples of active-semantic analysis showing the neutrality of the preposition 'in' with respect to the morphological details of the container. We therefore propose that the active-semantic effect of 'in' is to trigger routines that transform individual objects and their composition in a scene in the following manner:

- The container LM ('box', 'bowl', 'cage') is closed by adding virtual structures to complete its missing parts: the 'bowl' becomes a sphere, the 'cage' a continuous surface of similar shape, etc. Generally, this type of *closing* process replaces objects by continuous spherical or blob-like occupation domains.
- The contained object TR ('ball', 'fruit', 'bird') expands, for example by a contour diffusion process, irrespective of its detailed shape until it collides with the boundaries of the closed LM. In fact, these two phases of the 'in' routine, the closing of LM and the expansion of TR, can run in parallel.

- Both TR and LM expand simultaneously until they encounter each other's boundaries. In this case, the expansion of LM naturally creates its own closing while at the same time providing an obstacle to the expansion of TR.

Therefore, we propose the following active-semantic definition of the prototypical "containment" schema (corresponding to the 'in$_1$-in$_2$' cluster of examples (1) and their closely related variants):

A domain A lies 'in' a domain B if an isotropic expansion of A is stopped by the boundaries of B's closing or own isotropic expansion.

Although simple, the double expansion process that we propose for the "containment" schema is not trivial and leads us to introduce the following general principles of active-semantic morphodynamics:

(i) each object has a global tendency to occupy the whole space,
(ii) objects are obstacles to each other's expansion.

Through the action of structuring routines, the common space shared by the objects is divided into *influence zones* in an isotropic or anisotropic fashion. Image elements cooperate to propagate activity across the field and inhibit activity from other sources. Thus, the preposition 'in' involves an isotropic obstacle, whereas for example 'over' involves an obstacle in the lower half of the scene preventing a vertical expansion of the TR from reaching the bottom edge of the visual field.

In summary, within the general framework that all perceptual objects share the same surrounding space, we suggest that spatial interrelations between objects are inferred from the boundaries created by their territories (wave front collision curves and singular points on these curves) and the dynamical evolution of these boundaries. A directly measurable consequence of this definition is the fact that no TR-induced activity will be detected on the boundaries of the visual field. The final boundary between the domains is approximately equal to the skeleton of the *complementary space* between the objects, also called *skeleton by influence zones* or SKIZ (see Prewitt [316] and Serra [341]). In the case of the "containment" schema, the SKIZ surrounds TR and no region on the boundaries of the image will be encompassed in TR's expanded domain. The sheer absence of TR activity on the boundaries is a robust feature that contributes to categorize the scene as 'in'. It illustrates a typical morphodynamical routine at the basis of our perceptual-semantic classifier.

◇ *Shape-neutrality → singularities*

Other examples by Talmy clearly show that salient aspects of shapes are simply not taken into account by grammatical elements. In the following sentences (Talmy [374], p. 27):

(8) a. I zigzagged through the woods.
b. I circled through the woods.
c. I dashed through the woods.

the actual trajectory of the moving object TR (the first person), whether made of segments, curves or a single straight line, ultimately connects two opposite (and possibly virtual) sides of the extended domain LM (the woods). In our active-semantic view, this shows that the element 'through' creates a drastic schematization of TR in two steps:

- the trajectory is "convexified" or "tubified", again by outward expansion (yet, here, to a limited extent) and becomes what is called in geometry a *tubular neighborhood*; then
- the main axis of this tube is extracted through a *skeletonization* process that is basically a reverse inward expansion eroding the object down to its medial symmetry axis.

At the same time, it does the same with LM: the texture is first erased, then the domain is skeletonized. In short, the grammatical perspective enforced by the English preposition 'through' reduces the detailed original trajectory of TR and texture of LM to mere fluctuations around their medial axes. Finally, the characteristic feature of the 'through' schema lies at the intersection of the two skeletons.

As already mentioned above, skeleton transforms are widely used in computational vision and implemented using various algorithms. They offer a convenient way to simplify and schematize shapes by getting rid of unnecessary morphological details, while at the same time conserving their most important structural features. For this reason they are also well suited to the analysis of relations. We will further develop this point below (see Section 4).

In this example, the spatial relation between the objects is inferred from virtual *singularities* in the boundaries of the objects' territories. These singularities and their dynamical evolution are important clues that constitute the characteristic "signature" of the spatial relationship (Petitot [261]). Transformation routines considerably reduce the dimensionality of the input space, literally "boiling down" the input images to a few critical features. A key idea is that singularities encode a lot of the image's geometrical information in an extremely compact and localized manner.

In summary, after the morphodynamical routines have transformed the scene according to the expansion principles (i) and (ii), several types of characteristic features can be detected to contribute to the final categorization:

(iii) presence or absence of activity on the boundaries,
(iv) intersection of skeletons,
(v) singularities in the SKIZ boundary.

Our active-semantic approach proposes a link between the "things, relations, events" trilogy of cognitive grammar (Langacker [203]) and the "domains, singularities, bifurcations" trilogy of morphodynamical visual processing (Petitot [261]). We further suggest in Section 7 that the brain might rely on dynamical activity patterns of this kind to perform invariant spatial categorization.

3. What is "cognitive topology"?

The core invariance of spatial meaning reviewed above is sometimes referred to as the "linguistic form of topology" (Talmy [374]), or "cognitive topology" (CT) (Lakoff [199]), and is often compared to mathematical topology (MT). Section 2.2 showed that, just like MT, CT is magnitude- and shape-neutral. Yet, CT is also more abstract and less abstract than MT at the same time (Talmy [374], p. 30):

- In some areas, CT has a greater power of generalization than MT. For example, the invariant "containment" schema extracted from 'in the box', 'in the bowl' or 'in the cage' (see 2.2.2) shows that incomplete or disconnected parts can be equivalent to a single continuous surface through a closing process.
- In other areas, however, CT also preserves metric ratios and limits distortions in a stricter sense than MT. For example, the lexical elements 'cup' and 'plate' have distinct spatial semantic properties and uses, although a plate is nothing more than a flattened version of a cup (they are homeomorphic in the mathematical sense). As we will later see in Section 4, the grammatical element 'across' also exhibits subtle metric constraints in its applicability.

In reality, because of all these discrepancies, CT has actually little to do with true MT. In the mathematical sense, topology refers only to a certain level of structural description of spatial objects. Let us give a brief reminder of this topic.

In mathematics, structural levels go from the least constrained (sets) to the most constrained (Euclidean metric) and each level is associated to a particular class of mappings or structure morphisms preserving that level:

Lev 0 corresponds to pure *set* structures. Points are independent from each other and mappings can be any set applications.

Lev 1 is the *topological* level. Points are "glued together" through qualitative neighborhood relationships and mappings are continuous applications, which can be for example fractal.

Lev 2 is the *differentiable* level. Objects are "smooth" and can be approximated by tangent linear objects. Mappings are infinitely differentiable. Fractals are typical examples of objects from Level 1 that do not belong to Level 2.

Lev 3 is the *conformal* level and corresponds to the existence of "complex" structures, in the sense of complex numbers. Mappings are holomorphic, which means in the 2D case that they preserve angles.

Lev 4 is the *metric* level and introduces the concept of distance. Several subcategories can be distinguished here: In Riemannian spaces, metric is non-homogeneous (it is local and can vary from point to point); in Euclidean or non-Euclidean spaces (hyperbolic, Minkowskian, etc.) metric is homogeneous and mappings are isometric transformations.

Lev 5 is both metric and *linear.* This is the level of vectorial spaces with a norm (Euclidean spaces, Hilbert spaces).

We want to address again the question of a cognitive topology within this mathematical context. Levels 2 and 3, which lie between the malleable topological level and the rigid metrical level, are both "soft" in the sense that they allow stretching, and are at the same time more constrained than level 1. At first sight, these qualities would make levels 2 and 3 good candidates for the sought-after cognitive level of representation of CT. However, we think that this view is misleading. In our sense, if there is such a thing as a cognitive topology, it does not correspond to any of the levels (or intermediate levels) in the above mathematical hierarchy.

3.1. Convexification

In reality, the active semantics conception presented in Section 2.1 leads us to consider *inter-level* transformations, i.e., processes starting at one level and going up or down the hierarchy. For example, a *convexification* routine, which makes objects smoother and "rounder", is a true schematization in the sense that it discards a great amount of the objects' metric properties and yields only a small number of blob-like shape categories. To the limit of spherical convexification, all objects become equivalent to a unique 'sphere' class. This makes convexification appear to go down the level hierarchy, towards the "soft" topological levels. Yet, at the same time, the convex shapes it creates are more "rigid" than the original objects because, precisely, mappings acting within these shape classes are constrained to preserve convexity. Therefore, convex shapes somehow also belong to the higher metric levels.

This example illustrates the subtle and complex interplay between a process of impoverishment of details and a process of rigidification of structures. On the one hand, CT is "neutral" toward certain structural aspects (magnitude, continuity, shape, bulk, etc., see 2.2), which seems to imply that it lies in the lower range of the hierarchy of mathematical levels. However, on the other hand, active-semantic or morphodynamical routines essentially transform objects into *new* objects that are categorical prototypes and, precisely because they are prototypical, by nature more rigid and metrically constrained. For example, by transforming the topological and differentiable "amoeboid" forms of levels 1 and 2 into regular, prototypical spheres we are in fact going up the level hierarchy, towards the metric levels.

In short, one could say that schematization and categorization replace "soft complex" forms with "rigid simple" ones. Here lies the puzzling apparent paradox of cognitive topology.

3.2. Skeletonization

Another remarkable example of this phenomena is *skeletonization* (seen in Section 2.2.2). Like convexification, skeletonization performs a drastic simplification of objects, yet it is also highly sensitive to their metric details. An interesting problem is then to find out what transformations preserve the topology of the skeleton. If we denote by $\mathrm{Sk}(X)$ the skeleton of a shape X, the skeleton-based relation of equivalence $\cong$ between two shapes

$$X \cong Y \iff \mathrm{Sk}(X) \simeq \mathrm{Sk}(Y) \tag{1}$$

where $\simeq$, is much weaker than an isometry. Yet, it is also much more rigid than conformal isomorphisms and, a fortiori, diffeomorphisms and homeomorphisms.

In summary, active semantics is first and foremost a matter of processes, transformations, representations and re-coding. It does not correspond to any of the elementary levels of mathematical structures. As we will see below, it nevertheless corresponds to a family of operations that can be called morphological and it is in that sense that one may, somehow incorrectly, speak of a "morphological level" for cognitive topology.

> **Thesis**: *What is generally called "topology" in Gestalt semantics and cognitive grammar actually corresponds to the action of morphological operators.*

4. Operations on schemata: the 'across' puzzle

In order to illustrate the richness and difficulty of the problem raised by perceptual-semantic schemata, even leaving out their metonymic or metaphorical extensions and restricting ourselves to invariant perceptual kernels (see Section 1.2), we want to study here the example of the preposition 'across' in greater detail.

4.1. Invariant of transversality

The main geometrical concept underlying all uses of 'across' is the concept of "transversality", It is a natural and intuitive concept that was clarified formally only recently within the framework of modern differential geometry, in particular by the works of Hassler Whitney and René Thom. Let us start with a few simple considerations:

- Intuitively, two distinct lines L_1 and L_2 in the plane $\mathbb{R}^2$ are transverse if they intersect, i.e., $L_1 \neq L_2$ and $L_1 \cap L_2 \neq \emptyset$. This is a stable or "generic" situation in the sense that small perturbations cannot make transverse lines parallel.

- By extension, two (smooth) curves C_1 and C_2 in the plane $\mathbb{R}^2$ are transverse if they intersect at points where their tangent lines L_1 and L_2 are transverse.
- Two lines L_1 and L_2 in the space $\mathbb{R}^3$ are never transverse when they are coplanar, because this would not be a generic situation: any infinitesimal perturbation could bring them apart.
- On the other hand, a line L and a plane P such that $L \not\subset P$ in the space $\mathbb{R}^3$ are transverse if they intersect.
- By extension, a (smooth) curve C and a (smooth) surface S in the space $\mathbb{R}^3$ are transverse if they intersect at points where their tangents, respectively a line L and a plane P, are transverse in the previous sense.
- Two distinct planes P_1 and P_2 of the space $\mathbb{R}^3$ are transverse if they intersect.
- By extension, two (smooth) surfaces S_1 and S_2 are transverse at one of their intersection points if their tangent planes P_1 and P_2 at that point are transverse.

Extracting an invariant kernel from these various instances of transversality, one comes up with the following formal definition:

- Two (smooth) subspaces U and V of an ambient space M, for example $\mathbb{R}^2$ or $\mathbb{R}^3$, are said to be transverse at one of their intersection points x if their tangent spaces at x, $T_x(U)$ and $T_x(V)$, generate the whole tangent space of M at x, i.e., if $T_x(U) + T_x(V) = T_x(M)$.

4.2. Variants of transversality

Several archetypical schemata of transversality can be developed around this general conceptual kernel. One of them is the rather simple "crossing" schema, in the sense of road-crossing, and involves two 1D curves intersecting in a plane. Another schema, more complex, corresponds to the English preposition 'across' and will be examined in the rest of this section. It involves a domain D (the LM) delimited by boundaries, and a path C (the TR) traversing this domain. More precisely, the 'across' schema is in fact characterized by a double transversality and can be tentatively described as follows:

> **Across 1**: A path C is going across a domain D if (a) it enters D at a point x_1 where it intersects the boundary ∂D of D transversally, (b) it lies inside D, and (c) it exits D at a point x_2 on an "opposite" segment of ∂D where it intersects ∂D transversally again

Therefore, the particular perceptual-semantic schema conveyed by 'across' includes a notion of transversality between a 1D curve and a 2D domain of the plane, itself defined as a double transversality between the curve and the domain's boundary. The rigid prototypical form of this schema corresponds to a line intersecting a rectangle perpendicularly (Figure 1).

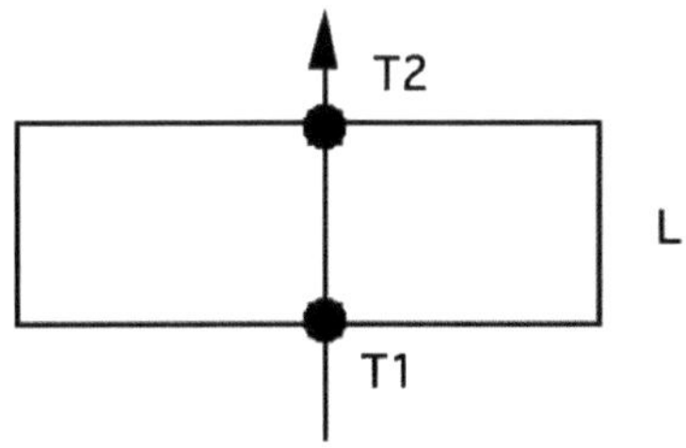

FIGURE 1. The 'across' prototype: a line intersecting a rectangle perpendicularly.

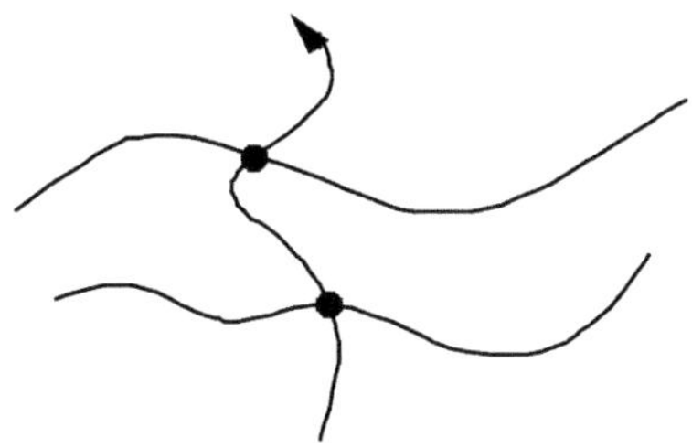

FIGURE 2. The 'across' prototype is plastic and can be greatly deformed up to a certain limit without losing its meaning.

4.3. Plasticity of perceptual-semantic schemata

Talmy identified a great number of different uses of the 'across' schema. These examples are typical of the high degree of plasticity displayed by percepto-semantic schemata.

4.3.1. *Topological plasticity.* First, the path C and the boundary ∂D can be considerably deformed without affecting the 'across' property, as long as the three criteria in the above **Across 1** definition are preserved (Figure 2).

However, this necessary condition is not sufficient. It is easy to show that, although they can be deformed to a great extent, neither C nor ∂D can be completely *arbitrarily* deformed as they would be in a purely topological sense, while preserving the 'across' property. In fact, identifying these limits precisely constitutes the core difficulty of the 'across' puzzle, as we will see below. For example, Figure 3 is not a good instance of 'across', because x_2 is not on the boundary "opposite" to x_1—but, again, what is an "opposite" boundary?

4.3.2. *The concrete/virtual dialectics.* Talmy gives numerous examples showing that the schema does not need to be complete and that one or more of its components can be virtual or missing. They span an entire range between two extreme cases:

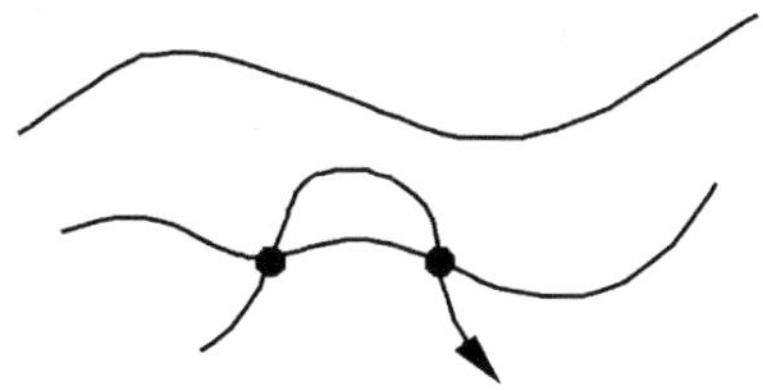

FIGURE 3. A bad instance of 'across': the exit point x_2 is not on the boundary "opposite" to the entry point x_1.

(9) a. The tumble weeds are swept across the prairie.
b. The snail crawled across the car.

In (9a) the boundary ∂D of the 'prairie' LM domain is virtual, and therefore x_1 and x_2 too, whereas in (9b) it is the path C of the 'snail' TR that is virtual.

4.3.3. *The 2D/3D ambiguity.* Talmy also gives adaptive examples of applicability of schemata. The following pair of sentences:

(10) a. He flew across/over the plateau.
b. He walked across/through the wheat field.

In (10a), 'across' takes the sense of 'over': the 'plateau' domain is 2D and the condition of transversality applies to two curves in a plane, the path C and the boundary ∂D. In (10b), by contrast, 'across' takes the sense of 'through': the 'wheat field' domain is now a 3D volume and its boundary a 2D surface. Here, transversality occurs between a curve C and a surface in a 3D ambient space.

4.3.4. *Constraints on distortion.* We just saw that there were additional constraints to the plasticity of a schema and that it cannot be of a purely topological nature, allowing free-range deformations. To further appreciate the problem in all its subtlety, let us take the example of a lake:

- A path that penetrates the domain of the lake only slightly, within its outer margin and then follows the shore to eventually exit on the same side (Figure 4a) does not really constitute a situation of transversality. It is rather an approximatively "tangential" situation.
- A path that penetrates the domain of the lake, goes almost all the way through to the other side but then turns around and comes back to exit on the entrance side poses a more delicate problem (Figure 4b). One can argue that it is not a good case of transversality either, because the second crossing x_2 is not on the "other side" of x_1—which again raises the question of a definition of the "other side".

These difficulties are characteristic of the morphological nature of perception and cannot be simply resolved by using some kind of metric template matching (levels 4 and 5 of the mathematical hierarchy evoked in Section 3)

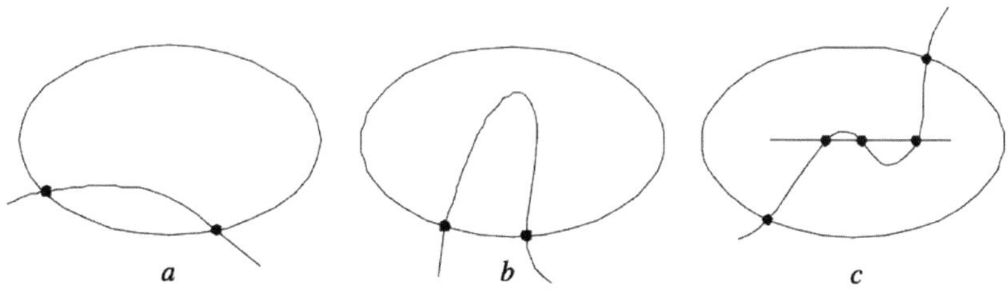

FIGURE 4. The difficulty of the 'across' puzzle and a proposal for a solution. (a) A path that is too "tangential" to the boundary cannot be described by 'across'. (b) A path that is sufficiently transverse to the boundary but exits on the "same" side cannot be described by 'across' either. (c) To be a good example of 'across', a path must intersect transversally the medial axis (cut locus) an odd number of times (see below).

or, on the other end of the spectrum, by using free-range topological deformations (levels 1 and 2). On the one hand, template matching would be an unpractical task because of the sheer number of templates necessary to cover the combinatorial complexity of typical and subtypical configurations. On the other hand, a more flexible template matching allowing greater deformations would in fact accept bad configurations and yield a lot of false positives.

4.4. Virtual structures

To find a way out of the misleading topological vs. metric dilemma, the solution we propose involves creating virtual structures *that embody metric information without being themselves metric.* The 'across' schema contains an implicit partitioning of the domain D into 'this side' and 'that side' (see 2.2.1), equivalent to the partitioning performed by 'here' and 'there'. This qualitative difference can be geometrically implemented by a virtual intermediate boundary C_D approximately located in the middle of the domain. Before the boundary C_D, we are on 'this side' of the domain and after C_D we are on 'that side.'

An interesting property of the virtual boundary C_D is that it is both of a metric nature, since it lies in the "middle" of D, and at the same time of a qualitative nature, since its only purpose is to divide D into two subdomains. The position of C_D is quantitatively precise (at maximal distance from the borders of D) but its function is really to discretize the underlying continuum. This is similar to the origin point on a coordinate axis: both precisely located and qualitatively arbitrary, present only to separate two half-lines.

The virtual structure C_D we need naturally corresponds to the "cut locus" (or "skeleton", "symmetry axis", etc.) already mentioned in Section 2.2 about the magnitude and morphology invariance of grammatical elements. A simple skeletonization of the domain D solves the problem rather easily, albeit partially, and the above criterion can be modified in the following manner:

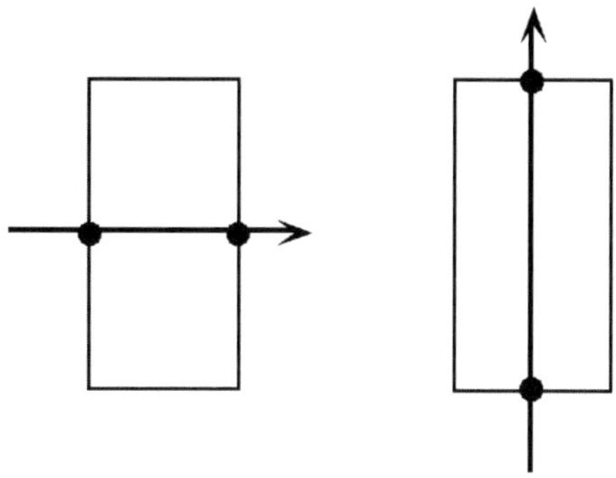

FIGURE 5. Ambiguity between 'across' and 'along.' Left: 'across' the swimming pool lane. Right: 'along' the swimming pool lane.

> **Across 2**: A path C is going across a domain D if (a) it lies inside D, (b) is transverse to the cut locus C_D of the domain in one or an odd number of points, and (c) intersects the boundary ∂D of the domain in two points: x_1 on one side of C_D when C enters D, and x_2 on the other side of C_D when C exits D (Figure 4c).

Therefore, we find again the pure invariant of transversality applied this time to the intersection between the cut locus C_D and the path C. The transversality archetype also operates on transformed shapes, giving the schema all its plasticity. For example, the path C can zigzag around the cut locus as long as the net result is a crossing from one side to the other.

In summary, this example clearly illustrates the fact that obtaining the right type of constraints, in order to characterize perceptual-semantic schemata, requires taking into account additional virtual structures.

4.4.1. *Across vs. along.* Another compelling demonstration of the idea that morphological analyzers such as the cut locus can help solve cognitive-topological puzzles is given by the following two sentences (Figure 5):

(11) a. I swam across the swimming pool lane.
b. I swam along the swimming pool lane.

Looking at Figure 5, the conditions for using 'across' seem to be satisfied in both cases according to the first definition **Across 1** above. Yet, in the situation depicted by (11b) 'across' is obviously incorrect in English and should be replaced by 'along'. Why is this the case? To resolve this issue, one could try to amend **Across 1** by introducing an additional *ad hoc* metric constraint that can discriminate between (11a) and (11b). This constraint could rely on the *ratio* of the lane edges with respect to the direction of the path: the edge parallel to the path should be shorter than the edge perpendicular to the path, i.e., their

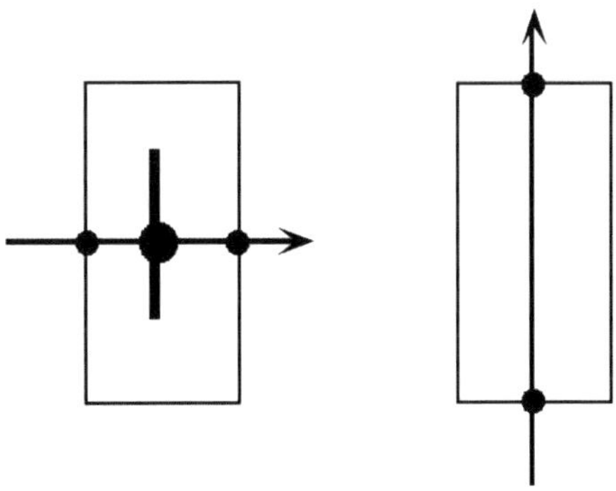

FIGURE 6. A solution to the 'across'/'along' ambiguity. Left: 'across' applies because the path is transverse to the cut locus. Right: 'along' applies because the path is not transverse to the cut locus (here, parallel).

ratio should be less than 1. But this kind of metric criterion is not satisfactory because it is in fact too metrical (Euclidean). It is applicable only to a small number of well-defined configurations, i.e., regular rectangle pool lanes in which the swimmer's path can be "parallel" to one edge and "perpendicular" to the other. It also loses the link with the topological invariant of transversality.

Instead, by introducing the cut locus C_D of the domain and applying the modified definition **Across 2**, the discrimination between (11a) and (11b) becomes natural and immediate. Whereas in case (11a) the path is *transverse* to C_D, in case (11b) it is not (here, parallel to C_D) (Figure 6).

However, this rather simple and elegant criterion also has limits in special cases: for example, if the path in (11a) was too close to one of the short edges of the pool lane, it would not intersect C_D anymore. Nevertheless, we believe that this kind of criteria involving basic virtual structures are able to bridge the gap between perception and semantics.

4.5. Other examples of virtual structures: fictive motion

The introduction of virtual, fictive, or subjective structures is crucial to Gestalt semantics. Len Talmy uncovered and categorized a wealth of linguistic instances in which fictive structures are grammatically embodied in language. In particular, his unified system of "fictive motion" (Talmy [373], [374], chap.2), a semantic phenomenon creating a sense of motion without physical realization, displays one of the most overt types of such structures. To illustrate this point, we give here a short sample of fictive motion categories identified by Talmy.

◇ *Emanation paths.* A finger pointing at an object draws a virtual line through space linking it to the object, and beyond:

(12) I pointed him toward/past/away from the lobby.

◇ *Shadow paths.* An object's shadow is perceived as moving from the object to a background surface:

(13) a. The tree cast its shadow down into the valley.
b. The pillar's shadow fell onto the wall.

◇ *Virtual processes.* In many cases, a static situation is construed as the result of a virtual dynamical process:

(14) a. The palm trees clustered together around the oasis.
b. Mountains are scattered all over this region.

◇ *Virtual fronts.* Any object defines a "front" comprising a virtual plane and a path emanating perpendicularly from this plane. This virtual structure is present in demonstrative paths, sensory paths, and radiation paths.

5. Modeling principles and algorithms

5.1. Gestalt computation

The family of problems that we examined in the previous sections constitute Gestalt-type problems in the strong technical sense of the term. They involve global structures, virtual organizer elements, dialectics between the quantitative and the qualitative, the metric and the categorical, along with properties of topological and morphological plasticity. They amplify the classical Gestalt phenomena (such as virtual contours or the *Prägnanz* of right shapes). At the algorithmic level, they also raise an arduous local/global dilemma: how to strike a balance between global Gestalts on the one hand, and local computation on the other hand.

The main issue is the following: what kind of computational procedures are able to generate (through a "scanning" process) the aforementioned virtual organizing structures? The core difficulty is that global (holistic) elements containing infinite information have to be processed by purely *local* algorithms only using informationally finite procedures. From this perspective, cognitive grammars are not satisfactory: they rely on rather mysterious and holistic properties of vision without attempting to create explicit models and work out algorithms that could explain these properties.

Moreover, those local procedures must derive from principles that are general enough to successfully explain how any unsupervised *learning* of the closed grammatical classes, with all their fascinating perceptual-semantic nuances, can even be possible.

5.2. Spreading activation

In fact, the structuring operations that we sketched in Section 2 all point toward a specific type of process. To be able to generate global structures using local mechanisms, the only possible solution is to trigger *diffusion-propagation* or *expansion processes* from the salient features of the original image, in particular the contours. These contours become active (excitatory or inhibitory) and trigger "activity waves". In turn, these waves create wave front singularities and, with multiple sources, wave front collisions.

To recapitulate, the routines that we gathered in our exploration of numerous linguistic examples are: convexification, tubification, skeletonization, cut-locus, triggering contours and obstacle contours, etc. Drawing from these observations, we propose the two following key ideas toward the foundation of an active morphological semantics:

> **Key idea 1**: Contours are active elements that either trigger processes of diffusion-propagation activity, or inhibit such processes coming from other sources.
>
> **Key idea 2**: Diffusion-propagation processes generate in turn new geometrical entities, in particular singularities, that act as virtual structuring elements.

5.3. Links with other works

5.3.1. *Terry Regier.* In his 1988 article [320] *Recognizing Image-schemata Using Programmable Networks* (see also Harris [144]), Terry Regier also used a contour diffusion routine for analyzing the cognitive (perceptual-semantic) content of prepositions such as 'in', 'above', etc. He starts from the hypothesis that

> recognizing an image-schema for a given preposition amounts to recognizing that the particular spatial configuration named by the preposition holds between the entities focused on in the current field of view. (p. 315)

In order to organize the scene, he uses Ullman's local visual routines, and then algorithms of "Bounded Spreading Activation", that is, contour diffusion.

> The basic idea is that activation spreads out from the copies of object A that were placed in the working image nets, and stops when it encounters units that correspond to the borders of the object in the B-net. (p. 318)

After this first model, Terry Regier refined his theory and, with Laura Carlson [51], introduced the role of *attention*: when processing spatial relations, geometrical information is mixed through attention with functional information provided by the mereological decomposition of objects into functional parts.[4] Taking into account mereology allowed him to develop a finer model, the AVS

4 See above, Section 1.2, the example of the spoon in the cup. It is the functional part "spoon's head" (not the handle) that is in the cup.

(Attentional Vector-Sum) model [322] in which projective prepositions such as 'above' are

> grounded in the process of attention and a vector-sum coding of overall direction.

The idea is to look at the functional parts LM_i of the LM, to weight the subrelations LM_i/TR by the attentional focus, and to take the weighted sum.

5.3.2. *Seibert and Waxman.* In their article [338] 'Spreading Activation Layers, Visual Saccades, and Invariant Representations for Neural Pattern Recognition Systems", Michael Seibert and Allen Waxman have built a model, the NADEL model—"Neural Analog Diffusion-Enhancement Layer"—for bottom-up pattern recognition. They have shown that a spreading activation network—a 2D diffusion—including a detection of local maxima

> can quickly perform a large number of early vision tasks (...) in a completely data-driven manner. (p. 9)

One of the main interest of their work is to have introduced a feed-back of the detection of critical points (maxima) on the diffusion itself.

> This feed back pattern provides a short-term memory reverberation between the levels, sharper peaks, and automatic gain control. (p. 14)

5.4. Morphological algorithms

The above hypotheses about Gestalt semantics must be empirically verified by computational synthesis. However, practical technical difficulties arise when trying to implement the standard differential equations of non-linear systems, such as reaction-diffusion or traveling waves. These equations require a significant amount of computing power and parameter tuning, therefore they are not the fastest way to model the elementary mechanisms of expansions and obstacles at work in perceptual semantics.We will come back to "dynamical systems" type of implementations in Sections 6 and 7, through cellular automata and spiking neurons, but for now we adopt a "functionalist" approach (still within the framework of dynamical models) and model these physicalist algorithms with more abstract counterparts found in *mathematical morphology* (MM). This domain of mathematics was developed in the 1970's at Ecole des Mines de Fontainebleau, France, by Georges Matheron and his colleagues Jean Serra [341], M. Schmitt, J. Mattioli and others. MM is a branch of image processing that focuses on the geometric structure of shapes and textures. We present here a review of the main concepts developed by MM (for an introduction see also Dougherty [87]) and draw a link to Gestalt semantics. Next, in Section 6, we present a series of numerical simulations of MM-based active semantics implemented in *cellular automata* (CA).

Lastly, in Section 7, we explore the deeper, physicalist, microstructure of active semantics. Specifically, we will establish a link between MM and *spiking*

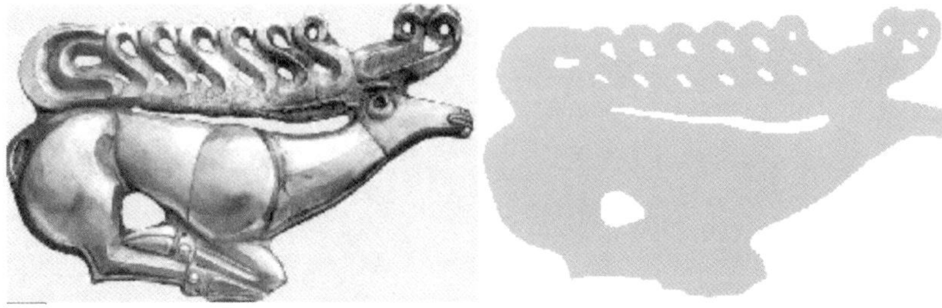

FIGURE 7. To illustrate the algorithms of mathematical morphology we use this picture of the celebrated Scythian Deer from the State Hermitage Museum in Saint Petersburg, Russia. Left: the original image. Right: its binary outline.

neural networks, which can play the role of excitable media supporting complex waveform dynamics. Dynamical events of an intrinsically *spatio-temporal* nature that are critical for the emergence of virtual structures and singularities cannot be captured by MM and can be realized only at this finer level of resolution.

5.4.1. *Morphological operations on binary images.* The main idea is to analyze images by using *structuring elements* and Boolean set operations. Let X be a 2D binary image defined on a window W of $\mathbb{R}^2$ centered around the origin O. X is defined by a characteristic function $\varphi_X : W \to \{0,1\}$, such that the figure corresponds to the inverse image $X = \varphi_X^{-1}(1)$ and the background against which it is profiled to $\varphi_X^{-1}(0)$ (Figure 7).

We now introduce a basic structuring element B in the form of a small disk of radius r centered around O. Element B should be viewed not just as a geometric element but also a dynamical element resulting from a "binary diffusion" of O during a characteristic period of time. However, an important difference between binary diffusion and real PDE-based diffusion is that the element B remains black and white without any "smoothing" of its boundaries. It remains discrete on the range of image values and is defined by its characteristic function φ_B: $a \in B \iff \varphi_B(a) = 1$. Binary diffusion is a simplified version of real diffusion especially well suited to binary morphological analysis.

◇ *Dilation*

Considering each point of the image X as an elementary source of binary diffusion, we can define the *dilation* of X by B as follows (Figure 8):

$$X \oplus B = \{u \mid u = x + a,\ x \in X,\ a \in B\}. \tag{2}$$

The effect of the dilation of X by disk B is to regularize the image. It connects components inside X that are sufficiently close, fills up small gaps ("close" and "small" with respect to B's size) and globally increases the linear size of X by B's radius, r. The exact result depends of course on B's shape and size, and on the various ways of implementing this continuous formalism in a discrete

FIGURE 8. A dilation of the deer image after 20 steps. Holes are gradually filled.

lattice, e.g., the type of neighborhood (4-neighbor square, 8-neighbor square, 6-neighbor hexagonal, etc.).

If we denote by B_u the translation of B by u, we get

$$B_u = \{v \,|\, v = a + u,\ a \in B\}, \tag{3}$$

and we can then define the operation of dilation as follows:

$$X \oplus B = \{u \,|\, B_u \cap X \neq \emptyset\} = \bigcup_{x \in X} B_x = \bigcup_{a \in B} X_a. \tag{4}$$

Morphological dilation has interesting algebraic properties, such as commutativity and associativity:

$$\begin{aligned} X \oplus B &= B \oplus X \\ (X \oplus B) \oplus B' &= X \oplus (B \oplus B'), \end{aligned} \tag{5}$$

and partial distributivity with respect to set operations:

$$\begin{aligned} X \oplus (B \cup B') &= (X \oplus B) \cup (X \oplus B') \\ X \oplus (B \cap B') &\subset (X \oplus B) \cap (X \oplus B'). \end{aligned} \tag{6}$$

It is also an increasing function:

$$X \subset X' \Rightarrow (X \oplus B) \subset (X' \oplus B), \tag{7}$$

and is translation invariant:

$$X_u \oplus B = (X \oplus B)_u. \tag{8}$$

Starting from dilation, we can now define other basic morphological routines, such as erosion, closing and opening.

◇ *Erosion*

The erosion of X by B is the dual operation of dilation by set complementation. If the complement of X is denoted by

$$X^c = W - X = \{u \,|\, u \notin X\}, \tag{9}$$

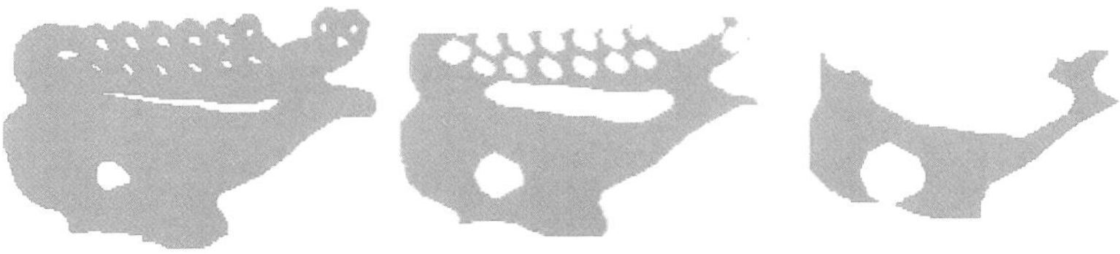

FIGURE 9. Two stages (20 and 50 steps) during the erosion of the deer image. Holes gradually disappear, while the shape contracts toward its skeleton.

then erosion corresponds to the dilation of the complement in the following manner (Figure 9):

$$X \ominus B = (X^c \oplus (-B))^c . \tag{10}$$

Image erosion obviously has the inverse effect of image dilation: it disconnects components linked by narrow strips, erases small spots (again, "narrow" and "small" with respect to B's size) and globally decreases the size of X by B's radius, r.

As in (4) it is easy to deduce alternative definitions of erosion based on translated sets:

$$\begin{aligned} X \ominus B &= \{u \,|\, u + a \in X,\ a \in B\} \\ &= \{u \,|\, B_u \subset X\} && (11) \\ &= \bigcap_{a \in B} X_{-a}, && (12) \end{aligned}$$

including the following algebraic property:

$$X \ominus (B \oplus B') = (X \ominus B) \ominus B', \tag{13}$$

and set-theoretic properties:

$$\begin{aligned} X \ominus (B \cap B') &\supset (X \ominus B) \cup (X \ominus B') \\ X \ominus (B \cup B') &= (X \ominus B) \cap (X \ominus B'). && (14) \end{aligned}$$

Like dilation, erosion is an increasing function:

$$X \subset X' \Rightarrow (X \ominus B) \subset (X' \ominus B), \tag{15}$$

and is translation invariant:

$$\begin{aligned} X_u \ominus B &= (X \ominus B)_u \\ X \ominus B_u &= (X \ominus B)_{-u}. && (16) \end{aligned}$$

◇ *Closing*

The closing of X by B consists of a dilation followed by an erosion:

$$X \bullet B = (X \oplus B) \ominus B. \tag{17}$$

FIGURE 10. Closing the deer image (30 steps).

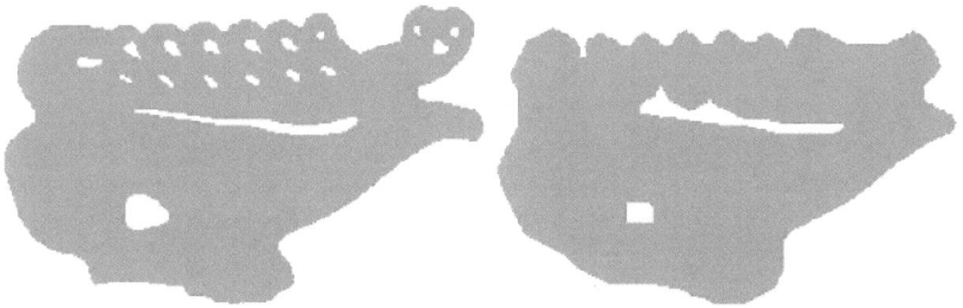

FIGURE 11. Opening the deer image (30 steps).

Like dilation, closing connects segments and fills-in gaps but it also renormalizes the size of X and therefore has a global effect of regularizing or "smoothing" the image without changing its size (Figure 10).

Closing is a natural match for continuity-neutral schemata like the preposition 'in' (see 2.2.2): it transforms a group of disconnected segments into a continuous curve (or surface in 3D). Note that closing is idempotent, i.e., it does not further modify an already closed shape:

$$(X \bullet B) \bullet B = X \bullet B \tag{18}$$

◇ *Opening*

The opening of X by B is the dual operation of closing and consists of an erosion followed by a dilation (Figure 11):

$$X \circ B = (X \ominus B) \oplus B. \tag{19}$$

Just like erosion corresponds to the dilation of the complement, opening corresponds to the closing of the complement:

$$X \circ B = (X^c \bullet (-B))^c, \tag{20}$$

and, like closing, is also idempotent:

$$(X \circ B) \circ B = X \circ B. \tag{21}$$

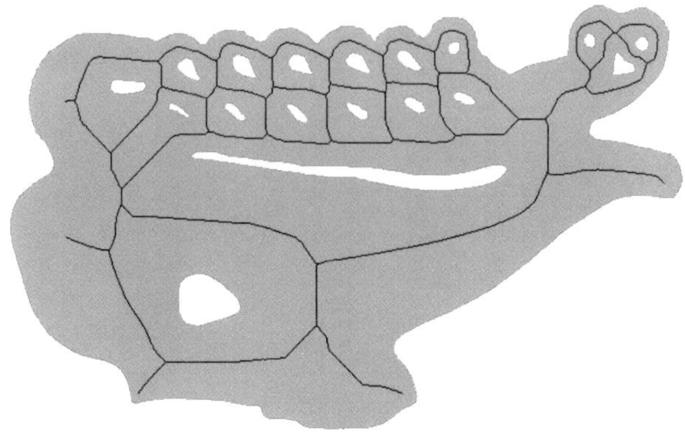

FIGURE 12. The morphological skeleton of the Scythian deer.

5.4.2. *Skeletonization of binary images.*

◇ *Medial axis skeleton*

Using morphological routines it becomes easy to compute the skeleton Sk(X) of a shape X, an otherwise complicated task when using partial differential equations. As we saw in Chapter 2, Sk(X) is made of the centers of all the maximal disks included in X (Figure 12).

Let us formalize this notion. For any subset X, we denote by $\overset{\circ}{X}$ its topological interior (the union of all open sets contained in X) and $\overline{X}$ its topological closure (the intersection of all closed sets containing X).[5] To avoid pathological cases, we assume in the remainder of this section that X is already an *open* set, is connected, has an arc-wise differentiable boundary ∂X, and that its topological closure $\overline{X}$ is compact. We denote by $B(x,r)$ the open disk of center x and radius r, and $\overline{B}(x,r)$ its topological closure:

$$\begin{aligned} B(x,r) &= \{u \mid d(x,u) < r\} \\ \overline{B}(x,r) &= \{u \mid d(x,u) \leq r\}, \end{aligned} \tag{22}$$

where d is the distance in the ambient space. The disks centered around the origin are denoted $B(r) = B(0,r)$ and $\overline{B}(r) = \overline{B}(0,r)$. A disk $B(x,r)$ is said to be *maximal* in X if it verifies

$$\forall x' \forall r' B(x,r) \subset B(x',r') \subset X \Rightarrow x' = x \text{ and } r' = r, \tag{23}$$

5 $\overset{\circ}{X}$ and $\overline{X}$ are classical topological objects and should not be confused with the morphological operations of opening, $X \circ B$, and closing, $X \bullet B$. Reminder in 1D: if X is the interval $[-1,1)$, including -1 but not 1, then $\overline{X} = [-1,1]$ and $\overset{\circ}{X} = (-1,1)$.

i.e., if no larger disk can be inserted between $B(x,r)$ and the boundary of X. If we now consider the erosion transformation of X by $B(r)$, we get:

$$
\begin{aligned}
X(-r) &= X \ominus \overline{B}(r) \\
\overline{X}(-r) &= X \ominus B(r)\,.
\end{aligned}
\tag{24}
$$

it is easy to prove that a disk $B(x,r)$ is maximal in X if and only if x belongs to $\overline{X}(-r)$ but not to any morphological opening of $\overline{X}(-r)$ by any closed disk $\overline{B'}(\varepsilon)$. In other terms, if we also introduce the intermediate concept of the r-skeleton $\mathrm{Sk}(X,r)$ of a shape X:

$$
\mathrm{Sk}(X,r) = \bigcap_{\varepsilon>0} \left(\overline{X}(-r) \left(\overline{X}(-r) \circ \overline{B'}(\varepsilon)\right)\right), \tag{25}
$$

then we can rewrite the above property as follows:

$$
B(x,r) \text{ maximal in } X \iff x \in \mathrm{Sk}(X,r). \tag{26}
$$

Finally, from the union of all r-skeletons we obtain the formula for the central skeleton $\mathrm{Sk}(X)$ (Lantuéjoul [208]):

$$
\begin{aligned}
\mathrm{Sk}(X) &= \bigcup_{r>0} \mathrm{Sk}(X,r) \\
&= \bigcup_{r>0} \bigcap_{\varepsilon>0} \left(\overline{X}(-r) - \left(\overline{X}(-r) \circ \overline{B'}(\varepsilon)\right)\right).
\end{aligned}
\tag{27}
$$

◇ *Shape reconstruction*

An important property of the skeleton $\mathrm{Sk}(X)$ is the ability to *reconstruct* the shape X and all its morphological transforms, if we also know how the radius varies along $\mathrm{Sk}(X)$.

- The shape X can be reconstructed from its r-skeletons by dilating them with corresponding disks of radius r:

$$
X = \bigcup_{r>0} \mathrm{Sk}(X,r) \oplus B(r). \tag{28}
$$

- Similarly, the skeleton of the eroded shape $X(-r')$ is made of the r-skeletons for $r > r'$:

$$
\mathrm{Sk}(X(-r')) = \bigcup_{r>r'} \mathrm{Sk}(X,r), \tag{29}
$$

  so that $X(-r')$ can be reconstructed from the r-skeletons with $r > r'$ by dilating them with corresponding disks of radius $r - r'$:

$$
X(-r') = \bigcup_{r>r'} \mathrm{Sk}(X,r) \oplus B(r-r'). \tag{30}
$$

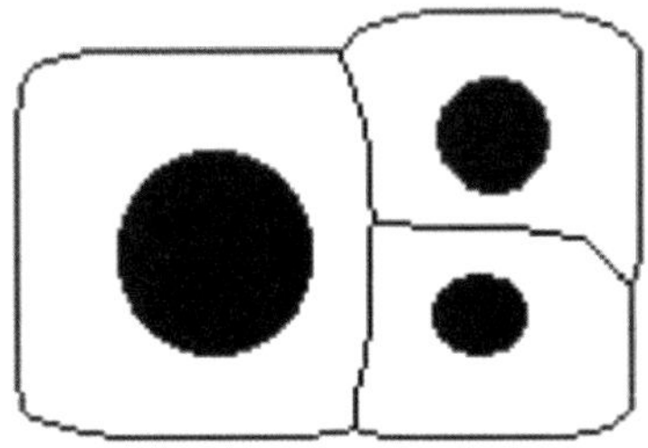

FIGURE 13. The SKIZ of a configuration of objects (3 black disks in a frame) is the skeleton of its complementary set (here in white). The external part of the SKIZ is induced by the frame.

- The same holds for dilations, if we define $X(r) := X \oplus B(r)$, then:

$$X(r') = \bigcup_{r>0} \mathrm{Sk}(X, r) \oplus B(r + r'). \tag{31}$$

- For openings, we get:

$$X \circ B(r') = \bigcup_{r>r'} \mathrm{Sk}(X, r) \oplus B(r). \tag{32}$$

Under the assumptions of regularity of X stated at the beginning of Section 5.4.2, it can be proved that the skeleton $\mathrm{Sk}(X)$ is a closed set of dimension 1, i.e., a *graph*, composed of regular curve segments that have termination points and are connected at a few independent vertices of finite order. Note that, on the other hand, the skeleton is also an unstable structure that is very sensitive to noise. As we have already emphasized in the previous chapter, small bumps in X's boundary suffice to create extra segments and vertices in $\mathrm{Sk}(X)$. This problem can be overcome by *multi-scale* skeletonization methods.

◇ *Skeleton by influence zones*

The idea of skeleton becomes especially interesting when applied to the complement $X^c = W - X$ of a configuration of objects $X = \{X_i\}$ in a domain W. The shapes X_i are the components of X, and we assume that they are closed, regular, well separated from each other and that X^c is connected. The skeleton $\mathrm{Sk}(X^c)$ is then pruned by removing secondary segments emanating from the singularities of X_i's boundaries. This yields the *skeleton by influence zones* or SKIZ (see 2.2.2), i.e., the subset of W containing the points bounding the X_i's influence zones (Figure 13).

Formally, the influence zone of X_i denoted by $Z(X_i)$ is the set of points that are closer to X_i than to any other X_j for $j \neq i$:

$$Z(X_i) = \{x \in W \mid d(x, X_i) < d(x, X - X_i)\}, \tag{33}$$

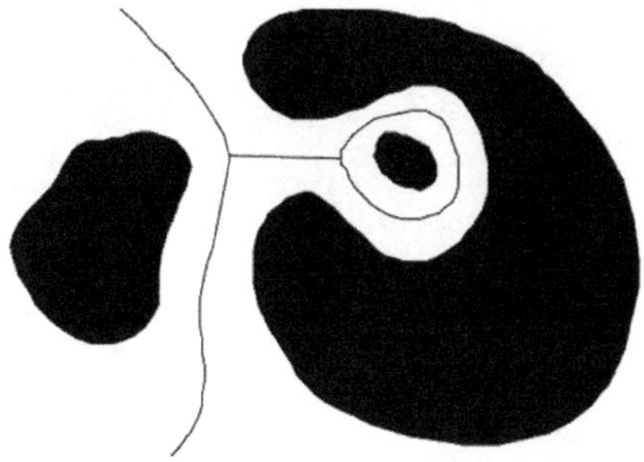

FIGURE 14. In this example of an external SKIZ, the triple point to the left codes the fact that there exist locally three parts in interaction. But these three parts are not the three objects: two parts of them are parts of the same non-convex object, while the third object that is 'inside' has no influence on the triple point. The existence of a loop in the right branch of *Sk* codes the relation of "insideness" and explains the global relational structure of the configuration.

and the SKIZ of X denoted by $\mathcal{Z}(X)$ is the critical set of boundaries where these zones meet, i.e., the complement of the union of all influence zones:

$$\mathcal{Z}(X) = W - \bigcup_i Z(X_i) = W - Z(X) = Z(X)^c.$$

Note that when the components X_i are points, the SKIZ corresponds to the well-known *Voronoi diagrams.*

The concept of SKIZ ascribes a morphological content to the gestaltic notion of relations, as exposed at the beginning of this chapter (see Section 1). The SKIZ offers a morphological coding of the background X^c and, therefore, of the spatial relations between the X_i components. Relations emerge from the global pattern in the same space as the objects. And such a genesis of relations is valid for complex relations, as it is clear from Figure 14.

Finally, it is easy to see the relation of inclusion between the medial axis skeleton and the skeleton by influence zones:

$$\mathcal{Z}(X) \subset \mathrm{Sk}(X^c), \tag{34}$$

since for any $x \in \mathcal{Z}(X)$ there exists i, j such that x is on the boundary between X_i and X_j, which means that $d(x, X_i) = d(x, X_j) = r$ and therefore that $B(x, r)$ is a maximal disk of X^c. However, the reverse relation does not hold because the medial axis skeleton $\mathrm{Sk}(X^c)$ contains centers of maximal disks that are in contact with only one of the X_i components.

5.4.3. *Implementation.* In Chapter 2, Section 9.2, we have seen Hugh Bellemare's implementation of the cut locus algorithm in a neural network. Figure 15 shows the SKIZ of a complex configuration of five objects as an external CL. We present a more sophisticated neural implementation in Section 7.

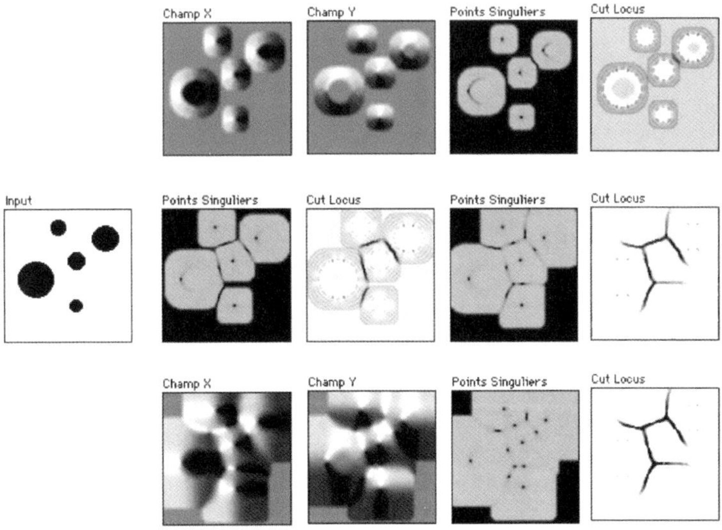

FIGURE 15. The SKIZ of a complex configuration of 5 objects as an external CL. Bellemare's implementation in a neural network with 5 layers (see Chapter 2, Section 9.2).

6. Numerical simulations based on cellular automata

To give a concrete illustration of the above principles and validate our hypotheses empirically we conducted a series of numerical experiments.

Our material consists of simple 100×100-pixel gray-level images, derived from artificial data. They contain various geometrical figures in the style of "contour sketches", e.g., circles, polygons, irregular curves, etc. We assume that they represent a typical outcome of the early stages of visual processing, essentially low-level segmentation and schematization processes. We also assume that each picture is composed of two objects that are already segmented and appropriately tagged by the system: the trajector (TR) and the landmark (LM), respectively acting as figure and background. In most cases, simple contour connectedness is enough to distinguish object domains from each other.

We now describe a perceptual-semantic machine, whose goal is to answer "yes" or "no" given one picture in input and one question about this picture, such as: "Is TR in LM?", or "Is TR above LM?", or "Does TR go across LM?", etc. As said above, the main idea is that a specific question triggers a specific chain of transformation routines, eventually yielding a categorical answer. The boundaries are active since they constitute the main triggering features of the routines. Generally, the TR plays the center role in the sense that it tends to expand, whereas the LM tends to block this expansion, either as a passive obstacle or by counter-expansion.

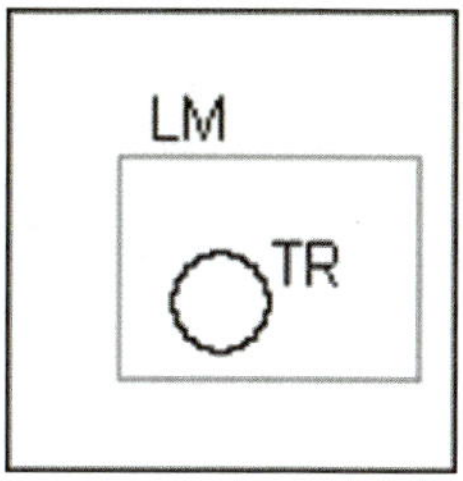

FIGURE 16. Schematic representation of "the ball in the box". TR (the circle) is the ball, LM (the rectangle), the box. We code TR in black and LM in gray. The image frame is a square of size 100×100.

6.1. Example 1: "the ball in the box"

The first example deals with the "containment" schema. Consider a very simple scene where TR and LM are two closed objects and TR is enclosed inside LM. For example, Figure 16 is a sketch representing "the ball in the box".

6.1.1. *Heat diffusion.* According to our morphological interpretation of 'in', asking whether "TR is in LM" means checking whether LM blocks TR's diffusion, i.e., whether activity from TR can spread out of LM and reach the boundary ∂W (the horizon) or the visual frame W[6] in a time $t \leq t_{max}$ where t_{max} is the characteristic travel time across W.

A first possibility is to rely on a simple diffusion process such as the heat equation

$$-\mu \frac{\partial a}{\partial t} = \Delta a, \tag{35}$$

where $a(x, y, t)$ is the activity rate at point (x, y) at time t, and μ a small coefficient. In the absence of specific boundary conditions, this diffusion process relaxes towards a flat landscape $a = a_0$, where a_0 is an arbitrary constant. In our case, we clamp the activity on the contours of TR and LM at two different constant values. At any time t, we set: $a(\partial LM, t) = 0$ and $a(\partial TR, t) = 1$, where ∂LM and ∂TR represent the set of points (x, y) belonging to LM's and TR's contours, respectively. To simplify, the initial state is also set to 0 everywhere else, although this has no influence on the final equilibrium state. Metaphorically, TR plays the role of an expanding "heat source" (spreading around positive values) and LM a "cold obstacle" (maintaining zero values).

Thus, we postulate that the question "Is TR in LM?" activates the boundaries of TR and LM in a specific way. As a result, the activity landscape relaxes towards a smooth gradient stretching between TR and LM (decreasing from

6 The definition of "visual frame" depends on the context. Generally, it is the visual field but it can also be larger. We will not address this issue here.

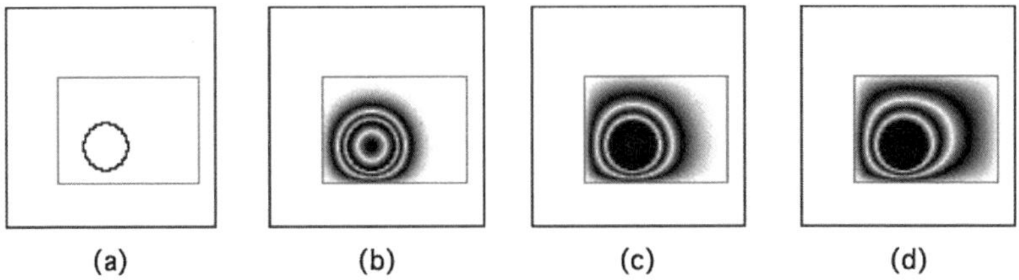

FIGURE 17. Heat diffusion process applied to "the ball in the box." (a) Initial state $t = 0$; (b) iteration $t = 100$; (c) iteration $t = 300$; (d) iteration $t = 2000$. With activity values ranging from 0 to 1, the gray-level code is periodic to make diffusion fronts visible (values $0, .4, .8$ coded in white; values $.2, .6, 1$ coded in black). The contour of TR (the ball) is clamped at 1 and the contour of LM (the box) at 0 (exceptionally coded in gray to keep it visible). The interior of the ball rapidly converges towards 1, the gap between the ball and the box is filled with a gradient (decreasing values from 1 to 0), and the outside of the box remains uniformly 0.

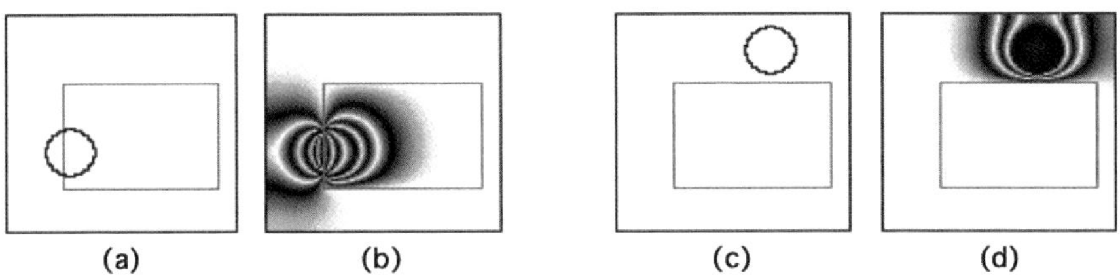

FIGURE 18. Heat diffusion process applied to counter-examples of "the ball in the box". (a) "crossing" example at $t = 0$ and (b) iteration $t = 500$; (c) "outside" example at $t = 0$ and (d) iteration $t = 300$. Boundary conditions and periodic gray-level code are the same as in the previous figure. In both cases, the activity on the top border of the image frame is rapidly increasing from zero to positive values.

1 to 0) and, since TR is entirely included in LM, the activity outside of LM remains uniformly 0 (Figure 17).

Thus, the fact that no activity transpires outside of LM's domain and reaches the image boundary ∂W can be considered as a characteristic categorical clue that TR is 'in' LM. Two counter-examples are shown in Figure 18: the ball is not in the box, whether crossing it or outside of it, and the diffusion process soon reaches the top border of the image frame, where it creates positive activity values.

Again, irrespective of the metric shapes of TR and LM, this type of analytical process reduces the applicability of 'in' to a purely binary (i.e., discrete) categorical criterion: is there or not noticeable activity on the boundary ∂W before t_{max}?

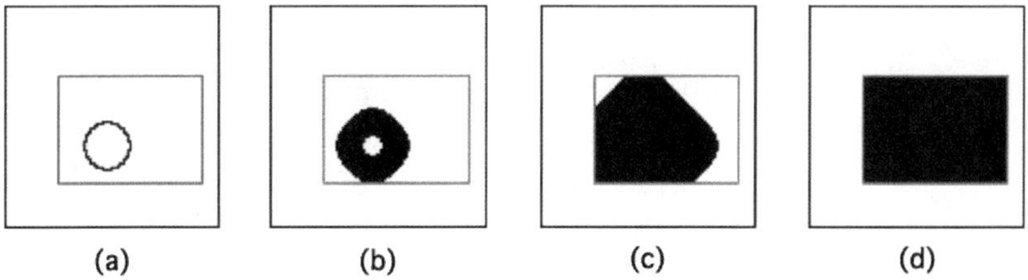

FIGURE 19. Dilation routine applied to "the ball in the box". (a) Initial state $t = 0$; (b) iteration $t = 6$; (c) iteration $t = 24$; (d) stopping at iteration $t = 60$. Black pixels aggregate onto TR's contour, i.e., wherever at least one of their 4 nearest neighbors is black. LM's contour is clamped on white (although made visible in gray) and eventually stops TR's expansion after a small number of iterations. TR's activity remains entirely confined in LM.

6.1.2. *Morphological transformations.* As mentioned above, diffusion processes governed by PDEs are rather computationally expensive and not well adapted to practical applications. For example, concretely, it is difficult to estimate how long they take to reach a satisfying state of equilibrium.

We saw in Section 5.4 that more convenient algorithms can be derived from mathematical morphology (MM). The two basic processes of dilation and erosion offer a range of interesting transformations. Shape dilation is simply done by "aggregation": the shape is progressively padded with additional layers of non-zero pixels, as if each pixel emitted a small disk around itself. Shape erosion does the opposite: non-zero pixels are homogeneously removed from the border, layer after layer. Note that the erosion of a shape is equivalent to the dilation of its complementary. In the simplest version of these routines, images are binary (shape in black, background in white) and a black pixel is added to (or removed from) the shape domain if at least one of its 4 nearest neighbors is black (respectively, white). All pixels are visited once per iteration, so that the number of iterations is roughly equal to the increase (or decrease) in thickness.

Figure 19 shows the dilation process applied to the ball in the box, where LM, the "box", again acts as an obstacle. Similarly to the heat diffusion, the obstacle is created simply by clamping all its pixels on white (although we make it visible in gray for the figures).

Since pixels belonging to the obstacle ∂LM are not allowed to change to black, the expansion process eventually stops. The number of iterations required to yield an answer is finite and small, making this kind of transformation more practical than physical diffusion. In any case, both algorithms model the same phenomenon—the active expansion of an object against a passive obstacle. Here too, TR is said to be in LM if no activity can be detected on the image frame (categorical binary criterion).

6.1.3. *Influence zones and cut-locus.* An interesting alternative to the previous scenario is to let LM play an active role, too, and create a counter-expansion

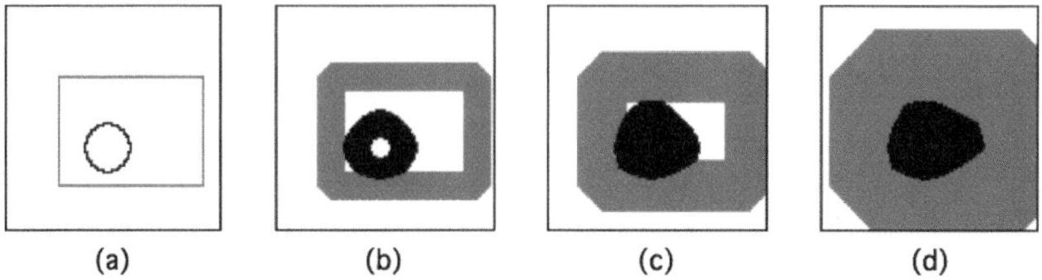

FIGURE 20. Cut-locus created by the simultaneous dilation of TR and LM, applied to "the ball in the box". (a) Initial state $t = 0$; (b) iteration $t = 6$; (c) iteration $t = 11$; (d) stopping after iteration $t = 21$. As in previous figures, TR is coded in black and LM in gray. Black pixels aggregate on TR only if they are not in LM's domain, and conversely. The routine stops when TR stops growing.

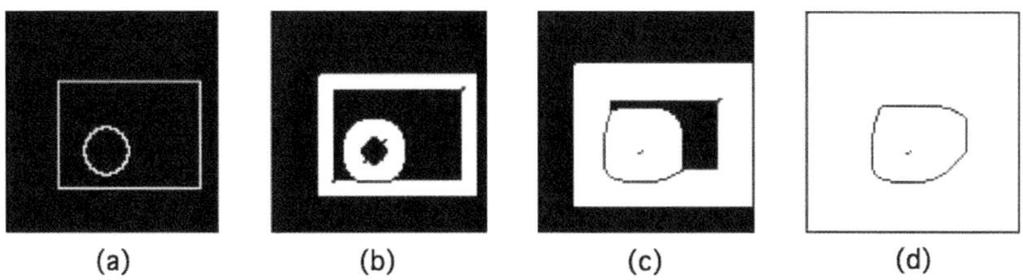

FIGURE 21. Skeletonization of the background external to TR and LM, applied to "the ball in the box". (a) Initial state $t = 0$; (b) iteration $t = 6$; (c) iteration $t = 16$; (d) stopping after iteration $t = 64$. Image values are reverted through $x \mapsto 1 - x$ with respect to the previous figure. Note the small internal cut-locus of the ball, reduced to a quasi-point (3-pixel domain).

toward TR. The resulting figure shows a boundary where the two expansion areas meet (Figure 20). The areas stretching on either side of this boundary are the influence zones of the objects. Here too, the "containment" criterion must distinguish between two types of activity, one coming from TR, the other from LM, and check that TR's activity is not detected on the image boundaries.

Note that TR's contour after expansion (the boundary) constitutes a good approximation of the *external cut-locus* of the scene, which is the generalized symmetry axis of the background (the complementary zone between the ball and the box). This cut-locus (here, looping back on itself) can also be revealed by a skeletonization of the background (Figure 21).

Technically, skeletonization algorithms gradually erode the black domain while preserving local pixel configurations of thickness 1. In a cellular automaton implementation, each one of the 256 possible neighborhoods of 3×3 pixels around a black pixel is associated with an update rule: removing (i.e., changing to white) or not the center pixel. Additionally, white pixels always remain white, independently of their neighborhood. On the other hand, the true cut-locus contains all the points that are equidistant from TR and LM (Blum [38]).

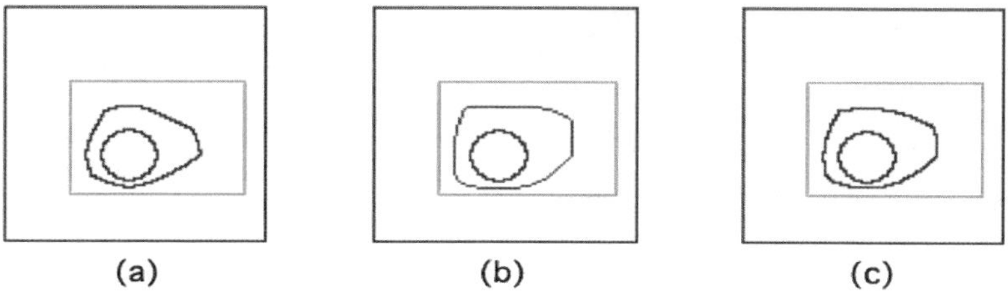

FIGURE 22. The boundary between TR and LM (shown in black) can be computed in three different ways, applied to "the ball in the box." (a) Approximation obtained by double dilation. (b) Approximation obtained by skeletonization of the complementary area. (c) Exact cut-locus computed by equidistance.

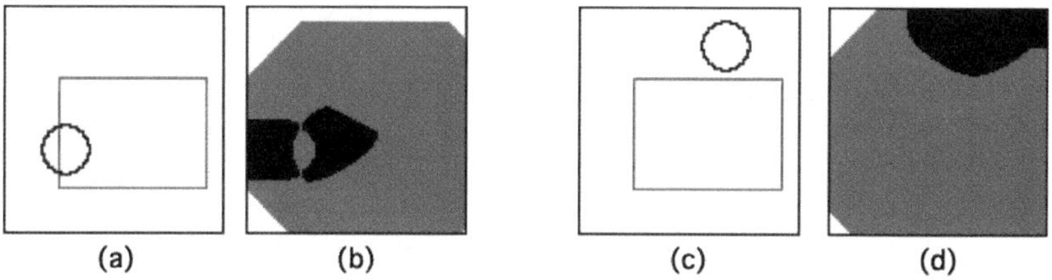

FIGURE 23. Double dilation applied to two counter-examples of "the ball in the box". TR's activity (in black) reaches the image boundary after 25 and 34 time steps, in (b) and (d) respectively.

Figure 22 compares three different methods generating a boundary between T and LM: double dilation, skeletonization, and exact cut-locus.

Finally, we will favor the double dilation method as it is faster and less *ad hoc* than the other two. It also naturally corresponds to the expansion phenomena that we are emphasizing. Figure 23 shows an application of this routine to the two counter-examples of "the ball in the box" seen previously (Figure 18).

6.2. Example 2: "the bird in the cage"

Consider now a more complicated case, "the bird in the cage", in which the container LM (the cage) has openings. We saw in Section 2 that the English preposition 'in' is neutral with respect to this type of modification. Within the framework of active semantics, such an invariance property is not given but is *constructed* through a preprocessing of LM.

6.2.1. *Preprocessing of LM.* The difficulty here is that the LM's contour is discontinuous. One solution is to perform a preliminary closing of LM before expanding TR (whether by diffusion or dilation). The simplest way to treat

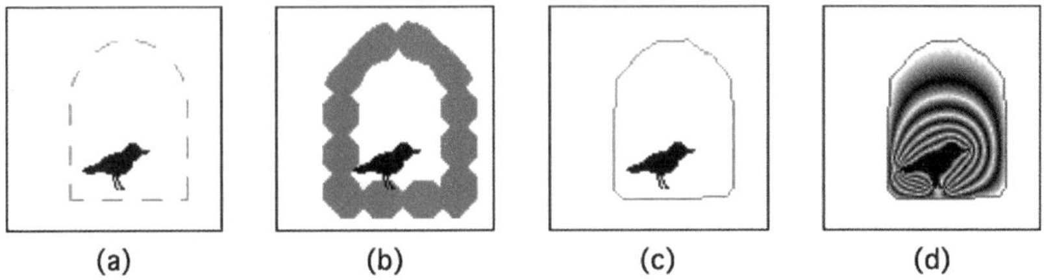

FIGURE 24. Detection of the "containment" schema by preliminary closing of a lacunary LM, applied to "the bird in the cage". (a) Initial state $t = 0$; (b) dilation of LM after iteration $t = 8$; (c) skeletonization of (b) in 18 additional iterations; (d) expansion of TR (here, by diffusion).

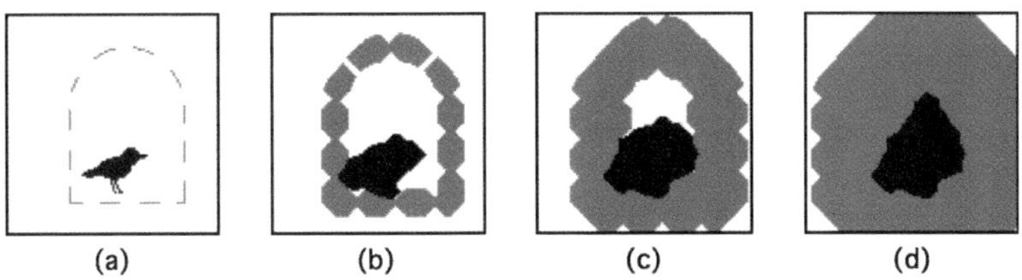

FIGURE 25. Detection of the "containment" schema by double dilation, applied to "the bird in the cage." (a) Initial state $t = 0$; (b) iteration $t = 6$; (c) iteration $t = 14$; (d) iteration $t = 27$.

any arbitrary contour is to compute a dilation followed by a skeletonization, instead of an erosion (Figure 24).

6.2.2. *Global processing by SKIZ.* However, here too, it appears simpler to trigger a double expansion of both TR and LM in parallel, which will reveal the external cut-locus (Figure 25).

Note that, as desired, morphological details of TR have only little influence on the outcome of the dilation. They are erased within a few iterations and, given any original shape, the dilated domain quickly reaches an undifferentiated "blob".

6.3. Remarks

To develop a complete morphological analysis of a preposition seemingly as simple as 'in', we would need much finer and more efficient algorithms than the ones just outlined. For example, to take into account situations where LM is not only lacunary but is also not completely surrounding TR, e.g., as in "the fruit in the bowl", a spherical completion LM^S of LM can be used, or a "solid" expansion of TR that blocks as soon as the diffusion front reaches LM^S.

In even more complex cases, e.g., metonymic cases such as "the spoon in the cup" (where only the spoon head is in the cup, while the spoon handle sticks out; see Section 1.2), higher-level cognitive preprocessing must intervene—here, a stored mereological decomposition of the spoon into two functionally different parts. As the head is considered more central than the handle, 'in' is applied preferentially and metonymically to the head. The reverse situation, in which the handle is inside and the head outside, is a much less good example of "the spoon in the cup."

6.4. Example 3: "the lamp above the table"

Let us consider now a different example: the English preposition 'above'.

6.4.1. *Preprocessing of LM.* The 'above' localizer fundamentally involves an up-down gradient. Its semantics is not only geometric (positional) but, in a qualitative sense, physical. At first, the morphological analysis of the spatial "superiority" schema could thus use an *anisotropic* expansion of TR along the vertical axis, along with detection of activity on the inferior image boundary ∂W (Figure 26a-b). If the TR finds itself 'above' LM, then LM will block TR's expansion and no activity should reach the bottom (this idea was also developed by Regier [320]). However, to properly handle problematic cases of translation (Figure 26c-d), a preliminary horizontal expansion of LM can be necessary (Figure 26e-g).

6.4.2. *Global processing by SKIZ.* Here again, the simultaneous and isotropic expansion of both TR and LM participants seems to offer a good solution in most situations. As before, the influence zones and their boundary Z are barely sensitive to variations around the prototypical configuration. In the case of 'above' the categorical binary criterion depends on the intersection of Z with the image boundary ∂W and the fact that this intersection should not be on the bottom border (Figure 27).

Generally, the intersection of Z with ∂W also provides critical information on the relative location of TR and LM. For example, in the case of 'above', it allows to distinguish among similar cases that different languages conceptualize by different grammatical elements (Figure 28).

6.4.3. *Toward a solution to cognitive linguistic challenges.* The previous example clearly shows the possibility of a *discrete* categorization on the basis of *continuously* varying spatial relations. They provide a solution to the deep perceptual-semantic problems posed by cognitive linguistics.[7] Another illustration of these principles is given by Figure 29, in which the TR is a disk hovering above a triangle LM. As TR is continuously shifted from left to right, it creates a smooth transition between the concurrent schemata "TR above LM"

7 See, e.g., Berkeley's L_0 and NTL (Neural Theory of Language) projects launched in the 1990's by Jerome Feldman and George Lakoff. See [104].

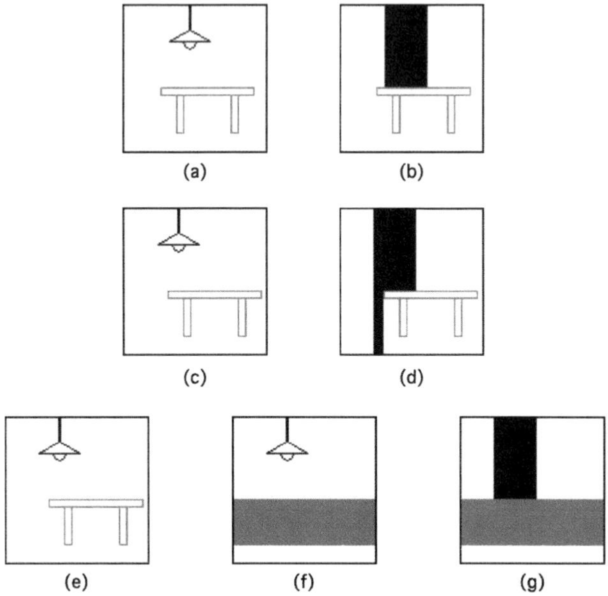

FIGURE 26. Detection of the "superiority" schema by sequential anisotropic expansion, applied to "the lamp above the table". (a)-(b) Prototypical case of TR 'above' LM. (c)-(d) Marginal case involving a problem due to translation. (e)-(g) Solving the translation problem by preliminary horizontal expansion of LM.

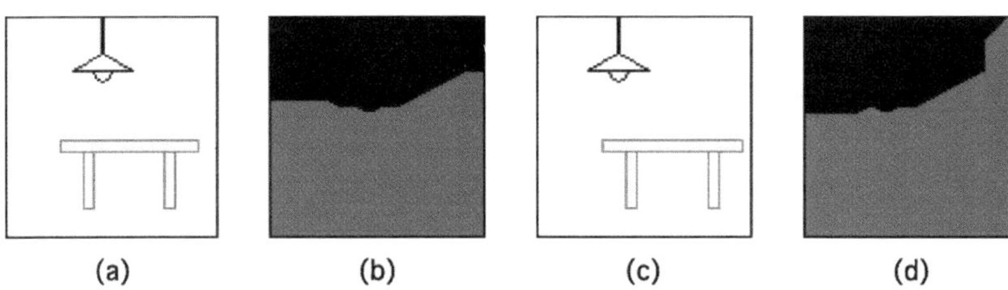

FIGURE 27. Detection of the "superiority" schema by double isotropic expansion, applied to "the lamp above the table." (a)-(b) Standard configuration. (c)-(d) Shifted configuration.

and "TR beside LM". Once more, the SKIZ boundary Z proves to be useful to extract a *categorical* criterion from the morphological analysis of this family of scenes. In the 'above' case, Z intersects the image boundary ∂W only on the top-top border (or, possibly, left-top, top-right, or even left-right), whereas in the 'beside' situation the intersection points are top-bottom (or, possibly,

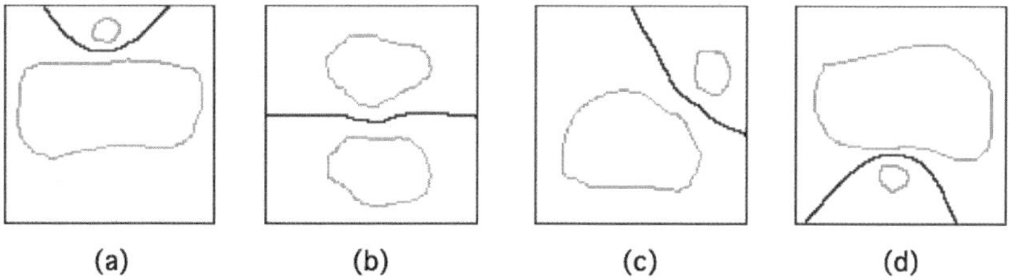

FIGURE 28. Different subtypes of "TR above LM", distinguished by the shape of the boundary Z and the location of its two intersection points with the image boundary ∂W. The two blobs represent TR and LM from top to bottom, respectively. The SKIZ boundary Z was computed exactly by equidistance. (a) Small TR above big LM: Z shows a positive curvature of paraboloid type. (b) TR and LM have similar size: Z is approximately flat. (c) TR is shifted to the side of LM: Z is paraboloid again but tilted so it also intersects a vertical side of the frame. (d) Big TR above small LM: Z has a negative curvature (corresponding to '*par-dessus*' in French).

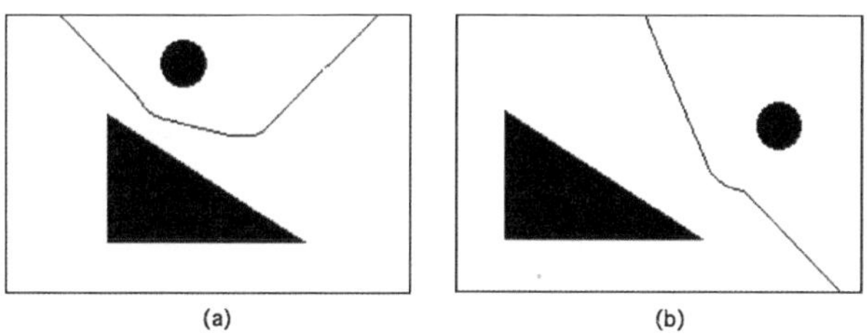

FIGURE 29. Continuous transition from (a) 'above' to (b) 'beside.' The bifurcation happens when the SKIZ Z touches the bottom boundary of W.

left-left, or right-right).[8] Therefore, the *bifurcation* from one class to another can be detected when Z reaches (or leaves) the bottom boundary of the image.

6.5. Links with Kosslyn's works

This "above vs. aside" example shows how a morphological approach can solve difficult problems in the relation between perception and language. One of these problems is the difference between *categorical* and *metric* processing of spatial relations. It has been thoroughly investigated by Stephen Kosslyn [191], who showed that they correspond to *hemispheric lateralization*. In a special issue of *Neuropsychologia* (44, 2006) dedicated to "New insights in Categorical and Coordinate Processing of Spatial Relations", S. Kosslyn [193] summarizes his theory concerning the "division of labor" of complex cognitive tasks into many simple sub-tasks. The occipital retinotopic areas implement a

8 In the case where ∂W is circular (closer to the real visual field), it can still be divided into top, right, bottom, and left by considering an embedded square. These four directions find natural cues in our universal sense of gravitation and bilaterality.

visual buffer where edge detection and segmentation are processed (see Chapter 2). The ventral "What" pathway goes from the occipital lobe to the inferior temporal cortex and processes the properties of objects. The dorsal "Where" pathway goes from the occipital lobe to the posterior parietal cortex and processes the properties of location. Additionally, a long-term associative memory processes spatial relations between objects and parts of objects. Now, just as the adumbrations of a single object can be infinitely many, the spatial relations in a configuration of objects can vary in a continuous manner, are also infinitely many, and must therefore be *categorized*. Hence, the necessity of a categorical processing of their continuous information.

In a series of papers, David Kemmerer [178] (see also [176], [177]) took into account cross-linguistic studies over more than 6,000 languages and looked for

> neuroanatomical correlates of linguistically encoded categorical spatial relations.

He found that such correlates exist and are located in the left supramarginal and angular gyri.

> Representations of coordinate spatial relations involve precise metric specifications of distance, orientation, and size; they are useful for the efficient visuomotor control of object-directed actions such as grasping a cup; and they are processed predominantly in the right hemisphere. In contrast, representations of categorical spatial relations involve groupings of locations that are treated as equivalence classes; they serve a variety of perceptual functions, such as registering the rough positions of objects in both egocentric and allocentric frames of reference; and they are processed predominantly in the left hemisphere.

Figure 30, from [178], shows (F = figure and G = ground)

> results from a PET study in which English speakers viewed drawings of spatial relations between objects (e.g., a cup on a table) and performed two tasks: naming F, and naming the spatial relation between F and G with an appropriate preposition. When the condition of naming objects was subtracted from that of naming spatial relations, the largest and strongest area of activation was in the left supramarginal gyrus.

Our morphological model can explain how a categorical perception of spatial relations can emerge from continuous variations: the cut locus of the configuration can bifurcate from one criterium ('above') to another ('aside'), and bifurcation is categorical.

6.6. Example 4: "zigzagging across the woods"

As our last example, we return to 'across', which was already examined in Section 4. In the case of the "transversality" schema, the morphological analysis has different requirements. As we saw previously, it mostly consists of a tubification of the trajectory of TR (dilation followed by a skeletonization), while extracting the cut-locus of LM, irrespective of the complexity of its texture

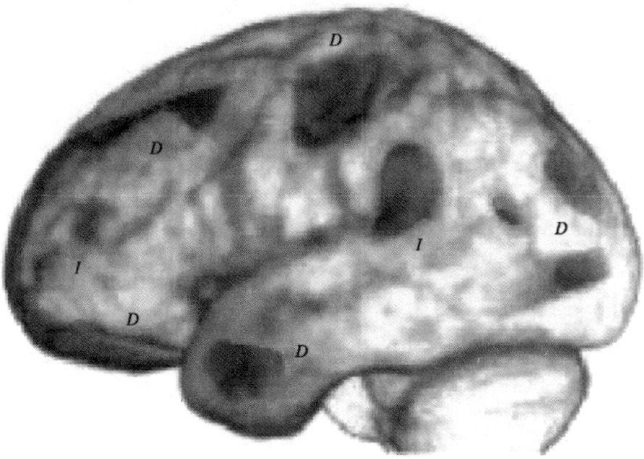

FIGURE 30. Kemmerer's and Damasio's results on the localization of the processing of categorical spatial relations in the left supramarginal gyrus. The letter I (resp. D) indicate blood flow increases (resp. decreases). (From [178], adapted from Damasio et al. [77])

(Figure 31). The latter can also be carried out by dilation and skeletonization. What remains is to detect the presence of a *quadruple singularity* point, instead of mere activity on the boundaries. This point signals somewhere on the image the crossing between two boundaries of *different origins* (TR and LM)[9].

Note the drastic simplification of the data performed by morphological schematization and how the virtual structures they create allow to extract the "transversality" invariant. Morphological algorithms reveal the typical crossing pattern $\times$, which can later be easily accessed to produce the desired categorical criterion. Our thesis is that the "TR across LM" *predication* becomes possible only on the basis of this preliminary morphological schematization.

7. Wave dynamics in spiking neural networks

7.1. Dynamic pattern formation in excitable media

Elaborating upon this first morphodynamical model, we now establish a parallel with neural modeling. Our main hypothesis is that the transition from analog to symbolic representations of space might be neurally implemented by traveling waves in a large-scale recurrent network of coupled spiking units, via the expansion processes discussed above (see Figure 33 for a preview). There is a vast cross-disciplinary literature, revived by Winfree [412], on the emergence of ordered patterns in excitable media and coupled oscillators. Traveling or kinematic waves are a frequent phenomenon in non-linear chemical reactions or multicellular structures (see, e.g., Swinney-Krinsky [366]), such as slime mold

9 This condition is important. Quadruple points of intersection between $\mathrm{Sk}(TR)$ and $\mathrm{Sk}(LM)$ must be distinguished from mere bifurcation points inside $\mathrm{Sk}(TR)$ and $\mathrm{Sk}(LM)$ respectively (these are generically triple points).

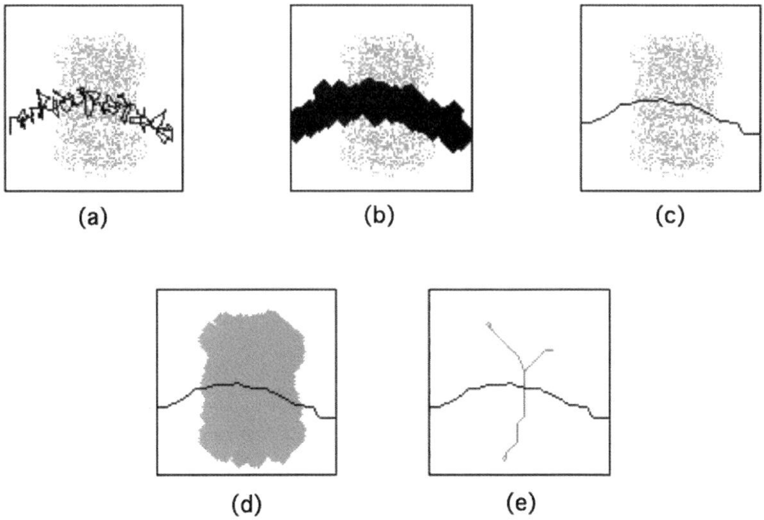

FIGURE 31. The "transversality" schema illustrated by "zigzagging across the woods'.' (a) The input image; (b)-(c) preprocessing of TR by dilation and skeletonization; (d)-(e) similar preprocessing of LM. Notice in (e) the quadruple intersection point between Sk(TR) and Sk(LM), typical of 'across'.

aggregation, heart tissue activity, or embryonic pattern formation. Across various dynamics and architectures, these systems have in common the ability to reach a critical state from which they can rapidly bifurcate between randomness or chaos and ordered activity. To this extent, they can be compared to "sensitive plates", as certain external patterns of initial conditions (chemical concentrations, food, electrical stimuli) can quickly trigger internal patterns of collective response from the units.

We explore the same idea in the case of an input image impinging on a layer of neurons and draw a link between the produced response and categorical invariance. In the framework proposed here, a visual input is classified by the qualitative behavior of the system, i.e., the presence or absence of certain singularities in the response patterns.

7.2. Spatio-temporal patterns in neural networks

During the past two decades, a growing number of neurophysiological recordings have revealed precise and reproducible temporal correlations among neural signals and linked them with behavior (see, e.g., Abeles [2], Bialek et al. [33], O'Keefe-Recce [247]). *Temporal coding* in the sense introduced by von der

Malsburg [397] is now recognized as a major mode of neural transmission, together with average rate coding. In particular, quick onsets of transitory phase locking have been shown to play a role in the communication among cortical neurons engaged in a perceptual task (Gray et al. [131]).

While most experiments and models involving neural synchronization were based on zero-phase locking among coupled oscillators (e.g., Campbell-Wang [49], König-Schillen [186]), *delayed* correlations have also been observed (Abeles [2], [3]). These nonzero-phase locking modes of activity correspond to reproducible rhythms, or waves, and could be supported by connection structures called *synfire chains* (Abeles [2], Ikegaya et al. [165]). Elie Bienenstock [35] construes synfire chains as the physical basis of elementary "building blocks" that compose complex cognitive representations: synfire patterns exhibit *compositional* properties (Abeles et al. [4], Bienenstock [36]), as two waves propagating on two chains in parallel can lock and merge into one single wave by growing cross-connections between the chains (in a "zipper" fashion). In this theory, spatio-temporal patterns in long synfire chains would thus be analogous to proteins that fold and bind through conformational interactions.

In the present work, we construe wave patterns differently: we look at their emergence on regular 2D lattices of coupled oscillators to implement the expansion dynamics of our morphodynamical spatial transformations. Compared to the traditional blocks of synchronization, i.e., the phase plateaus often used in segmentation models, we are also more interested in traveling waves, i.e., phase *gradients*.

7.3. Wave propagation and morphodynamical routines

A possible neural implementation of the morphodynamical engine at the core of our model relies on a network of individual spiking neurons, or local groups of spiking neurons (e.g., columns), arranged in a 2D lattice similar to topographic cortical areas, with nearest-neighbor coupling. Each unit obeys a system of differential equations that exhibit regular oscillatory behavior in a certain region of parameter space. Various combinations of oscillatory dynamics (relaxation, stochastic, reaction-diffusion, pulse-coupled, etc.) and parameters (frequency distribution, coupling strength, etc.) are able to produce waveform activity, however it is beyond the scope of the present work to discuss their respective merits. We want here to point out the generality of the wave propagation phenomenon, rather than its dependence on a specific model.

For practical purposes, we use Bonhoeffer-van der Pol (BvP) relaxation oscillators (FitzHugh [108]). Each unit i is located on a lattice point x_i and described by a pair of variables (u_i, v_i). Unit i is locally coupled to neighbor units j within a small radius r and may also receive an input I_i:

$$\text{(36)} \qquad \begin{cases} \dot{u_i} = c(u_i - \frac{u_i^3}{3} + v_i + z) + \eta + k\sum_j (u_j - u_i) + I_i \\ \dot{v_i} = \frac{1}{c}(a - u_i - bv_i) + \eta \end{cases}$$

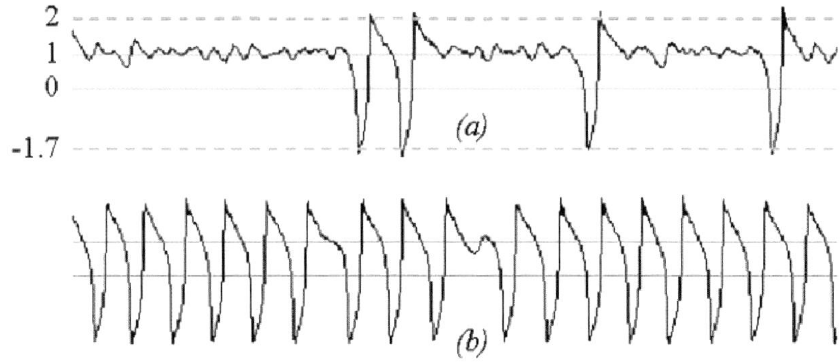

FIGURE 32. Typical firing modes of a stochastic BvP relaxation oscillator. Plots show $u(t)$ with $a = .7$, $b = .8$, $c = 3$, $\eta > 0$, $k = 0$, $I = 0$. (a) Sparse stochastic firing at $z = -.2$ (spikes are upside-down). (b) Quasi-periodic firing at $z = -.4$. At critical value $z_C = -.3465$ without noise, there is a bifurcation from a stable fixed point $u \approx 1$ to a limit cycle.

where η is a Gaussian noise, $d(x_i, x_j) < r$ and $I_i = 0$ or a constant I. Parameters are tuned as in Figure 32, so that individual units are close to a bifurcation in phase space between a fixed point and a limit cycle, i.e., one spike emission. They are *excitable* in the sense that a small stimulus causes them to jump out of the fixed point and orbit the limit cycle, during which they cannot be disturbed again.

Figure 33b shows waves of excitation in a network of coupled BvP units created by the schematic scene "a small blob above a large blob". Block impulses of spikes trigger wave fronts of activity that propagate away from the object contours and collide at the SKIZ boundary between the objects. These fronts are "grassfire" traveling waves, i.e., single-spike bands followed by refraction and reproducing only as long as the input is applied. Under the non-linear dynamics, waves annihilate when they meet instead of adding up. Figure 33a shows the same influence zones obtained by mutual expansion in a cellular automaton, as seen in Section 6.4. Again, there is convincing perceptual and neural evidence for the significant role played by this virtual SKIZ structure and propagation in vision (Kimia [179]).

7.4. Two wave categorization models

We now show how wave dynamics can support the categorization of spatial schemata by proposing two models based on the principles discussed previously. In both models, waves implement the expansion-based transformations stated in principles (i)-(ii) (see 2.2.2), then the detection of global activity or of singularities created by the wave collisions is based on principles (iii)-(v). One wave model implements the boundary detection principle (iii) used in the

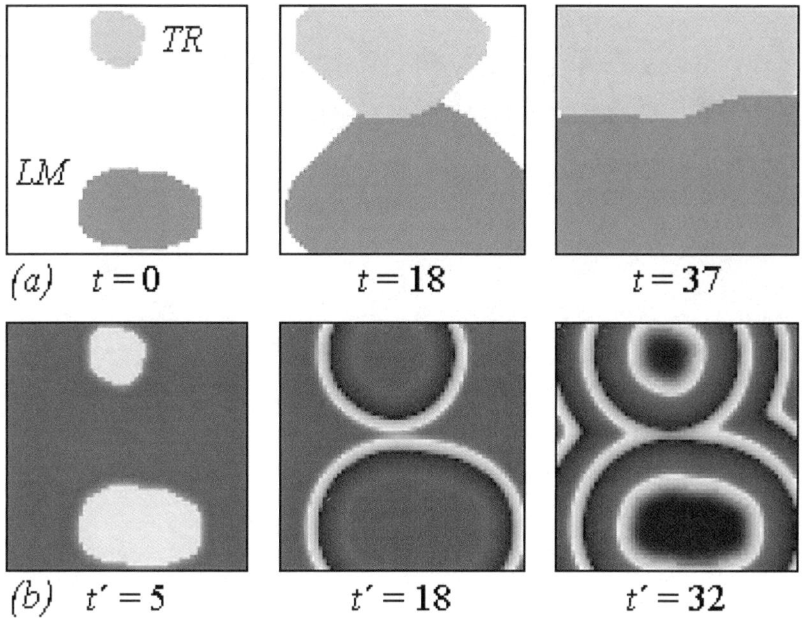

FIGURE 33. Realizing morphodynamical routines in a spiking neural network. (a) SKIZ obtained by diffusion in a 64×64, 3-state cellular automaton. (b) Same SKIZ obtained by traveling waves on a 64×64 lattice of coupled BvP oscillators in the sparse stochastic regime with $\eta = 0$, connectivity radius $r = 2.3$ and coupling strength $k = .04$. Activity u is shown in gray levels, brighter for lower values, i.e., spikes $u < 0$. Starting with uniform resting potentials $u \approx 1$ (or weak stochastic firing with noise $\eta > 0$), an input image is continuously applied with amplitude $I = -.44$ in both TR and LM domains. This amounts to shifting z toward a subcritical value $z = -.3467 < z_c$, thus throwing the BvP oscillators into a quasi-periodic firing mode. This in turn creates traveling waves in the rest of the network.

"containment" and "superiority" schemata. The second model focuses on the SKIZ singularities and "signature" detection principle (v), which can be used as a complement or alternative to boundary detection. In this case, we illustrate SKIZ detection with the same 'above' schema as in boundary detection.

7.4.1. *Boundary detection with cross-coupled lattices.* Detecting the presence or absence of TR activity on the boundaries of the image, as for 'in' or 'above', is not possible in the single lattice of Figure 33b because the waves triggered by TR and LM cannot be distinguished from each other. However, as discussed above and in Chapter 2, a number of models have shown that lattices of coupled oscillators can also carry out *segmentation from contiguity* by exploiting a simpler form of temporal organization in the lattice: zero-phase synchronization or "temporal tagging". We take here these results as the starting point of our simulation and assume that the original input layer is now split into

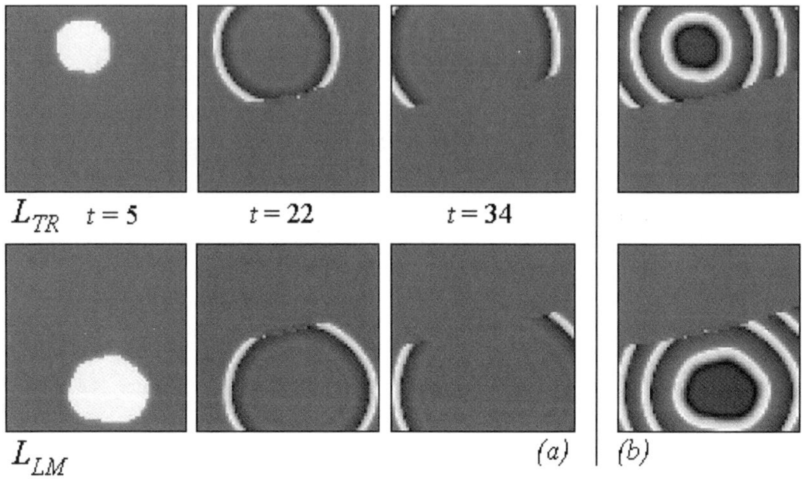

FIGURE 34. Detection of the 'above' schema by mutually inhibiting waves. Two 64 × 64 lattices of BvP units, L_{TR} and L_{LM}, are internally coupled *and cross-coupled* with $r = r' = 2.3$, $\eta = 0$, $k = .03$, $k' = -.03$. (a) Single wave fronts obtained by injecting a pulse input $I = -.44$ in TR and LM for $0 \leq t < 2$ (10 time steps $dt = .2$). (b) Multiple wave fronts obtained by applying the same input amplitude indefinitely. In both cases, no spike reaches the bottom of L_{TR}.

two distinct sublayers, each holding one component of the scene. Thus, after a preliminary segmentation phase (not presented here), TR and LM are forwarded to layers L_{TR} and L_{LM}, where they generate wave fronts separately (Figure 34). Mutual wave interference and collision is then recreated by *cross-coupling* the layers: unit i in layer L_{TR} is not only connected to units j in L_{TR} inside a neighborhood of radius r, but also to units j' in L_{LM} inside a neighborhood of radius r'. The modified dynamics is therefore

$$(37)\quad \dot{u_i}^{TR} = F(u_i^{TR}) + k\sum_j (u_j^{TR} - u_i^{TR}) + k'\sum_{j'} (u_{j'}^{LM} - u_i^{TR}) + I_i^{TR}$$

where $F(u)$ is the right hand side of Eq. (36) without k and I terms. A symmetrical relation holds for u_i^{LM}, swapping TR and LM. Variables v_i are not coupled and obey equation (36) as before. The net effect is shown in Figure 34: the spiking wave fronts created by TR are canceled by LM's wave fronts and never reach the bottom border of L_{TR}, while hitting the top and partly the sides. This can be easily detected by external cells receiving afferents from the border units and linked to an 'above' response (not shown here). Again, the invisible collision boundary line is the SKIZ, which we now examine more closely in the next network model.

7.4.2. *SKIZ signature detection with complex cells.* As previously mentioned in Section 6.4.2, detection of boundary activity provides a simple categorization mechanism but is generally not sufficient. Alone, it does not allow to distinguish among similar but non-overlapping proto-schemata, such as the ones in Figure 28. This is where the properties of the SKIZ can help: for example, the concave or convex shape of the SKIZ is able to separate Figure 28a and Figure 28d. As we already emphasized in Chapter 2, Section 9.3, a cut locus is a dynamical structure, and one should also take into account the *flow velocity* along the SKIZ. Indeed, the dynamics of coupled spiking units (Figure 33b) is richer than the morphological model (Figure 33a) because it contains specific patterns of activity that are absent from a static geometric line. In particular, the wave fronts highlight a secondary flow of propagation *along* the SKIZ line, which travels away from the focal shock point with decreasing speed on either branch. The focal point (where the bright band is at its thinnest in Figure 33b, $t' = 32$) is the closest point to both objects and constitutes a *local optimum* along the SKIZ. While a great variety of object pairs produce the same static SKIZ, the speed and direction of flow along the SKIZ vary with the objects' relative curvature and proximity to each other. For example, a vertical SKIZ segment between a pair of bracketed contours resembling (|) flows *inward* when facing their concave sides, whereas it flows *outward* when facing the convex sides of reversed brackets)|(. This refined information can be revealed by wave propagation.

In order to detect the focal points and flow characteristics (speed, direction) of the SKIZ, we propose in this model to introduce additional layers of detector neurons similar to the so-called "complex cells" of the visual system. These cells receive afferent contacts from local fields in the input layer and respond to segments of moving wave bands, with selectivity to their orientation and direction. More precisely, the spiking neural network presented in Figure 35 is a three-tier architecture comprising: (a) two input layers, (b) two middle layers of orientation and direction-selective "D" cells, and (c) four top layers of coincidence "C" cells responding to specific pairwise combinations of D cells. (These are not literally cortical layers but could correspond to functionally distinct cortical areas.) As in the previous model, TR and LM are separated on two independent layers (Figure 35a). In this particular setup, however, there is no cross-coupling between layers and the waves created by TR and LM do not actually interfere or collide. Rather, the regions where wave fronts "coincide vertically" (viewing layers L_{TR} and L_{LM} superimposed) are captured by higher feature cells in two direction-selective layers, D_{TR} and D_{LM} (Figure 35b), and four coincidence-detection layers, $C_{1...4}$ (Figure 35c). Layer D_{TR} receives afferents only from L_{TR}, and D_{LM} only from L_{LM}. Layers C_i are connected to both D_{TR} and D_{LM} through split receptive fields: half of the afferent connections of a C cell originate from a half-disc in D_{TR} and the other half from its complementary half-disc in D_{LM}.

In the intermediate D layers, each point contains a family of cells or "jet" similar to multi-scale Gabor filters. Viewing the traveling waves in layers L as

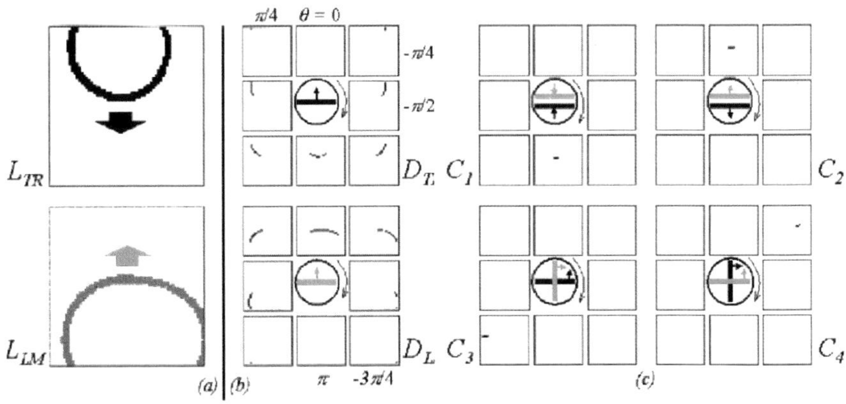

FIGURE 35. SKIZ detection in a three-tier spiking neural network architecture. (a) Input layers show single traveling waves as in the previous figure, except that cross-coupling between TR and LM is removed and potentials are thresholded to retain only spikes $u < 0$. (b) Orientation and direction-selective cells. Each D layer is shown as 8 sublayers (smaller squares) of cells selective to orientation $\theta = k\pi/4$, with $k = 8 \ldots 1$ from the top, clockwise. (c) Pairwise coincidence cells. Each cell in $C_1(\theta)$ is connected to a half-disc neighborhood in $D_{TR}(\theta)$ and its complementary half-disc in $D_{LM}(\theta - \pi)$, rotated at angle θ. For example: a cell in $C_1(0)$ (illustrated by the icon in center of C_1) receives afferents from a horizontal bar of $D_{TR}(0)$ cells in the lower half of its receptive field, and a bar of $D_{LM}(\pi)$ cells in the upper half. Same for $C_2(0)$, swapping TR vs. LM and upper vs. lower. Similarly, $C_3(\theta)$ cells are half-connected to $D_{TR}(\theta)$ and half to $D_{LM}(\theta - \pi/2)$, via orthogonal bars (swapping TR/LM for C_4). The net output is sparse activity confined to $C_1(\pi)$, $C_2(0)$, $C_3(3\pi/4)$ and $C_4(-\pi/4)$. Note that we use only global C_i activity: in reality, C_1 cells fire first when the two wave fronts meet, then C_2 cells when they separate again, and finally C_3 and C_4 cells in rapid succession when the arms cross. This precise rhythm of C_i spikes could also be exploited in a finer model.

moving bars, each D cell is sensitive to a combination of bar width λ, speed s, orientation θ and direction of movement φ. In the simple wave dynamics of the L layers, λ and s are approximately uniform. Therefore, a jet of D cells is in fact single-scale and indexed by one parameter $\theta = 0, \ldots, 2\pi$, with the convention that $\varphi = \theta + \pi/2$. Typically, 8 cells with orientations $\theta = k\pi/4$ are sufficient. Each sublayer $D_{TR}(\theta)$ thus detects a portion of the traveling wave in D_{TR} (same with LM). Realistic neurobiological architectures generally implement direction-selectivity using inhibitory cells and transmission delays. In our simplified model, a $D(\theta)$ cell is a "cylindrical" filter, i.e., a temporal stack of discs containing a moving bar at angle θ (Figure 35b, center of D layers): it sums potentials from afferent L-layer spikes spatially and temporally, and fires itself a spike above some threshold.

Among the four top layers (Figure 35c), C_1 detects converging parallel wave fronts, C_2 detects diverging parallel wave fronts, and C_3 and C_4 detect crossing perpendicular wave fronts. Like the D layers, each C_i layer is subdivided into 8 orientation sublayers $C_i(\theta)$. Each cell in $C_1(\theta)$ is connected to a half-disc neighborhood in $D_{TR}(\theta)$ and the complementary half-disc in $D_{LM}(\theta - \pi)$, where the half-disc separation is at angle θ. The net output of this hierarchical arrangement is a *signature* of coincidence detection features providing a *very sparse coding* of the original spatial scene. The input scene 'above' is eventually reduced to a handful of active cells in a single orientation sublayer $C_i(\theta)$ for each C_i: $C_1(\pi)$, $C_2(0)$, $C_3(3\pi/4)$ and $C_4(-\pi/4)$ (Figure 35c).

In summary, the active cells in C_1 and C_2 reveal the focal point of the SKIZ, which is the primary information about the scene, while C_3 and C_4 reveal the outward flow on the SKIZ branches, which can be used to distinguish among similar but non-equivalent concepts. This sparse SKIZ signature is at the same time characteristic of the spatial relationship and largely insensitive to shape details. For example: 'below' yields $C_1(0)$ and $C_2(\pi)$; 'on top of' yields $C_2(0)$ like 'above' but no C_1 activity because TR and LM are contiguous (wave fronts can only separate at the contact point, not join); the French preposition 'par-dessus' with a convex SKIZ facing up (Figure 28d) yields $C_3(\pi)$ and $C_4(-\pi/2)$, etc. Note that the actual regions of $C_i(\theta)$ where cells are active (e.g., the location of the SKIZ branches in the south-west and south-east quadrants of Figure 28d) are sensitive to translation and therefore are not good invariant features.

CHAPTER 4

Processes: What is an "Attractor Syntax"?

1. Introduction

Developing dynamical models of cognitive processes raises fundamental issues, some of which have been addressed by connectionist models implementing dynamical systems. One of the most difficult questions is whether dynamical models implemented in connectionist networks have the capacity for adequately supporting syntactic constituency and constituent-structure, which are traditionally expressed in a symbolic way. We have already shown that, for perception, mereological dynamical models exist and can be used in a morphological approach to spatial relations. But in what concerns *syntax*, and especially actantial syntax[1], the difficulty is to model grammatical relations, semantic roles, constituency and compositionality in a purely dynamical fashion. In a nutshell, it can be formulated in the following manner: if terms of sentences are modeled by attractors of some underlying dynamics, what is the dynamical status of a "syntax" relating these attractors? *What could be an attractor syntax*?

2. Dynamical models of syntax

In the late 1980's and early 1990's, following the 1988 debate that opposed Jerry Fodor and Zenon Pylyshyn [113] to Paul Smolensky [355] (see also Smolensky et al. [359] and Fodor, McLaughlin [111]), a great number of works have been devoted to this question. In 1991 and 1992 two conferences on the *Compositionality in Cognition and Neural Networks* organized by Daniel Andler, Elie Bienenstock and Bernard Laks (COMPCOG I [64] and COMPCOG II [65]) were held in Royaumont, France. Bernard Victorri and Catherine Fuchs organized in 1992 at the University of Caen a nice conference on *The Continuum in Linguistics* [115]. We organized in 1995 another international conference *Topology and Dynamics in Cognition and Perception* [289] with Patrizia Violi in San Marino. We also edited a special issue of *Sémiotiques*, "Dynamical Models and Cognitive Semiotics" [290]. Another conference on *Dynamic Representations in Cognition* was organized by Robert Port and Tim van Gelder

1 We recall that, following the standard convention, we use the terms "actant", "actantial", "actantiality" to refer to semantic roles in the sense of case grammars and narrative grammars.

at Indiana University in 1992. They later edited at the MIT Press a reference book entitled *Mind as Motion* [312] on dynamical models in linguistics.

The particular status of dynamical explanations, contrasting with the deductive-nomological ones, has been stressed by Tim van Gelder [312]. The epistemological problems raised by a dynamical interpretation of basic theoretical concepts such as *structure*, *constituency* and *syntax*, have been carefully analyzed by Daniel Andler [312] and Yves-Marie Visetti [312].

What could be an attractor syntax? The problem is especially difficult for the following reasons.

2.1. Weak CN vs. strong CN

In the case of syntax—deep universal and formal syntax, not English or French morpho-syntax—we need to at least distinguish:

(i) between two syntactic (categorial) types: things or objects (terms) and relations, and
(ii) between two types of relations: static and dynamic[2].

A process described by a verb is a dynamical transformation of actantial relations. If, however, we represent terms by activity patterns that are the attractors of dynamical systems implemented in connectionist (CN) networks, the challenge becomes[3]: how can we represent these two differences "term vs. relation" and "static vs. dynamic"? It is clear that syntactic relations between attractors cannot be reduced to mere linear superpositions. We call *weak* CN a CN that models semantic entities of *different* syntactic types by attractors of the *same* dynamical type, without taking into account differences in their grammatical categories (this is a category mistake).[4] In order to work out a solution to this challenge, we need to strengthen weak CN into a *strong CN* that has the capacity to model different grammatical categories by mathematical entities of *different* types.

2.2. Elementary vs. non-elementary CN syntax

One could think that it would be trivial to elaborate a strong CN. One would only have to

2 We must distinguish three meanings of "dynamical": a physical phenomenon is dynamical when the causality of its temporal evolution is concerned; a relation is dynamical when it is temporal; a mathematical model is dynamical when it proceeds from the theory of dynamical systems (global analysis).

3 In the remainder of this chapter, CN stands for "connectionism" or "connectionist", CNC for "connectionist cognitivism", CL for "classic", and CLC for "classical cognitivism".

4 We will see below what is the technical sense of this "weakness": the lack of any interpretation of syntactic relations in terms of dynamical bifurcations.

(i) represent activity patterns (attractors) coding the terms by units belonging to some higher level layer (what is called in the connectionist literature a "localisation" through "grand-mother" neurons), and
(ii) represent the relations by connections between these units.

We call such a solution *elementary*. Yet, this solution does not work for the very reasons given by Fodor and Pylyshyn [113] in their criticism against CNC: it analogically projects the neuronal implementation onto the functional architecture and admits

> the implicit—and unwarranted—assumption that there ought to be similarity of structure among the different levels of organization of a computational system.

Static and dynamic relations between terms must be modeled by relationships between dynamical attractors, and these relations are of a completely different nature than the underlying connections implementing them. We call "non-elementary" a dynamical model that does justice to this principle.

Main problem: Under the initial hypothesis that contents can be modeled by attractors, can an "attractor syntax" be worked out in the framework of a strong non-elementary CN, that is, the theory of dynamical systems?

This problem is essentially theoretical and mathematical. We present here a few elements toward its resolution.

3. Theoretical strategy: using the "morphological turn"

As Domenico Parisi [113] pointed out, in this sort of dynamical modeling of "higher level capacities such as language",

> the most serious problem (...) is to free itself from the grip of concepts, definitions of problems, and even description of phenomena, that have been used for decades by symbolic accounts of these capacities.

We agree with this "categorical imperative" for dynamical modeling. Consequently, our strategy will be the following.[5]

3.1. The concept of "structure" and Morphodynamics

We already know that in the physical realm, theories of self-organization have shown how discrete constituent structures can emerge from continuous material substrates: they are essentially dependent on critical phenomena, on phenomena of symmetry-breaking that induce qualitative discontinuities. Thus there is a basic link between the concept of "structure" and morphodynamics.

---

5 We thank Bob Port for his remarks on the best way to summarize these rather technical theoretical currents.

3.2. Cognitive processing

To understand constituency dynamically at the cognitive level, we must also use the abovementioned theories showing how elementary macro-structures can emerge from the underlying (e.g., neural) complex micro-dynamics in which they are implemented.

3.3. The configurational definition of roles

We will see that in a dynamical theory of constituent structures, the difference between the semantic roles and the syntactic relations expressing events of interaction between these roles corresponds to the difference between attractors and *bifurcations* of attractors. In such models, semantic roles cannot be reduced to a mere assignment of labels. There exists an *embodiment* of their meaning. Bifurcation theory allows one to work out a *configurational definition* of the semantic roles in much the same way as, in the symbolic conception of formal grammars, syntagmatic trees yield a configurational definition of grammatical relations. For us, therefore, the main problem is not the symbolic binding between a role label and a filler term, but rather such a configurational definition.

3.4. The link with spatial cognition

Following René Thom, one of our main theses is that syntactic structures linking participants roles in verbal actions are organized by universals and invariants of a morphological nature. It is our dynamical interpretation of Talmy's, Langacker's and Lakoff's works concerning *Gestalts* and *image-schemata* in language. We have already developed it in the previous chapter concerning static relations.

3.5. The shift of mathematical level: "kernel sentences"

A dynamical approach to syntactic constituent structures upsets the classical conception of formalization because it shifts the level of mathematical modeling. Indeed, in the classical symbolic view, the fact that terms can be bounded by relations is taken for granted as a basic elementary fact that deserves no further explanation. Consequently, in this view, the only interesting structures are not elementary structures—the "atomic formulae" of logicians or the "kernel sentences" of Chomskyans—but complex structures. On the contrary, in the dynamical view, the concept of relation is in itself a hugely difficult problem. Moreover, supposing that it were solved, it would still be very difficult to formalize complex structures in this way.

Our purpose is to lay the foundations for a dynamical and physical theory of constituency and constituent structures. Such a theory already needs a lot of sophisticated mathematical tools to model even the most elementary structures—while complex structures are still intractable that way and will require higher levels of description.

Because these questions are very important, before proposing models we present a careful analysis of the criticism raised by Fodor and Pylyshyn against the very idea of a connectionist syntax. In 1988, Jerry Fodor and Zenon Pylyshyn published an important paper in *Cognition*: "Connectionism and Cognitive Architecture: A Critical Analysis". It was a radical criticism of the thesis held by Paul Smolensky in his paper "On the Proper Treatment of Connectionism", which appeared in *Behavioral and Brain Sciences.* A little later, in his talk on "Why connectionism is such a bad thing", given at the École Normale Supérieure in Paris, J. Fodor elaborated upon his objections.

In the following, we review his main points. Our own arguments will seek to show

(i) that Fodor and Pylyshyn's criticism is essentially valid if it concerns a very basic form of CN ("weak CN"); but that on the other hand
(ii) it is not admissible for a more elaborate form of CN that fully employs the mathematical resources of the theory of dynamical systems ("strong CN").

4. Connectionism and the theory of dynamical systems

4.1. The CNC main thesis and its precursors

We already know (see Chapter 1) that classical symbolic cognitivism (CLC) is analytic and constructivist and favors logical automatisms, conscious rules, calculus, and deductive inference, while connectionist (subsymbolic) cognitivism (CNC) is synthetic and associationist and favors dynamics of networks, intuitive performances, equilibrium states and induction. For CNC, the entities possessing a meaning are, on the micro-dynamical underlying level, global and complex distributed activation patterns of elementary meaningless local units. These scattered units are interconnected and process the data in parallel.

4.2. CN networks and dynamical systems

In its simplest expression, a formal neural network S consists of N units u_i whose instantaneous global activation state $\mathbf{x} = (x_i)_{i=1,\dots,N}$ varies within a given configuration space $M = E^N$. M plays the role of an internal space and in most cases $E = \{0,1\}$, $\{-1,1\}$, or $[0,1]$. Units u_i are interconnected through synaptic weights w_{ij} that determine the network's *computing program.* Positive weights represent excitatory connections and negative weights represent inhibitory connections. Reflexive weights are generally absent: $w_{ii} = 0$.

The network "computes" in the following way. Each neuron u_i receives incoming signals from its presynaptic neighbors, computes a new signal (i.e., modifies its internal state according to a transition law) and sends this signal via outgoing connections to its postsynaptic neighbors. The input to unit u_i is generally defined as a weighted sum of incoming signals:

(1) $$h_i = \sum_{j=1}^{j=N} w_{ij} x_j , \text{ or } \mathbf{h} = \mathbf{w}\mathbf{x} \, .$$

Neurons u_i are modeled as threshold automata, whose internal states are governed by a local transition rule of the following type:

(2) $$x_i(t+1) = g\left(h_i(t) - T_i\right), \text{ or } \mathbf{x}(t+1) = g\left(\mathbf{h}(t) - \mathbf{T}\right)$$

where T_i is a threshold and g a gain function. Typically:

- $g =$ Heaviside function for $E = \{0, 1\}$,
- $g =$ sign function for $E = \{-1, 1\}$,
- $g =$ sigmoidal function $1/(1 + e^{-x})$ for $E = [0, 1]$.

Weights w_{ij} and thresholds T_i vary in a control space W that plays the role of an external space. The global dynamics of the network results from all the local rules put together and executed iteratively. This dynamics characterizes the computing properties of the network.

In the limit of continuous time, one obtains a large set of differential equations, typically:

(3) $$\dot{\mathbf{x}} = -\mathbf{x} + g\left(\mathbf{w}\mathbf{x} - \mathbf{T}\right) \, .$$

In the limit of a spatial continuum of neuronal substrate, the system can be described by partial differential equations on density variables, such as:

(4) $$\frac{\partial x(r,t)}{\partial t} = -x(r,t) + g\left(\int \left(w\left(r,s\right) x\left(s,t\right) - T(r)\right) ds\right)$$

where $x(r,t)$ is the activity at time t of neuron $u(r)$ located at position r of the neural layer, $T(r)$ its threshold and $w\left(r,s\right)$ the synaptic weight between units $u(r)$ and $u(s)$.

In recurrent networks, where outputs feed back into inputs, the representative states of the system are its *asymptotic states*, in particular its *attractors*.[6] The local transition laws of an elementary unit from one state to another, considered as a function of the information that this unit receives from its immediate neighbors, define an endomorphism $\mathcal{T} : M \to M$ that associates the instantaneous global state $x \in M$ of S with the next state $\mathcal{T}(x)$. In general, it encodes a large amount of information. It is the *iteration* of $\mathcal{T}$ that defines the *internal dynamics* of the network S. The stable asymptotic states of $\mathcal{T}$ (its attractors) are the *internal states* of S. If x is an input of S (an instantaneous initial state), its trajectory $(\mathcal{T}^k(x))_{k \in \mathbb{N}}$ will tend in general towards an attractor A that will be the output (the response) of S defined by x. By varying the

---

6 See Amit [11].

local transition laws— for instance by changing the weights of the connections of S—$\mathcal{T}_w$ is modified.

The attractors define the internal states of the network. This kind of models are thus special cases of morphodynamical models. The basic dynamical phenomenon is the asymptotic capture by attractor A of a global instantaneous state $\mathbf{x}_0$ of the network. In this sense, neurally inspired networks compute in a way that is radically different from Turing-von Neumann machines. They are dynamical computers that bifurcate from attractor to attractor.

The types of dynamics obtained this way can be dauntingly complex. In general, attractors have *an internal structure* and a non-trivial topology. This fact is fundamental, for one can think that this topology encodes the internal intrinsic semantics[7] of the associated representations. But if the internal dynamics $\mathcal{T}$ is the gradient of an "energy" function, then the internal intrinsic semantics becomes trivial (in the same way as, in CLC, one represents a content by a symbol). This will be the case if $\mathcal{T}$ consists in minimizing an "energy" function $H : M \to \mathbb{R}$.

4.2.1. *Symmetric weights.* Even in this maximally simple case, the internal dynamics is rather complex. For example, in the case—unrealistic from a neurobiological viewpoint— where connections are *symmetric*, with $E = \{-1, +1\}$ and g = sign function, Hopfield remarked that the network's equations are equivalent to a system of interacting spins. The corresponding energy function minimized by the dynamics reads:

$$H = -\frac{1}{2}\sum_{i \neq j} w_{ij} x_i x_j + \sum_i T_i x_i \,. \tag{5}$$

Here, synaptic weights w_{ij} are the analog of coupling constants that can form an intricate mix of positive and negative values.

These systems exemplify the *simplest type* of artificial neural networks, yet at the same time the *most complex type* of spin systems called "spin glasses".[8] Their energy function displays a great number of local relative minima (metastable states), making ineffective the traditional methods of gradient descent used for searching global minima. It becomes necessary to use more sophisticated algorithms coming from statistical physics, such as *simulated annealing.*[9] Noise is introduced into the system via a computational "temperature" T. Starting from an initial configuration $\mathbf{x}_0$ with high T (which gives access to all basins of attraction), a neighboring configuration $\mathbf{x}_1$ is chosen at random. If the energy difference ΔH between $\mathbf{x}_1$ and $\mathbf{x}_0$ is negative, the system

7 The internal intrinsic semantics must not be confused with an "external" denotative semantics. Using the Fodorian opposition between "narrow" and "large" contents, we can say that it belongs to the narrow content of the associated representation.

8 See, e.g., Mézard et al. [233].

9 See, e.g., Azencott [24].

shifts to $\mathbf{x}_1$, otherwise, for $\Delta H > 0$, it shifts with probability

$$\frac{1}{1+e^{\frac{\Delta H}{T}}}, \tag{6}$$

which allows to raise the threshold values. The next steps are computed until reaching a local minimum, then temperature T is decreased and the same operation repeated.

More precisely, let us consider distributions of probabilities $P(x)$ over M. For a given mean energy C, that is for $\sum_{x \in M} P(x)H(x) = C$, the distribution $G_T(x)$ which maximizes the entropy is the well known Gibbs distribution:

$$G_T(x) = \frac{1}{Z_T} e^{-\frac{H(x)}{T}} \tag{7}$$

where Z_T is the partition function:

$$Z_T = \sum_{x \in M} e^{-\frac{H(x)}{T}} . \tag{8}$$

When the computational "temperature" $T \to 0$, G_T concentrates on the set $M_{\min}$ of absolute minima of H, which can have a very large cardinal. This is a well known physical result. The "simulated annealing" algorithm is based on the construction of sequences of random variables X_n and temperatures T_n such that:

$$\begin{aligned} \lim_{n\to\infty} P(X_n = x) &= G_T(x) \\ \lim_{n\to\infty} P(X_n \in M_{\min}) &= 1 . \end{aligned}$$

More precisely, as explained by Robert Azencott [24], we are given:

(i) An exploration matrix $Q = (q_{x,y})_{x,y \in M}$ satisfying the condition that, for all $x, y \in M$, there exists a chain (x_k) linking x and y and such that $q_{x_k,x_{k+1}} > 0$ for all k. If we call $V_x = \{y \mid q_{x,y} > 0\}$ the set of neighbors of x, this condition means that two configurations x and y are always linkable by a chain of neighboring configurations.

(ii) A sequence T_n of "temperatures" such that $T_n \to 0$ (cooling schedule), the decrease of T_n being sufficiently "slow". This later condition is expressed by the fact that $\lim_{n\to\infty} T_n \operatorname{Log}(n) = R$ where R is a sufficiently large constant.

(iii) A Markov chain of random variables X_n on M such that the conditional probability:

$$P_n(x, y) = P(X_{n+1} = y \mid X_n = x) = q_{x,y} e^{-\frac{(H(y)-H(x))^+}{T_n}} \tag{9}$$

if $y \neq x$ (where $a^+ = a$ if $a \geq 0$ and $= 0$ else), and

$$P_n(x, x) = 1 - \sum_{y \neq x} P_n(x, y) . \tag{10}$$

A configuration x is a *local* minimum of H ($x \in M_{\min}$) if $H(x) \leq H(y)$ for all $y \in V_x$. Its depth $D(x)$ is then defined as the minimal height of the thresholds (the saddles) that limit its basin of attraction.

A theorem due to Bruce Hajek says that:

$$\lim_{n \to \infty} P\left(X_n \in M_{\min}\right) = 1$$

iff $\sum_{n=1}^{n=\infty} e^{\frac{D}{T_n}} = \infty$, where $D = \operatorname{Sup}\{D(x) \mid x \in M_{\text{locmin}} - M_{\min}\}$.

4.2.2. *Asymmetric weights.* When synaptic weights are *asymmetric*, the energy function is undefined, and the dynamics can become extremely complex. Steve Renals and Richard Rohwer considered the following model:

$$x_i\,(t+1) = g\left(r \sum_{j=1}^{j=N} w_{ij} x_j(t)\right) \tag{11}$$

where r represents the slope of the sigmoid. They analyzed its Fourier spectrum

$$P_i(k) = \frac{1}{T}\left(\sum_{t=0}^{t=T-1} x_i\,(t)\, e^{-\frac{2i\pi kt}{T}}\right)^2 \tag{12}$$

and studied the bifurcations generated by the behavior of the x_i activity states when r varies. They found numerous classical route-to-chaos scenarios, in particular for $r \in [12, 14]$ the period-doubling route and subharmonic cascade of Coullet-Feigenbaum-Tresser. They also recalculated Feigenbaum's universal constant δ by recursion:

$$r_n = r_\infty - \text{const}\,.\delta^{-n}\ ,\ \text{i.e.}\ \frac{r_n - r_{n-1}}{r_{n+1} - r_n} = \delta \tag{13}$$

and found $\delta = 4.67 \pm 0.04$, which is in excellent agreement with the standard value $\delta = 4.6692016091029909$.

Haim Sompolinsky, Manuel Samuelides and Brunello Tirozzi also studied such systems in the limit of very large N for randomly distributed asymmetrical weights w_{ij} (e.g., Gaussian) of mean zero and variance w^2/N.[10] For a critical value $rw = 1$, they found a phase transition from a convergent to a chaotic regime. Samuelides especially studied the route to chaos in diluted systems (i.e., almost all $w_{ij} = 0$), with thresholds T_i and where random variables w_{ij} are not centered anymore.

Numerous other results of this kind show that it is now possible to reestablish the Thomian thesis (mentioned in Section 2.2) on a rigorous mathematical footing, in which mental contents are attractors of dynamical systems implemented in neural networks. Therefore, cognitive functions can be naturally

10 See Sompolinsky, Crisanti, Sommers [362], Doyon, Cessac, Quoy, Samuelides [93] and Tirozzi, Tsodyks [386].

rephrased within the framework of neural morphodynamics and thermodynamics.

In fact, connectionist models have rediscovered other ideas first proposed by René Thom. For example, the process of *categorization* is construed as resulting from the partition of an internal space M in a number of attractor basins of *prototypes*. What psychologists (e.g., Eleonor Rosch) call "typicality gradients" can thus be reinterpreted as Lyapunov functions on these basins. The typical vs. atypical (i.e., generic vs. special) opposition can also be made geometrical if M is stratified by the stable manifolds of the internal dynamics. These dynamical categorization phenomena allow neural networks to function as an associative memory, i.e., a content-addressable memory, distinct from the traditional computer RAM.

Actually, in the dynamical explanation of categorization, René Thom went significantly farther than the connectionist models that were developed after him. He also considered processes of categorization in *external* spaces induced by the unfolding of the singularities of the internal dynamics. Such external categorization processes arise naturally whenever the internal states depend on control parameters. For example, this is the case in phonetics: the internal dynamics are acoustic and the external controls are acoustic cues, such as voicing or articulation cues.[11]

In neural network theory, external space dynamics only play a role during *learning*. Learning can be viewed as the *inverse* problem of finding the attractors $X_{\mathbf{w}}$ given the matrix of synaptic weights $\mathbf{w}$: it consists of finding $\mathbf{w}$ on the basis of a desired set of attractors. A few algorithms were developed toward this goal, in particular the *backpropagation* of errors, which decreases the discrepancy between the desired attractors $X_{\mathbf{w}}$ and the attractors $X_{\mathbf{w}_0}$ calculated starting from an initial matrix $\mathbf{w}_0$. Such methods define *slow* external dynamics in the external spaces W of synaptic weights.

However, here too the morphodynamical viewpoint remains ahead of connectionist theories. It stresses the fact that in slow/fast systems there exists a *stratification* of W that categorizes the attractors $X_{\mathbf{w}}$ into different qualitative types. This stratification is a partitioning of W by a closed set K (what is called a catastrophe set). Algorithms such as backpropagation provide external dynamics that are defined only in open strata (the connected components of $W - K$, i.e., the categories of $X_{\mathbf{w}}$). But generally, the specificity of a learning process is to radically transform the qualitative type of $X_{\mathbf{w}}$. It remains to be understood how backpropagating dynamics defined on the different open strata are glued along K. This problem is still unresolved.[12]

---

11 For a morphodynamical analysis of phonetic categorization, see Petitot [269].

12 One of the few experts interested by the instabilities of learning algorithms caused by bifurcations contained in the system is Kenji Doya [92].

4.3. Harmony theory

In his harmony theory, Paul Smolensky [352] has applied these ideas to a certain number of subsymbolic processes and has stressed the importance of going beyond the von Neumann computational conception towards a dynamical conception of information processing. His aim was to link up the higher levels of cognition with the lower levels of perception. For example, in the interpretation of a visual scene we can suppose that the cognitive system undergoes the following operations:

(i) Let $(r_i)_{i \in I}$ be a set of representational features which "constitute the cognitive system's representation of possible states of the environment with which it deals". A representational state is thus a vector r with components the values of the r_i (± 1, for example). The cognitive system interprets its environment using "knowledge atoms". Each of these atoms α is characterized by a knowledge vector k_α that attributes values to each feature r_i. It is or it is not activated (we introduce an activation variable $a_\alpha = 0$ or 1). The atoms α encode sub-patterns of values of the features occurring in the environment. Their frequency is encoded into their force σ_α.

(ii) Let (r, a) be the state of the cognitive system and let K be the knowledge base defined by k_α and σ_α. We define H_k by:

$$H_k(r, a) = \sum_\alpha \sigma_\alpha q_\alpha h_k(r, k_\alpha) \tag{14}$$

with

$$h_k(r, k_\alpha) = \frac{r.k_\alpha}{|k_\alpha|^{-k}}\,. \tag{15}$$

(iii) We now apply the general thermodynamical method sketched above to this particular case. This allows us to interpret the visual scene (i.e., the vector r) by completion, that is, by optimizing the *global coherence* (the consistency) of the (local) partial interpretations α: for a given representational state r and for an activation state a of the cognitive system, H is a sum including a term for each knowledge atom α, each term being wheighted by the force σ_α of r and each weight σ_α multiplying the self-consistancy between r and α (i.e., the similarity between r and k_α).

(iv) This inference and decision process is then indentified with the result of a parallel and distributed stochastic process driven by H_k.

In their "harmonic" framework, Paul Smolensky and Géraldine Legendre [358] developed a "grammar-based strategy for connectionnist language research" where connectionist models are used to construct grammatical formalisms for "abstract, higher-level descriptions". The strategy is the following:

a. Choose a mathematically precise formulation of the PDP principles.
b. Derive from these principles a precise but general grammar formalism (or grammatical theory).
c. To evaluate the proposed formalism, choose a particular class of target empirical generalizations concerning human language.
d. Apply the grammar formalism to language data instantiating the target generalizations, defining a formal 'account' of the target phenomena.
e. Compare the degree of explanation of the target generalizations that can be achieved with the new account with that achieved with previous grammar formalisms. ('Explanation' here means deduction of generalizations and particular data from the principles defining the proposed account.)

In harmonic grammar, distributed connectionist representations are tensorial products of roles and fillers. We return to this point later in Section 7.1.

4.4. The morphodynamical and PTC agendas

We have seen that morphodynamics raises a number of hard epistemological questions, the most important feature of the "morphological turn" consisting in a radical questioning of the logical-combinatorial formalist point of view on perception, cognition, and language.

Very similar questions have been tackled by Paul Smolensky. Adopting a dynamical point of view in semantics and an emergentist point of view in syntax, he has clearly presented their bases, their characteristics and their epistemological consequences. Let us recall the essential points of his 1988 article "On the Proper Treatment of Connectionism" (PTC) ([355]):

(i) The CN level is neither the conceptual and symbolic CL level nor the neuronal one. It does not concern the implementation of cognitive algorithms in massively parallel machines, but rather the structure, the architecture and the dynamical behavior of the cognitive processes themselves.
(ii) We cannot model the performances of intuitive knowledge on the basis of the assumption that the intuitive processor applies, simply in an unconscious way, sequential programs of formal rules. The processes at work cannot be adequately modeled in the framework of the CL symbolic paradigm, where symbols denote external objects (denotative semantics) and are operated upon syntactically by the application of rules.
(iii) In a dynamical CN model, the units possessing a semantics are complex patterns of activity distributed over numerous elementary units. This conception of semantics is characteristic of the CN approach.
(iv) Whereas in the symbolic formal models, all the processing levels are of the same type, in the subsymbolic semantic models there is a semantic shift:

Unlike symbolic explanations subsymbolic explanations rely crucially on a semantic shift (...) from the conceptual to the subconceptual levels.

(v) In order to get a unified conception of cognition, it is necessary to combine the CN and CL approaches. The rules consciously applied by the interpreter will be then interpreted as structurally stable emergent regularities:

patterns of activity that are stable for relatively long periods of time (of the order of 100 ms.) determine the contents of consciousness.

(vi) As far as the linguistic rules are concerned, this presupposes in particular the possibility of representing subsymbolically and sub-conceptually, in CN dynamical models, the propositional structures of language. This is certainly very difficult, but it is necessary. Now, according to Smolensky, a constituent structure can be obtained for the patterns of activity possessing a conceptual semantics when considering them as *superpositions* of subpatterns (constituents). As we shall see, this is the Achille's heel of the PTC agenda, and the criticism of Fodor and Pylyshyn focuses on this very point.

(vii) The mathematical universe of CN models is not the universe of formal languages and Turing machines. It is the universe of dynamical systems, that is, the qualitative theory of differential equations (global analysis). The subsymbolic computation is continuous, geometrical and differential. Inference here is a statistical inference that optimizes the fit with the input (harmony theory):

macro-inference is not a process of fixing a symbolic production but rather of qualitative state change in a dynamical system, such as phase transition.

(viii) The CN dynamical systems are suited for the elaboration of a theory of schemata, prototypicality, and categorization.

5. Fodor and Pylyshyn's arguments against connectionism

Let us turn now to Fodor and Pylyshyn's (F-P) arguments against CN.

5.1. The general structure of the F-P arguments

A.1. Both classical (CL) and connectionist (CN) cognitivists are *representationalists.* They both admit mental states encoding the properties of higher level cognitive processes and, therefore, the CLC/CNC debate is internal to cognitivism. It bears upon a precise issue:

the architecture of representational states and processes.[13]

The CNC paradigm is thus committed to showing that it can provide a good theory of the cognitive architecture, i.e.,

---

13 In this section, all quotations are from Fodor, Pylyshyn [113].

> processes which operate on the representational state of an organism.

A 2. Now, there is a fundamental difference between CL and CN paradigms. CL cognitivists assign semantic content to expressions, and admit causal relations, but also *structural* relations, between semantically evaluable entities. They consider it to be characteristic and essential:

(i) that the mental representations share a combinatorial syntax and semantics, and
(ii) that the mental processes *are dependent on this structure* ("structure sensitive"): operations operate on the mental representations as a function of their combinatorial structure, i.e., their syntactic form.

On the contrary, according to the authors, CNC would assign semantic content only to holistic entities, without internal combinatorial structure (labelled nodes symbolizing activity patterns of the network). Furthermore, they would only admit *causal* relations between the semantically evaluated entities. Only CL cognitivists

> are committed to a symbol-level of representation, or to a "language of thought", i.e. to representational states that have combinatorial syntactic and semantic structure.

Contrary to CNC, CLC emphasizes the fact that computational operations act on the syntactic structure of complex symbols and that, in so far as the syntactic relations are parallel to the semantic ones,

> it may be possible to construct a syntactically driven machine whose state transitions satisfy *semantical* criteria of coherence.

This is

> the foundational hypothesis of Classical cognitive science.

A.3. According to the authors, the limits of CNC are rather dramatic since the mental representations *must* possess an internal syntactic-semantic combinatorial structure if they have to be able to explain four fundamental aspects of cognition:

(i) *Productivity and Generativity.* As all natural languages, the "language of thought" shares the capacity of generating an indefinite number of expressions from finite means. Consequently, there must exist rules of generation, and this presupposes an internal structure of the expressions.
(ii) *Systematicity.* Even if we put into question productivity and generativity of the cognitive capacities, we cannot reasonably put into question their systematicity, that is, the intrinsic links relating the comprehension and the production of certain expressions with those of certain other expressions. Systematicity is explicable only if there exists an internal structure of expressions allowing us to define well-formedness rules and to structurally relate different expressions.
(iii) *Compositionality.* There are *semantic* transformations (a "covariance") between systematically related expressions (such as "John loves Mary"

and "Mary loves John", or "being a brown cow", "being brown" and "being a cow", etc.). The principle of compositionality according to which the semantic properties of constituents are independent from the context can only be understood if there exists a syntactico-semantic constituent structure.

(iv) *Inferential coherence.* The relations of logical similarity between different inferences presuppose the same conditions.

A.4. In a word, it is only if we posit an internal structure of representations, that we can speak of representations of the *same* structure, *similar* structures, or structures that are *related* to each other in different ways. But, according to Fodor and Pylyshyn, an essential feature of CNC would be to reject such a structure. For CNC, cognitive systems are systems

> that can exhibit intelligent behavior without storing, retrieving, or otherwise operating on structured symbolic expressions.

Surely, the labels marking the semantically evaluable holistic entities have in general a constituent-structure, but the internal dynamics of the system is *not* determined causally

> by the structure—including the constituent-structure—of the symbol arrays that the machines transform.

CN graphs are *not* structural descriptions of mental representations, but specifications of purely causal relations.

> The intended interpretation of the links as causal connections is intrinsic to the theory. (...)
> A network diagram is not a specification of the internal structure of a complex mental representation. Rather, it is a specification of a pattern of causal dependencies among the states of activation of nodes.

On the other hand, the fact that the mental representations are distributed over micro-features derived from learning, and extracted by multivariational analysis from the statistical regularities of the stimuli samples, does not imply that these representations are structured. Actually,

> you have constituent-structure when (and only when) the parts of semantically evaluable entities are themselves semantically evaluable. (...)
> Complex spatially-distributed implementation in no way implies constituent-structure.

The main error of CNC, its "major misfortune", would then be to have confused a componential analysis of micro-features with a combinatorial structure.

> The question whether a representational system has real-constituency is independent of the question of micro-feature analysis. (...)
> It really is very important not to confuse the semantic distinction between primitive expressions and defined expressions with the syntactic distinction between atomic symbols and complex symbols.

In short, as far as the semantically evaluable entities (nodes, activation-patterns, etc.) are conceived of as atomic and holistic Gestalts that are related only by causal relations, it becomes impossible to account for the fundamental features of cognition, namely productivity, generativity, systematicity, compositionality, and inferential coherence (cf. A.3.):

> The connectionist architecture (...) has no mechanism to enforce the requirement that logically homogeneous inferences should be executed by correspondingly homogeneous computational processes.

CNC presupposes a systematic organization of cognition. But it should also be able to explain it.

> It is not enough for a connectionist to agree that all minds are systematic; he must also explain *how nature contrives to produce only systematic minds.*

Hence the final verdict:

> The only mechanism that is known to be able to produce pervasive systematicity is classical architecture. And (...) classical architecture is not compatible with connectionism since it requires internally structured representations.

A.5. Moreover, according to the authors, CNC's main criticism against CLC is not acceptable. It claims that, in CLC, the behavioral regularities must come from explicitly encoded rules. But this is false. In CLC, several functions can be encoded implicitly (for example, as part of the hardware). What should be explicit are only the data structures that the cognitive machines transform, not the rules (the grammar) of transformations.

A.6. As a consequence, the CN perspective should be rejected as a *cognitive* theory. It relies upon a "bad" associationist psychology against which one can repeat the well-known rationalist criticisms formulated since Kant.

A.7. Fodor and Pylyshyn conclude that the only real interest of CNC is to provide an alternative theory of *implementation* for the classical functional architecture. They stress the fact that most of the arguments put forward by CNC bear only on the limitations imposed on competence by the concrete constraints of performance. According to them, the material limits of performance result from an interaction between an unbounded formal competence (unlimited but finitely describable by generative rules (see A.3.(i)) on the one hand and limited resources on the other. Adopting a functionalist perspective radically opposed to the emergentist CN one, they sharply separate the functional architecture (the software algorithms) from its implementation (the hardware). This is, for them, a "principled distinction". The (micro-level) models of implementation are neutral with respect to the nature of cognitive (macro-level) processes and to deny this fact is to confuse structure and function.

Such a confusion leads to unfortunate consequences. For example,

(i) from the evident efficiency of neural networks one tends to conclude to an associationist psychology (networks of representations), or

(ii) from the equally evident anatomic distributivity of neurons one tends to conclude to a functional distributivity of the mental representations themselves (componential analysis in micro-features), or

(iii) from the reinforcement of the connection between two neurons by their co-activation one tends to conclude to associationist statistical models of learning, or still

(iv) in the other direction, from a functional locality (position of a symbol in an expression, for example) one tends to conclude to a physical localisation in instanciation.

The "brain style" of CNC is definitely a dramatic epistemological error: it makes

> the implicit—and unwarranted—assumption that there ought to be similarity of structure among the different levels of organisation of a computational system.

It projects the neuronal level onto the cognitive one, and so doing, it reactivates

> the worst of Hume and Berkeley.

A.8. Thus, the CN stance may only provide

> an account of the neural (or "abstract neurological") structures in which classical cognitive architecture is implemented.

The symbolic structures of CLC are of course physical ones. They are neurally encoded and instantiated and it is their physical implementation which cause the operational behavior of the cognitive system. The CN arguments therefore become valid if we interpret them as supporting a physical implementation in massively parallel networks. For example, the fact that cognitive processes are fast, whereas neuronal phenomena are slow, or the fact that a considerable amount of forms (words, faces, etc.) stored in memories can be quickly recognized, or the continuity, fuzziness, approximation and structural stability properties of cognitive processes, all these facts support the thesis of a CN implementation of the algorithms constituting the CL functional architecture. But if the CN models are rather to be seen as concerning only implementation, then they should waive their cognitive pretensions. They should in particular refuse to assign

> a representational content to the units (and/or aggregates) that they postulate.

A.9. A key argument which is not made explicit by the authors is that "structural" necessarily means "formal-symbolic". If mental representations possess a combinatorial syntax and semantics then they are ipso facto "symbol systems". As we will see, it is this formalist dogma—the dogma of logical form—which, in turn, makes the Achilles' heel of all their arguments.[14]

14 For a philosophical criticism of this dogma, see Mulligan et al. [241].

5.2. Comments: the problem of a dynamical structuralism

The F-P arguments seem well constructed and forceful. However, we can question their effective validity at different levels.

C.1. The arguments A.1. and A.2. (as regards the characterization of CLC) A.3., A.5. and A.8. (except for its conclusion) are, we think, excellent and, furthermore, irrefutable. But they *do not* at all imply a rejection of CN as a cognitive theory. They simply impose on it certain constraints and additional requirements (as explained in A.1.): to be able to develop what we shall call the *structural hypothesis.*

C.2. The presentation and characterization of CNC given in A.2. and A.4. are caricatures. They only comment on the following "syllogism":

(i) a "good" CNC should be able to develop the structural hypothesis;
(ii) for intrinsic reasons, the caricature of CNC presented in A.2. and A.4. does not fulfill this requirement;
(iii) therefore CNC, whatever can be its further developments is *de jure* a "bad" cognitive theory.

But of course, nothing proves that the caricature can be identified with the full theoretical power of CNC. It is why their argument is dogmatic.

Let us continue to call "dynamical" the CN cognitivism. The central question is the following :

Question: *just as it is possible, using appropriate formal theories, to develop a symbolic structuralism, is it also possible, using the mathematical theories of dynamical systems, to develop a dynamical structuralism*?

If we reduce a priori all possible CN models to graphs of causal relations between holistic units lacking internal structure, then the reponse is of course trivially negative. *But these elementary models are only a very tiny part of the mathematical theory of dynamical systems.* We shall return to this point later. It is essential.

C.3. Even if we could prove that, contrary to the authors' assertion, it is possible to elaborate an authentic dynamical structuralism, this would not lead to transforming the CL/CN opposition into a Manichean alternative. There are certainly higher processing levels of the cognitive system which are of symbolic nature. But this does not entail that there are no lower levels that are of a dynamical nature. Associationist processes are certainly not sufficient to explain the structure of cognition, but they can nevertheless be necessary. Logical-symbolic superstructures can possess associationist infrastructures. The question is not whether CNC should replace CLC (or if the later should excommunicate the former), but to find out whether *the structural hypothesis can or cannot be already elaborated at the dynamical level of cognitive processes.* Such a dynamical structuralism must be clearly distinguished from the formal symbolic one. It must be:

(i) an authentic structuralism,

(ii) a *proto*-symbolic structuralism compatible with the symbolic level.

If one still wants to criticize it, one must develop more refined arguments than the F-P ones (see C6 below).

C.4. Fodor and Pylyshyn seem unaware of the true nature of the *emergence* of a macro-level from a micro-level. By separating the functional level of algorithms from the level of implementation, they disregard what is really the central issue in the point of view they are attacking. However, the physical parallel that they suggest should have incited them to more circumspection.

> The point is that the structure of 'higher levels' of a system is rarely isomorphic, or, even similar, to the structure of 'lower levels' of a system. No one expects the theory of protons to look very much like the theory of rocks and rivers, even though, to be sure, it is protons and the like that rocks and rivers are 'implemented in'.

The argument is fallacious. In physics, the relation between micro-levels and macro-levels is a matter of emergence. No physicist would separate the levels and postulate, as the authors do, that micro-levels are "neutral" in relation to macro-levels and that the latter are thus independent from their "implementation". The "form/matter" opposition, which is of Aristotelian origin, has been eliminated—and even eradicated—in modern science. The very physical problem here is to understand how an emergent—and therefore *non*-independent—macro-level can nonetheless present a certain *autonomy* of structure. That two levels are of different nature does not imply that they are independent and "neutral" in relation to each other. To assert this is to seriously misunderstand the epistemology of emergence. CN cognitivists are thus right when they distinguish the problem of implementation from the *intra-cognitive* problem of the emergence of a symbolic level from a dynamical subsymbolic level. But we repeat that the CL cognitivists are right when they assert that this dynamical level, in order to be considered cognitive, should be *structural.*

C.5. The point mentioned is the crux of the problem. The authors attack CNC because it assumes the systematic organization of cognition without explaining it (A.4.). But essentially the same argument can also be opposed to them. For they themselves do not *explain* this systematicity. They only describe it formally. By reducing the performance constraints to the material concreteness of implementation, by separating the levels, according to their functionalist perspective, and by autonomizing competence, they can surreptitiously identify a formal logical-combinatorial description of competence with the development of the structural hypothesis. But this identification is possible only if we assume the thesis A.9. according to which "structural" means "symbolic". However, if we admit this equivalence, then the argument becomes trivial: CNC is not symbolic (by definition), "hence" it is not structural, "hence" it cannot account for the structural character of cognitive processes.

A formal symbolic description of mental representations and mental processes is clearly possible. But as such, it should not be confused with an explanation. To achieve an explanation, we must:

(i) model the semantically evaluable entities by *mathematical* structures—perhaps very sophisticated—of a certain type, that is, belonging to a certain mathematical universe;
(ii) show that a theory of structures can be developed within this universe.

The question of CNC then becomes (see C.2 and C.3):

Question: *if semantically evaluable entities are modeled by attractors of dynamical systems, is it or not possible, within the framework of the theory of dynamical systems, to develop a theory of structure?*

C.6. If CL cognitivists satisfy themselves with a formal symbolic description, it is because for them the explanation of cognitive structures must be of an innatist nature. Behind the CL/CN controversy and the conflict of arguments, behind the rationalist criticism of empiricist associationism, there lies, in fact, an epistemological alternative. It was brilliantly sketched by Massimo Piattelli-Palmarini ([309]) in his paper *"Evolution, Selection and Cognition : from 'Learning' to Parameter Fixation in Biology and in the Study of Mind"*. The argument is as follows.

In all biological domains, one progressed from *instructivist* theories ("Lamarckian") to *selective* ("Darwinian") ones. In every case, both experimentally and theoretically, one arrived at the conclusion that there cannot be a transfer from the structure of the environment to the organism, and that only mechanisms of internal selection can be mechanisms of learning. This internal selection involves filtering and fixation of parameters that selectively stabilize certain possibilities among a very rich universe of genetically predetermined possibilities. For the instructivist point of view, the genetic constraints are poor and structuration comes from general capacities, such as adaptation, resolution of problems by trial and error, etc. For the selective point of view, on the contrary, the genetic constraints are essential, the structuration is strongly innate and modular, and adaptation is replaced by "exaptation", that is, by the fact that the characters can be selected independently of all adaptive value, even if, later on, they acquire such a value. For the selective thesis, the impossibility for an organism to assimilate external structures is a nomological one: it is nomologically improbable that

> structures external to the organism might possibly be 'internalized' through a 'learning' process;

but it is, however, nomologically very probable that

> a process of selection, of triggering and parameter-fixation, acting on a vast, profligate and highly articulated repertoire of innate structures may prove to be the most productive explanatory hypothesis. (Piattelli-Palmarini, [309], p. 23)

It is such an innatist and selective point of view that is now further developed in the cognitive science domain, in syntax as well as in semantics. Hence the radical criticism undertaken by Chomsky, Fodor and their colleagues against empiricist theories of learning by imitation, association, assimilation, induction, problem-solving, etc. Many results seem to indicate that there exists a rich syntactic-semantic architecture of language whose universality is of genetic origin:

> our species innately possesses a rich, specific, modular and highly articulate capacity for language, organized around certain universal 'principles'.

This cognitive capacity would be independent of perception and action. It would manifest

> a very intricate and closely inter-dependent process, full of 'deductive' consequences that are known to each of us in a totally unconscious way.

That is why, genetic constraints being contingent, a formal description can amount to an explanation.

It is this formalist dogma that we are criticizing. For these arguments are relevant only at the symbolic level. They do not imply at all that the innate symbolic form of the cognitive system exhausts its structure. By contrast, it is of course perfectly legitimate to assume

(i) that there is an objective content on which this form operates;
(ii) that a *dynamic* functional architecture can also be innately constrained.

5.3. The main point of the F-P argument

Let us come now to the central argument of Jerry Fodor and Zenon Pylyshyn. They consider the way in which certain CN cognitivists (Hinton, McClelland, Rumelhart) have treated a sentence like "John loves Mary". The problem is evidently

> the role relations that traditionally get coded by constituent-structure.

They accept therefore with fair-play the conception of syntax that is the least symbolic and the most akin to CN sensitivity, namely *case grammars*. But they stress the fact that, to be acceptable and amenable, CNC must provide a good CN account of the semantic roles that select cases. This is the main problem: modeling in a CN framework what European linguistics and semiotics call *actantial relations*.

The CN cognitivists mentioned above represent actantial relations by a set of activated units such as John-subject, +loves, +Mary-object, where the descriptors *J-S*, *L*, *M-O* are labels of holistic units *without* internal syntactic structure and *without* structured inter-relations. These descriptors combine an identity (an actant *J*, *M*) and an actantial role (*S*, *O*) and allow the representation of the syntactic structure of the sentence in a set-theoretic manner. Fodor and Pylyshyn can easily show that such a representation immediately leads to a series of inescapable difficulties, which can be solved only by what they call a "grotesque" proliferation of the number of descriptors:

> The idea that we should capture role relations by allowing features like John-subject thus turns out to be bankrupt. (...)
> It is of course, no accident that the connectionist proposal for dealing with role relations runs into these sorts of problems. Subject, object and the rest are classically defined *with respect to the geometry of constituent-structure trees.* And the connectionist representations don't have constituents.

If we just additively *superpose* the activated holistic entities in order to account for the sentences, then it becomes impossible to account for the relation between *J-S* + *L* + *M-O* and *M-S* + *L* + *J-O* (argument of systematicity, see A.3.(ii)).

> This consequence (...) offers a particularly clear example of how failure to postulate internal structure in representations leads to failure to capture the systematicity of representational systems.

Furthermore, in the case of a conjunction of sentences, it becomes impossible to retrieve the initial structures. The superposition leads to an irreversible destructuring (see the "superposition catastrophe" evoked in Section 10.2 of Chapter 2).

This is really the key point:

> when representations express concepts that belong to the same proposition, they are not merely simultaneously active, but also *in construction with each other.*

And to be in a relation of "construction"—that is, to be related by dependence relations—, representations should be constituents of more complex representations (see the arguments A.2 and A.3).

> Representations that are 'in construction' form parts of a geometrical whole, *where the geometrical relations are themselves semantically significant.*

The main problem is therefore to build what we have called *a configurational definition* of case roles.

Of course, for the CL paradigm, the problem of a configurational definition of actantial relations is *a priori* solved by using formal and combinatorial symbolic structures. But this does not imply at all that every such configurational definition *must* be, for *de jure* reasons, of a symbolic nature.

5.4. Towards a geometry of syntax

We see that Fodor and Pylyshyn's criticism against the possibility of working out a connectionist theory of high-level cognitive abilities—and in particular a theory of syntactic constituency and compositionality— are based on two theses.

(a) A thesis concerning the internal constituent structures—that is the internal form—of mental representations. The structures they consider are principally case-structures where semantic roles (the "actantial relations") are "in construction with each other". These semantic roles "classically defined *with respect to the geometry of constituent-structure trees*" are geometrical *and* syntactical relations which constitute the roots of constituency. Here, constituency is more basic than combinatorial compositionality.

(b) A thesis concerning the necessary symbolicity of constituent structures and the reducibility of constituency to combinatorial compositionality. For Fodor and Pylyshyn, representations can manifest internal constituent-structures if and only if they are *symbolic* representations. Their internal constituency is therefore a *combinatorial* one, analogous to the constituency found in formal languages. The "geometrical" form of constituency is therefore reduced to a pure combinatorial one.

Now, the point is that thesis (a) *does not* entail thesis (b). Actually, there exist many natural structures that are not symbolic but present nevertheless constituent-structures. The most evident case are atoms and molecules in quantum physics: electronic orbitals provide a typical example of constituents that are not symbolic in nature and can be dynamically modeled as solutions of partial differential equations. Therefore, if constituent-structures are *natural* structures they need not be necessarily symbolic.

If we admit thesis (a) without admitting thesis (b), however, then the problem raised by the F-P arguments changes its nature. It becomes about whether the main concepts of relation, structure, constituency and compositionality can be mathematically interpreted in a purely dynamical framework. Is it possible to work out a "syntactic geometry" of structures in the CN context?

6. Refutation of the F-P argument: the main problem

R.1. The central F-P argument is valid only as far as it is applied to a *weak* CN. Remember that we call "weak CN" a CN that models uniformly semantically evaluable entities of *different* syntactic types by mathematical structures of the *same* type, without taking into account their different grammatical categories. On the contrary, we call "strong CN" a CN that has the capacity of modeling the differences and the relations between different grammatical categories.

R.2. Let us clarify this a little further. The F-P argument points out a category mistake. Its "syllogism" is as follows:

Syllogism S1

(i) Let A_i $(i = 1, ..., n)$ be the actants of a sentence and V the verb organizing the actantial interactions. Let us *model* the actants A_i by means of mathematical structures $\mathfrak{A}_i$ of a certain type (for example activity patterns) on which is defined an abelian (associative, commutative, with neutral element and inverse elements) operation of composition $\oplus$, that is, an abelian group law (for example, the superposition of activity patterns).

(ii) Let us model the verb V by a structure $\mathfrak{V}$ of the *same* type as the $\mathfrak{A}_i$.

(iii) Model the actantial interaction V of the A_i by the sum $\mathfrak{V} = \oplus_{i=1}^{i=n}\mathfrak{A}_i$.

(iv) Experimental observation: such a modeling strategy runs into unavoidable difficulties.
(v) Conclusion: the modeling of the actants A_i by the structures $\mathfrak{A}_i$ should be rejected since it is experimentally refutable.

Furthermore, the authors oppose this syllogism to another one, aimed at showing the superiority of CLC.

Syllogism S2

(i) Let us symbolize the actants A_i by symbols A_i^*.
(ii) Let us symbolize the verb V by a symbol V^*.
(iii) Let us symbolize the actantial interaction V of the A_i by syntagmatic relations linking the A_i^* and V^* (for example, by a syntagmatic tree of a generative or constituent-structure grammar).
(iv) Empirical observation: such a symbolization is "good".
(v) Conclusion: it is to be accepted since it is experimentally valid.

The problem is that the first syllogism (S1) is fallacious and the second (S2) tautological.

S1 is fallacious.

In fact, it is equivalent to say that the (logical-combinatorial) structures of syntagmatic-tree type being non-associative and non-commutative (and therefore non-abelian), they cannot be modeled adequately by abelian algebraic structures of group type.

Let us mimic the F-P argument in another theory, for instance a physical theory (overly simplistic and therefore fictional) of elementary particles (e.p.).

(i) Let us model the free e.p.'s P_i $(i = 1, ..., n)$ by irreducible representations G_i of the Poincaré group in a Hilbert space.
(ii) Let us model the concept of an interaction V between the P_i by another irreducible representation F.
(iii) Let us model the interaction of the P_i by the sum $F = \oplus_{i=1}^{i=n} G_i$.
(iv) Empirical observation: such a modeling strategy runs into unavoiable difficulties and is experimentally refutable.
(v) Conclusion: the modeling of the P_i by the G_i must be rejected.

In such a case the fallacy is striking. There was a category mistake in confusing the concept of interaction in (ii) and (iii) with an additional free e.p. (iv) is trivial since an interaction of P_i is not the same thing as the system of the free P_i to which V has been added. The inference $(iv) \rightarrow (v)$ is illegitimate.

It is the same case with the F-P argument. It also points out a category mistake: an interaction of actants is modeled by a mathematical structure $\mathfrak{V}$ of the *same type* as those $\mathfrak{A}_i$ which are used to model the actants themselves. Fodor and Pylyshyn are then right to denounce such an error in weak CN, but nevertheless their drastic conclusion is fallacious. The only correct conclusion is that, if actants are dynamically modeled by attractors, then verbs expressing interactions of actants *cannot* be modeled by attractors of the same type.

S2 is tautological.

It is clear that if one symbolizes constituents by means of formal symbols, then one can symbolize their structural relations *a priori* by means of formal relations. But, as we have already seen, such a formalization does not at all *explain* the relations.

R.3. It is thus necessary to clarify and elaborate upon F-P's central argument. This can be done in the following way.

(i) First of all, we must be aware of the distance that separates a true mathematical modeling from a mere formal symbolization (see Section 5.3 of Chapter 1). Modeling a certain class of natural phenomena is interpreting them by sophisticated mathematical theories that allow to reconstruct their properties mathematically; whereas symbolizing them, in contrast, is only representing such properties formally. The requirement of modeling has nothing to do with symbolization: for instance a mathematical physics of elementary particle interactions has nothing to do with symbolic representations of the type $V(P_i)$ where V is an n-ary relation. The main limitation of the symbolic-formalist point of view in cognition is when it confuses a formal description with a mathematical explanation (see Petitot [260], [261], [262]). We see here the consequences of argument A.9 criticized in C.5.

(ii) So far, CNC constitutes the most decisive attempt to move from formal symbolization to mathematical modeling in cognitive science. And, as far as it aims to provide an explanation for *proto*-symbolic structural phenomena only, the lack of formalism cannot be attributed to it.

(iii) We can refute F-P's arguments if we can answer positively the question whether, in the case of syntactic structures, there are *two* structural levels that correspond respectively to the dynamical and symbolic levels. In a number of works, we have tried to show that it is in fact the case. Underlying the strictly grammatical level of grammatical relations (which are quite adequately described in terms of symbolic structures: syntagmatic trees, etc.) there does exist, in fact, a level of actantial relations where the actants are defined by their semantic (casual) roles and where the verbs express the actantial interactions. The differences between formal grammars and case grammars are well known.

(iv) Now, we notice that F-P's arguments are *neutral* with respect to this difference of levels. They concern as well the actantial roles and only refer to the "geometry" of structures where "the geometric relations are themselves semantically significant". Thus we may apply the results of R.2. regarding actantial syntax. From there, the question:

Main question: *If the actants A_i of a process are modeled by attractors $\mathfrak{A}_i$ of a dynamical system, is it possible, within the framework of the mathematical theory of dynamical systems, to elaborate a theory of actantial interactions—that is, a theory of the verb?*

(v) Let us further develop this question. In many dynamical models, the situation can be greatly simplified if one makes the hypothesis that the dynamics X defining the attractors $\mathfrak{A}_i$ admits a global Lyapunov function[15] or, even more simply, that X is a gradient: $X = -\operatorname{grad} f$, f being a function on the space underlying X. The $\mathfrak{A}_i$ are then the minima m_i of the potential function f.

Simplified main question: *If the actants A_i of a process are modeled by the minima m_i of a potential function, is it possible, within the framework of the dynamical theory of potential functions, to elaborate a theory of actantial interactions—that is, a theory of the verb?*

The mathematical challenge is therefore to develop a theory of *interactions of attractors*—what we call an *attractor syntax.* We shall see later that *bifurcation theory* provides adequate tools to address this challenge.

7. Connectionist binding and configurational roles

Since 1988, many CN cognitivists have proposed strategies to take up Fodor and Pylyshyn's challenge. One of the most interesting proposals, but still insufficient, has been Smolensky's idea of using the *tensorial product* operation.[16]

7.1. Smolensky's tensorial product

Smolensky's main idea is to take for granted the CL finitist and combinatorial view of symbolic structures and to *represent* them in a CN way—in much the same way as one represents abstract groups in linear groups in the well known group representation theory. To do this, he first adopts a case conception of syntax and thinks of syntactic structures as compounded by three types of entities:

(i) semantic case-roles r_i;
(ii) fillers f_j;
(iii) binding relations between roles and fillers.

He then supposes that the roles and the fillers are already represented in a CN way and solves the problem of representing the binding relations using the linear device of *tensorial product.*

Suppose that the roles r_i (resp. the fillers f_j) are vectors belonging to the vector space V_R (resp. V_F) of the global states of a network R (resp. F). Let u_ρ (resp. v_φ) be the units of R (resp. F). One connects R and F using

15 See Section 3.9 of Chapter 5 for details.
16 See Smolensky [352], [353], [354], [356].

connections $u_\rho \Leftrightarrow v_\varphi$ with Hebbian weights $w_{\rho,\varphi} = \Sigma_i r_{i,\rho}.f_{i,\varphi}$ where $r_{i,\rho}$ (resp. $f_{i,\varphi}$) is the activity of the unit u_ρ (resp. v_φ) in the global activity pattern of R (resp. F) representing r_i (resp. f_j). The tensorial product device consists in introducing *new* units $b_{\rho,\varphi}$ between R and F, $b_{\rho,\varphi}$ being connected by two weights $= 1$ to u_ρ and v_φ and having $w_{\rho,\varphi}$ as activity.

(16)

$$\begin{array}{ccccc}
\begin{array}{c} u_\rho \\ \text{activity } r_{i,\rho} \end{array} & \oplus & \xrightarrow{w_{\rho,\varphi}} & \oplus & \begin{array}{c} v_\varphi \\ \text{activity } f_{j,\varphi} \end{array} \\
 & | & & | & \\
\begin{array}{c} u_\tau \\ \text{activity } r_{i,\tau} \end{array} & \oplus & \xrightarrow{w_{\tau,\psi}} & \oplus & \begin{array}{c} v_\psi \\ \text{activity } f_{j,\psi} \end{array} \\
\left\{\begin{array}{c} R = \textit{Roles} \\ \text{Pattern} \\ \text{of activity} \\ \text{encoding} \\ r_i = \sum_\rho r_{i,\rho} u_\rho \end{array}\right. & & & & \left\{\begin{array}{c} F = \textit{Fillers} \\ \text{Pattern} \\ \text{of activity} \\ \text{encoding} \\ f_j = \sum_\varphi f_{i,\varphi} v_\varphi \end{array}\right. \\
\begin{array}{c} u_\rho \\ \text{activity } r_{i,\rho} \end{array} & \oplus \xrightarrow{1} & \begin{array}{c} b_{\rho,\varphi} \\ \bigotimes \\ \text{activity } w_{\rho,\varphi} \end{array} & \xleftarrow{1} \oplus & \begin{array}{c} v_\varphi \\ \text{activity } f_{j,\varphi} \end{array}
\end{array}$$

It is easy to see that we get that way a CN implementation $R \otimes L$ of the tensorial product $V_R \otimes V_L$ with basis $b_{\rho,\varphi} = u_\rho \otimes v_\varphi$. With $w_{\rho,\varphi} = \Sigma_i r_{i,\rho}.f_{i,\varphi}$, the state of $R \otimes L$ becomes:

$$\sum_{\rho,\varphi} w_{\rho,\varphi} b_{\rho,\varphi} = \sum_{\rho,\varphi} \left(\sum_i r_{i,\rho}.f_{i,\varphi} \right) u_\rho \otimes v_\varphi \tag{17}$$

$$= \sum_i \left(\left(\sum_\rho r_{i,\rho} u_\rho \right) \otimes \left(\sum_\varphi f_{i,\varphi} v_\varphi \right) \right) \tag{18}$$

$$= \sum_i r_i \otimes f_i \, . \tag{19}$$

We therefore get a *representation* $\Psi : S \to V$ of a set S of structures in the state space of a network, where a representation is defined in the following manner. We suppose that there exists a *role decomposition* $F|R$ of S, that is, a truth function assigning to each pair (f_j, r_i) the truth value of the predicate $f_j|r_i$ on S: "f_j fills the role r_i in $s \in S$". We suppose as given a CN representation of the fillers/roles bindings $\Psi_{\text{bind}} : \{f_j|r_i\} \to V$ and we define the representation of the role decomposition $F|R$ by the map:

$$\begin{aligned} \Psi : \; & S \to V \\ & s \to \sum_{\{(f_j,r_i) \mid (f_j|r_i)(s)\}} \Psi(f_j|r_i) \, . \end{aligned} \tag{20}$$

In a tensorial product, r_i (resp. f_j) is identified with an activity pattern $\sum\limits_{\rho} r_{i,\rho}u_\rho$ (resp. $\sum\limits_{\varphi} f_{i,\varphi}v_\varphi$) and $f_j|r_i$ is identified with $r_i \otimes f_j$. As far as a structure s is a conjunction of $f_i|r_i$ and a conjunction is represented by addition, we finally get:

$$\Psi(s) = \sum_i r_i \otimes f_i \ . \tag{21}$$

Paul Smolensky has shown in a detailed and convincing manner that this type of procedure allows in various ways to represent operations and transformations on symbolic structures. According to him ([353], p. 1), this shows that it is possible to integrate

> in an intimate collaboration, the discrete mathematics of symbolic computation and the continuous mathematics of connectionist computation.

This view is clearly anti-eliminativist. Smolensky does not want to reduce all symbolic structures and processes to CN ones. He wants to *represent* in a CN way these symbolic descriptions in order to explain "higher thought processes".

For instance, in a joint work with G. Legendre and Y. Miyata (Smolensky et al. [359]), he applies to binary trees this strategy of understanding

> how symbolic computation can arise naturally as a higher-level virtual machine realized in appropriately designed lower level connectionist networks.

Let r_{x_i} be the *positional* roles in a binary tree with nodes x_i. A tree s with atom f_i at node x_i is represented by the tensorial product $s = \sum_i r_{x_i} \otimes f_i$. This representation is in fact a recursive one. The x can be coded by binary strings using the code 0 = "left child" and 1 = "right child". Let $r_{x_0} = r_x \otimes r_0$ and $r_{x_1} = r_x \otimes r_1$. Using such a binary coding, a tree s can be represented by a vector of the vector space

$$\bigoplus_{k=1}^{k=\infty} V_R^{\otimes k} \otimes V_F \tag{22}$$

where V_R is generated by r_0 and r_1. Smolensky shows how to implement a programming language such as LISP in this framework. He also gives a CN representation of Context Free Grammar theory.

Harmonic grammar already mentioned in Section 4.3 applies harmony theory to these tensorial connectionist representations of symbolic structures. One can add constraints of well-formedness of sentences, which are hierarchized by different specific ranking orders in different languages.

7.2. Dynamical binding

With regard to the implementation of the binding relations between roles and fillers, we also want to mention the *dynamical binding* by means of synchronized oscillatory neural groups independently and variously developed by Gerald Edelman, Christoph von der Malsburg, Charles Gray, Wolf Singer, Elie Bienenstock, and Lokendra Shastri.[17] The key idea is that the terms of sentences (objects) are internally represented and encoded by rhythmic patterns—oscillators—and the bindings by processes of *synchronization* —phase locking—between such oscillators. In this framework (Shastri-Ajjanagadde [343]):

> reasoning is the transient but systematic propagation of a *rhythmic* pattern of activation, where each *phase* in the rhythmic pattern corresponds to an object in the dynamic or short-term memory, where bindings are represented as the *in-phase* or *synchronous* firings of appropriate nodes, where long-term facts are subnetworks that act as temporal pattern matchers, and where rules are interconnection patterns that cause the propagation and transformation of rhythmic patterns of activation.

Using such oscillatory patterns and phase-locking processes, one is now able to embody dynamical bindings in the fine temporal structure of firing patterns in the brain. This idea is akin to the works of Christoph von der Malsburg and Elie Bienenstock concerning the role of fast synapses in neural networks, that is, the possibility for a neural network to change non-adiabatically its synaptic weights during its transient functioning.

We see, therefore, that the way in which one can bind a role label with a filler term raises fundamental issues. But these are not the central issues that led to take up F-P's challenge. Indeed, the main problem is the *configurational* definition of roles, which can substitute for the classical role labels. We will see in the next chapter that, in such a configurational definition, roles are identified with positions or "places" in configurations of positions. Of course, these places have to be filled by fillers, but the key difficulty is to elaborate a CN theory of such positional relations without taking for granted any prior CL representation of them.

How, according to this strategy, can roles be dynamically processed without being previously symbolically represented? According to Smolensky's slogan that "hardness emerges from softness", one must consider that in the CN stance, the challenge is to construct symbols "out of soft stuff".

As was strongly emphasized in Smolensky's article "On Variable Binding and the Representation of Symbolic Structures in Connectionist Systems" [353], CN must

---

17 See, e.g., Edelman [96] and Sporns, Tononi, Edelman [363], von der Malsburg-Bienenstock [398], Bienenstock [34], Shastri-Ajjanagadde [343]. As we have already seen in Section 10.2 of Chapter 2, synchronization of oscillatory neural groups is also used to model constituency relations (labeling hypothesis).

> find ways of naturally instantiating the sources of power of symbolic computation within fully connectionist systems.

Such a CN instantiation is much more than a mere CN implementation of a symbolic stuff. It is an "extended version of connectionist computation" which would

> naturally incorporate, without losing the virtues of connectionist computation, the ingredients essential to the power of symbolic computation.

This is the core of the main problem, because it is the "syntactic geometry" of the internal form of constituent structures that defines the semantic relations characterizing the roles. To solve this problem, it is not sufficient to bind noun fillers and case-frame slots of verbs, the roles being themselves tensorial product of semantic verb features and case roles. We must be radical and *instantiate*—not only implement or represent—the concept of role in a purely dynamical way.

7.3. The need for a configurational definition of roles

In their response to Smolensky's response [354], Fodor and McLaughlin [111], reconsider the systematicity problem and the fact that (p. 185)

> cognitive processes are causally sensitive to the constituent structure of mental representations.

They summarize their main point claiming that (p. 187)

> all we really need is that propositions have internal structure, and that characteristically, the internal structure of complex mental representations corresponds, in the appropriate way, to the internal structure of the propositions that they express.

More specifically, they introduce (p. 187) a condition (C) that

> expresses a *psychological law* that subsumes all systematic minds.
>
> (C): "If a proposition P can be expressed in a system of mental representations M, then M contains some complex mental representation (a "mental sentence") S, such that S expresses P and the (classical) constituents of S express (or refer to) the elements of P.

Condition (C) plus the fact

> that mental processes have access to constituent-structure of mental representations,

allows to explain the cognitive systematicity of the mind.

Supported by this theoretical background, Fodor and McLaughlin can provide an evaluation of Smolensky's tensorial product device. Their main criticism is that it is impossible to retrieve from tensorial product representations and from abelian operations of superposition a constituent structure whose constituents can have a *causal* status. Indeed, in a vector space the choice of a basis and hence of a vector decomposition is *not canonical.* Every vector decomposition is therefore *counterfactual* and the constituents (components) it generates cannot have causal efficiency.

We think that this negative argument is essentially right, even if it is overly categorical. For instance, it is true that there exists no canonical basis in a vector space V (in fact, V possesses a large transitive symmetry group, the linear group $GL(V)$). But nevertheless, the vector space V_R of the states of a network R *does* possess a distinguished basis, namely the basis defined by its units. In that case, vector decompositions are not counterfactual operations. Nevertheless, the criticism underlines a major difficulty that can be expressed in the following manner.

For Smolensky, the basic problem of a CN theory of symbolic structures are the binding relations between roles and fillers. He succeeded in solving this problem, but in a way which replies to only *half* of Fodor and Pylyshyn's challenge. Indeed, it says nothing about the possibility of reaching in the CN framework itself a configurational definition of actantial roles. Moreover, it takes for granted a symbolic pre-definition of the roles. As was pointed out by Yves-Marie Visetti ([395], p. 186), in the tensorial product approach,

> the associative conception of memory as a relaxation to a preferential state

disappears, together with

> the concept of attractor as an intrinsic meaningful state.

Now, the problem is not only to represent semantic roles as local or distributed activity patterns of some appropriate network, it is also to give a correct, purely CN account of the *relations of actantial interaction* which are involved in syntactic structures. These relations are not binding relations. They concern the roles independently of their fillers. The PTC[18] agenda, which according to Smolensky [64]

> consists in taking (the) cognitive principles and finding new ways to instantiate them in formal principles based on the mathematics of dynamical systems,

must *also* be applied to *the configurational definition of the actantial roles.*

In some sense, it is such a requisite that is stressed by Fodor and McLaughlin when they claim that in order to build a CN theory of constituency and systematicity one must:

(i) find "some property D, such that if a dynamical system has D its behavior is systematic";
(ii) "make clear what property D is"; and
(iii) "show that D is a property that CN systems can have by law" ([111], p. 201).

18 Reminder: PTC = Proper Treatment of Connectionism.

But suddenly, the authors become dogmatic. They state that such a requisite is impossible to satisfy. Even when Smolensky stresses the fact that constituent structures do exist in physics (for instance a molecule with its atomic nucleus and its electrons), they reply that

> since *being a representation* isn't a property in the domain of physical theory, the question whether mental representations have constituent-structure has no analog in physics. (p. 200)

Of course, it is then very easy for them to conclude

> that Fodor and Pylyshyn's challenge to connectionists has yet to be met. We still don't have even a suggestion of how to account for systematicity within the assumptions of connectionist cognitive architecture. (p. 204)

This claim subsumes some previous claims from their 1988 paper:

> So far as we know, there are no worked out attempts in the Connectionist literature to deal with the syntactic and semantical issues raised by relations of real-constituency. (p. 22)
>
> There doesn't seem to be any other way to get the force of structured symbols in a Connectionist architecture. Or, if there is, nobody has given any indication of how to do it. (p. 24)
>
> There are no serious proposals for incorporating syntactic structure in Connectionist architectures. (p. 67)

However, these peremptory judgements are not at all true. Indeed, as we will show, the concept of *bifurcation* of attractors precisely provides a real D property, and such a dynamical conception of syntactic structures had already been available *for over 20 years* when Fodor and Pylyshyn raised their criticism.

8. The link with Chalmers' criticism of F-P arguments

In his 1990 paper [54] "Why Fodor and Pylyshyn were wrong. The simplest refutation", David Chalmers also presented a criticism of Fodor and Pylyshyn, pointing out a contradiction concerning their

> underestimation of differences between localist[19] and distributed representations.

According to him, Fodor and Pylyshyn do not understand how distributed representations "can carry semantics".

> This is the fundamental flaw in F&P's argument: lack of imagination in considering the possible ways in which distributed representations can carry semantics.

They argue that

> connectionist models cannot admit of a compositional semantics...

Although this is true for localist networks, it is provably false for distributed representations, which

19 "Localist" in the connectionist sense: one neural node = one symbol content, the weights between the nodes being identified to associationist links between contents.

> can be used to support direct support-sensitive operations, in a manner quite unlike the classical approach.

In fact, connectionist implementations of Turing machines do exist.

D. Chalmers comments a finer point. Even if we admit a compositional semantics in distributed connectionist representations, this information could be intractable and functionally useless. However this is not the case:

> distributed representations of compositional structures *can* be operated on directly, without proceeding through an extraction stage (of components).

But it is certainly true that connectionist compositionality is not of the same nature as classical compositionality. In this chapter, we gave the example of attractor syntax, in which connectionist compositionality is implemented in the interactions between attractors.

9. The epistemology of the morphodynamical paradigm

To conclude this chapter, we summarize a few critical epistemological points.

Morphodynamics aims at explaining natural morphologies and iconic, schematic, Gestalt-like aspects of structures, whatever their underlying physical substrate may be, using the mathematical theory of dynamical systems. Syntactic structures can be treated as Gestalts and can be morphodynamically modeled.

One must carefully distinguish between the formal description of symbolic structures on the one hand and their dynamical explanation on the other. It is not because the former is correct that one is committed to a symbolic conception of mental states and processes. In morphodynamics, the conceptual contents of mental states are no longer identified with symbols. Their meaning is embodied in the cognitive processing itself. More precisely, it is identified with the topology of the complex attractors of the underlying neural dynamics, and the mental events are identified with sequences of bifurcations of such attractors. Symbolic structures are conceived of as macro-structures emerging from the underlying micro-neurodynamics. Information processing is therefore thought of not as an implemented symbolic processing but *as a dynamical process*.

The morphodynamical paradigm shares many epistemological views with Smolensky's theses. We emphasize in particular the following three convergences.

(i) The "structure/statistics dilemma", i.e., the central cognitive paradox opposing distributed statistical-numerical descriptions of real performance to symbolic rule-based descriptions of higher-level competence:

(ii) The necessity of grounding model-centered approaches on principle-centered meta-theoretical integrated ones.

(iii) The necessity of distinguishing, with David Marr, the three computational, algorithmic and implementational levels. To adopt a logico-combinatorial and symbolic point of view at the computational level *does not* commit one to also adopt such a point of view at the algorithmic level. The symbolic constituents that emerge at higher cognitive levels do not have a *causal* efficiency at the underlying algorithmic level. This later is not symbolic but dynamical.

Computational effectiveness does not have to be of a discrete logical nature. It can be dynamical *without* being identified with the physical level of implementation. We will call *dynamical functionalism* the thesis according to which dynamical structures are to a large extent independent from the particular physical substrate they are implemented in. Dynamical functionalism is the key of the naturalization of syntactic structures. It relies upon deep mathematical theorems whose importance for a morphodynamical theory of natural structures was first stressed by René Thom and Christopher Zeeman.

Therefore, we now turn to a presentation of this historical breakthrough, which must be considered as one of the most important precursors to CNC.

CHAPTER 5

From Morphodynamics to Attractor Syntax

1. Introduction

It was in the late 1960's and early 1970's that morphodynamics established the basis for a dynamical approach to higher level cognitive performance such as categorization and syntax. In this chapter we will give a short review of the principles, mathematical tools and results of these works. It will be only a crude summary but, even if the matter is very technical, it may help the reader to better understand the issues of morphodynamical modeling in cognitive sciences.

2. Christopher Zeeman's initial move

To our knowledge, it was Christopher Zeeman who introduced the first dynamical approach to explain the links between neurology and psychology. In his seminal 1965 article *Topology of the Brain* [416], he introduced the key idea that brain activity must be modeled by dynamical systems (flows) X_w on configuration spaces $M = I^N$, where $I = [0,1]$ is the range of activity of a neuron, N is the number of neurons of the system under consideration, and the flows X_w depend on control parameters w, micro-parameters such as synaptic weights and macro-parameters such as behavioral or psychological values. The main novelty was to identify mental states with attractors of the flows X_w, their content with the topological structure of the attractors, and the flux of consciousness with a "slow" temporal evolution of the X_w. Consequently, the strategy for explaining mental phenomena was to use the mathematical theory of dynamical systems (global analysis)—especially theorems concerning the general structure of the attractors and their bifurcations—for drawing empirical conclusions from this dynamical scheme *without knowing explicitly the* X_w.

This strategy was very clearly explained in Zeeman's 1976 article *Brain modelling*. Let us read a long quotation (Zeeman [418], p. 287):

> What is needed for the brain is a medium-scale theory. (...) The small-scale theory is neurology: the static structure is described by the histology of neurons and synapses, etc., and the dynamic behaviour is concerned with the electrochemical

> activity of the nerve impulse, etc. Meanwhile the large-scale theory is psychology: the static structure is described by instinct and memory, and the dynamic behaviour is concerned with thinking, feeling, observing, experiencing, responding, remembering, deciding, acting, etc. It is difficult to bridge the gap between large and small without some medium-scale link. Of course the static structure of the medium-scale is fairly well understood, and is described by the anatomy of the main organs and main pathways in the brain. (...) But what is strikingly absent is any well developed theory of the dynamic behaviour of the medium-scale. True, there have been several models concerned with groups of cells and computer simulations, but none have really matched up to the two main requirements of providing a framework for prediction and experiment, and providing a link between the large and small. On the one hand the network theories of neurons and the combinatorial theories of synapses seem unable to escape from the small-scale, and therefore appear to be unrelated to psychology. On the other hand the computer simulations of perception and problem-solving seem unable to escape from the large-scale, and therefore appear to be unrelated to neurology.
> Question: what type of mathematics therefore should we use to describe the medium-scale dynamic? Answer: the most obvious feature of the brain is its oscillatory nature, and so the most obvious tool to use is differential dynamical systems. In other words for each organ O in the brain we model the states of O by some very high dimensional manifold M and model the activity of O by a dynamic on M (that is a vector field or flow on M). Moreover since the brain contains several hierarchies of strongly connected organs, we should expect to have to use several hierarchies of strongly coupled dynamics. Such a model must necessarily remain implicit because it is much too large to measure, compute, or even describe quantitatively. Nevertheless such models are amenable in one important aspect, namely their discontinuities.

The fundamental trick was then to use the classification theorem of elementary catastrophes (a deep theorem, see Section 3 below) in the following manner.[1] If the mental states are modeled by attractors, then their significant changes during mental processes are modeled by "discontinuities", that is by bifurcations. These are empirically given as "catastrophes". They are observable in the control space W of the *relevant* control parameters (the relevance depends of course on the nature of the mental phenomena under consideration). We have, therefore, a dynamics X_w defined on the very high dimensional manifold $M \times W$ (which is the direct product of the "internal" manifold M by the "external" control space W). This dynamics X_w is "vertical", that is, compatible with the fibration (the canonical projection) $\pi : M \times W \to W$. This means that the vector of X_w at a point (x, w) of $M \times W$ has no "horizontal" component parallel to W: it is tangent to the fiber $M \times \{w\}$. X_w can therefore be considered as a family of dynamics defined on the internal space M and

1 Our notations are not Zeeman's own but those conventionally used and also employed below.

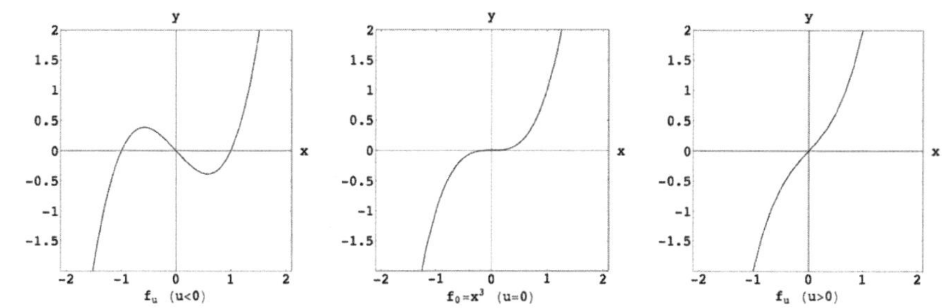

FIGURE 1. The potentials $f_u(x) = x^3 + ux$.

parametrized by the external space W. In W we observe a system of discontinuities K. Now, if the catastrophes are *elementary*, namely, if the internal dynamics is a gradient: $X_w = -\operatorname{grad} F_w$ where F_w is a potential function on M, if the family F_w is structurally stable and if $\dim W \leq 4$, we can, according to the Whitney-Thom classification theorem, generate K *locally* using only *elementary* models $\chi : \Sigma \to W$, with $\Sigma \subset \mathbb{R}^2 \times W$ (and even $\Sigma \subset \mathbb{R} \times W$). In such a model we consider:

(i) *potentials* (Lyapunov functions) $f_w(x)$ of two (or even one) real variables x which are parametrized by $w \in W$: they substitute for the F_w;
(ii) the critical points of the f_w, that is the points x where $\operatorname{grad}_x f_w = 0$;
(iii) the critical subset $\Sigma = \{(x, w) \mid x = \text{critical point of } f_w(x)\} \subset \mathbb{R}^2 \times W$;
(iv) the restriction χ to Σ of the canonical projection $\zeta : \mathbb{R}^2 \times W \to W$; the bifurcation set $K \subset W$ is then the *apparent contour* of χ, that is the projection of the set of points $x \in \Sigma$ where the tangent map $D_x\chi$ of χ at x is not of maximal rank ($= \dim W$)[2].

Let us consider a very simple example, the fold singularity where the potentials are

$$f_u(x) = x^3 + ux \ . \tag{1}$$

The graphs of the f_u are represented in Figure 1. For $u < 0$, $f_u(x)$ presents a minimum and a maximum. These two critical points merge for $u = 0$ into a flex point and then disappear: for $u > 0$, $f_u(x)$ has a monotonously decreasing graph.

The critical set Σ is the parabola in the (x, u) plane (i.e., $\mathbb{R} \times W$) of equation

$$f_u'(x) = 3x^2 + u = 0 \ . \tag{2}$$

For $u < 0$ there exists two critical points above u and no critical point for $u > 0$: the parabola is "folded" above the u-axis, hence the name of fold singularity given to x^3. The map χ is the vertical projection of Σ onto the u-axis and its

2 i.e., where χ is not a local diffeomorphism between Σ and W.

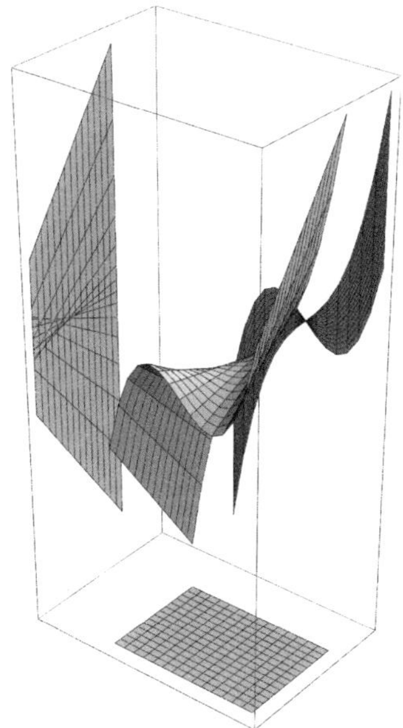

FIGURE 2. The fold singularity. The base space is the (x, u) plane and the y-axis (the values of $f_u(x)$) is vertical. We have also represented the projections of the surface $y = f_u(x)$ on the (y, u) and (y, x) planes.

apparent contour, that is, the catastrophe set K, is the point $u = 0$, $x = 0$ where the direction of projection is tangent to Σ. Figure 2 shows the complete structure of the singularity.

Let us return to Zeeman's proposal. We get two types of models that are equivalent in what concerns the observable discontinuities: one $\mathcal{M}_\pi$ coming from $\pi : M \times W \to W$ and the second $\mathcal{M}_\varsigma$ coming from $\zeta : \mathbb{R}^2 \times W \to W$. In the passage from $\mathcal{M}_\pi$ to $\mathcal{M}_\varsigma$ we meet a drastic *reduction of the dimension* of the internal space (from $\dim M$ to 2) which is very similar to what we meet in thermodynamics when we reduce an enormous number of degrees of freedom using what is called an *order parameter*. This drastic reduction is assimilated by Zeeman ([418], p. 290) to the passage from the dynamical meso-scale to the "psychological" macro-scale.

For instance in the celebrated Zeemanian model of the Lorenzian theory of animals' aggression (Zeeman [418]), the behavioral conflicting factors are "rage and "fear" and the controlled behaviors are "attack" and "flight". The conflict between simultaneously high values of the two controls induces an instability (bimodality or "double-bind") between the two behaviors and explains the "catastrophic" suddenness of the attack or flight of the animal. In such a model, the "rage" and "fear" controls are treated as intensive magnitudes measurable by a "degree" on a scale. The behaviors are modeled by attractors of some neural internal dynamics, and their "catastrophic" jumps (triggered by critical values of the controls) are modeled by bifurcations of these attractors. The fact that the same bifurcation scheme can be generated by a simple and typical "cusp" catastrophe (where the "rage" and "fear" scales correspond to the diagonal axis $u = -v$ and $u = v$ and the "attack" and "flight" behaviors to the two opposed stable sheets of S) sets up the desired link between

> the observed behavior of aggression (and) the underlying neural mechanism. Each point of the surface S represents an attractor of some huge dynamical system modelling the limbic activity of the brain, and the jumps occur when the stability of an attractor breaks down. The dynamical system remains implicit in the background; the only part that we need to make explicit for experimental prediction is the catastrophe model. (Zeeman [418], p. 13).

Moreover, to set up the link with the "psychological" large scale,

> We take S as an explicit model for the large-scale psychology. We label the various sheets of S with psychological words. (...) For coordinates in $\mathbb{R}^2$ we seek two psychological indices that correlate with the labelled sheets. Then Σ is an explicit quantitative psychological model for testing experimentally. (Ibid., p. 290)

Therefore, the dynamics X_w on M

> arise from the neuronal network, the synaptic connections and the metabolism of O (the brain organ under consideration), imparting a homeostatic tendency to certain oscillations, and thereby creating the attractors of the flows on M. (Ibid., p. 291)

In the "psychological" model $\chi : \Sigma \to W$, the generating potentials f_w are Lyapunov functions with no neural interpretation. But even if a dynamics X_w on M and a potential f_w on $\mathbb{R}^2$ are of a completely different nature, they can generate isomorphic qualitative behavior. It is this non-linear diffeomorphism that, according to Zeeman, sets up the qualitative

> connection between the neurological measurements and the psychological measurements. (Ibid., p. 291)

3. The general morphodynamical model

We will now introduce the mathematical concepts which are the main ingredients of a general morphodynamical model. This section is rather technical.

3.1. The internal dynamics and the internal states

Our setting will be as general as possible. Let us return to the general model already sketched in Section 3 of Chapter 2. Let S be a system, for instance a neural network. We suppose that S satisfies the following hypotheses.

The first hypothesis is that there exists an internal dynamical mechanism X which defines the *internal states* of S. More precisely:

(i) There exists for S—as for every physical system—a configuration space (or a phase space) M which is a differentiable manifold and whose points x represent the instantaneous transient states of S. M is the *internal space* of S. For instance, for a neural network S of N neurons u_i with activations $x_i \in I = [0,1]$, we get $M = I^N$ (see Section 4 of Chapter 4).

(ii) X is a flow on M, that is, a system of ordinary differential equations (ODE) $\dot{x} = X(x)$ (where $x = (x_1, ..., x_N)$ is the global state of S) which shares three properties: it is firstly complete (its trajectories are integrable from $t = -\infty$ to $t = +\infty$); secondly deterministic; and thirdly smooth relatively to the initial conditions. The smooth vector field X is the *internal dynamics* of S.

For the previous neural net, X can be, for instance, the vector field

(3) $$\dot{x} = -x + g(Wx - T)$$

where g is a sigmoidal gain function, W the $N \times N$ matrix of synaptic weights and T the N-vector of the thresholds of the u_i. Such a vector field corresponds therefore to the system of ODE

(4) $$\dot{x}_i = -x_i + g\left(\sum_j w_{ij}x_j - T_i\right),$$

which is the continuous approximation of the discrete system

(5) $$x_i(t+1) = g\left(\sum_j w_{ij}x_j(t) - T_i\right)$$

(see Section 4 of Chapter 4).

Let Γ_t be the application of the manifold M into itself which associates to every point $x \in M$ at time 0 its position at time t. It is easy to show that Γ_t is a *diffeomorphism* of M, that is a one-to-one bi-continuous and bi-differentiable map. Clearly, $\Gamma_{t'} \circ \Gamma_t = \Gamma_{t+t'}$ and $\Gamma_{-t} = (\Gamma_t)^{-1}$. $\Gamma_t : \mathbb{R} \to \mathrm{Diff}(M)$ is therefore a morphism of groups from the additive group of $\mathbb{R}$ to the group of diffeomorphisms of M—what is called a *one parameter subgroup* of diffeomorphisms of M. Γ is the *integral* version of the vector field X and is called its *flow*. The internal states of S are then the (asymptotically stable) *attractors* of X. They must not be confused with the instantaneous transient states $x \in M$.

In fact, for a general dynamical system, it is very difficult to define rigorously the notion of an attractor. We can give now a more precise definition than in Section 3.2 of Chapter 2. Let $\omega^+(a)$ be the positive limit set of $a \in M$, that is, the topological closure "at infinity" of the positive trajectory of a. It expresses the asymptotic dynamical behavior of a. Let $A \subset M$ be a subset of the internal space M. A is an attractor of the flow X if it is topologically closed, X-invariant (i.e., if $a \in A$ then $\Gamma_t(a) \in A$ for every $t \in \mathbb{R}$), minimal for

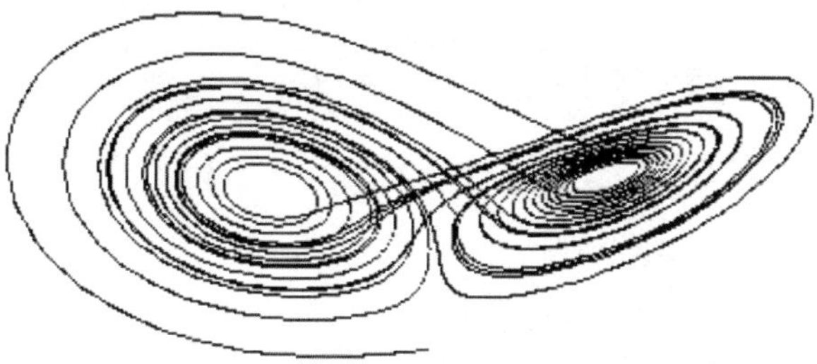

FIGURE 3. The Lorenz attractor. It is the attractor of a system of 3 coupled non-linear ordinary differential equations of degree 2 introduced by Edward Lorenz in 1963 as a very simplified 3D model for atmospheric convection. For certain values of the control parameters the asymptotic behavior is chaotic.

these properties (i.e., $A = \omega^+(a)$ for every $a \in A$), and if it attracts asymptotically every point x belonging to one of its neighborhoods U (i.e., there exists U s.t. $A = \omega^+(x)$ for every $x \in U$). A is *asymptotically stable* if in addition it confines the trajectories of its sufficiently neighboring points. If A is an attractor of X, its *basin* $B(A)$ is the set of points $x \in M$ which are attracted by A (i.e., s.t. $\omega^+(x) = A$). If U is an attracted neighborhood of A, we have of course $A = \bigcap_{t>0} \Gamma_t(U)$ and $B(A) = \bigcup_{t<0} \Gamma_t(U)$.

There is a hierarchy of complexity between attractors. The simplest case is that of fixed-points attractors. More complex is the case of attracting limit cycles (oscillatory motion). Still more complex is the case of attracting tori with quasi-periodic motion (the trajectories wind densely around the tori). And still more complex is the case of "strange" attractors where trajectories present a complex and even chaotic motion. Figures 3 and 4 show two celebrated examples of such strange attractors: the Lorenz attractor and the Rössler attractor.

3.2. The criterion of selection of the actual state

The second hypothesis is that there exists some criterion I (for instance a physical principle of minimization of energy) which selects from among its possible internal states the *actual* internal state of the system S.

3.3. The external control space

The third hypothesis is that the system S is controlled by control parameters varying in a control space W. W is the external space of S. The internal dynamics X is therefore a dynamics X_w which is parametrized by the external points $w \in W$ and varies smoothly relative to them. For the previous neural network, W is the space of the synaptic weights w_{ij} and of the thresholds

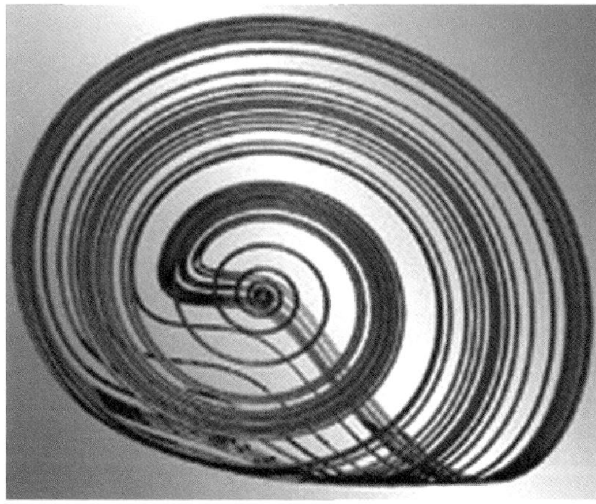

FIGURE 4. The Rössler attractor.

T_i. In a neurologically plausible model, many modules are strongly coupled and the external space of each module will in general be the output of some other modules (including the external stimuli in the case of perception). These complex systems are called dynamical "cascades".

3.4. The field of dynamics

Let $\mathcal{X}$ be the functional space of the smooth vector fields on the internal space M. $\mathcal{X}$ is the space of smooth sections of the tangent vector bundle TM of M. The possible dynamical behaviors of S are completely described by the *field of dynamics* $\sigma : W \to \mathcal{X}$ which associates X_w to $w \in W$. If another dynamics (an "external" one) drives the control w, then the controlled internal dynamics X_w drifts and can become unstable.

3.5. Stuctural stability

In this mathematical context we can better explain the Thomian (W, K) models already used in Section 3.1 of Chapter 2. The mathematical theory of *structural stability* is needed in order to account for the observable morphologies K_W. Indeed, the critical values $w \in K_W$ are those where, according to the criterion I, the actual state A_w of S bifurcates towards another actual state B_w. In general, such a bifurcation is forced by the fact that A_w becomes structurally unstable when w crosses K_W.

To define the concept of structural stability, we need two things.

(i) First a topology $\mathcal{T}$ on the functional space $\mathcal{X}$. In general, the chosen topology $\mathcal{T}$ is the Whitney C^∞-topology, which is the topology of uniform convergence of the vector fields and of all their partial derivatives on the compact sets of M, with equality "at infinity" (i.e., outside some compact set).[3]

(ii) Second, we need an equivalence relation on $\mathcal{X}$ which allows us to define the notion of "qualitative type". In general, if $\mathcal{F} = C^\infty(M,N)$ is the functional space of smooth maps between two manifolds M and N, two elements f and g of $\mathcal{F}$ are called C^∞-equivalent if there exist diffeomorphisms $\varphi \in \mathrm{Diff}(M)$ and $\psi \in \mathrm{Diff}(N)$ such that $g = \psi \circ f \circ \varphi^{-1}$, i.e., if f and g are conjugate by two changes of "global coordinates", one in the source space M and the other in the target space N. For the functional space $\mathcal{X}$ of vector fields on M this definition must be refined.

Now let $X \in \mathcal{X}$ be a vector field on M. Let $\widetilde{X}$ be its equivalence class. X is called *structurally stable* if $\widetilde{X}$ is (locally) $\mathcal{T}$-*open* at X, that is if there exists a neighborhood U of X for the topology $\mathcal{T}$ such that every $Y \in U$ is equivalent to X. If X is structurally stable, its qualitative structure "resists" to small perturbations.

3.6. Categorization

Let $K_\mathcal{X}$ be the subset of $\mathcal{X}$ consisting of the structurally *unstable* vector fields. The main fact to be stressed here is that $K_\mathcal{X}$ *categorizes* $\mathcal{X}$. The open subset $R_\mathcal{X}$ of structurally stable vector fields is partitioned in open connected components which are identifiable with "species" of vector fields (the structurally stable equivalence classes) and these components are glued together by $K_\mathcal{X}$. $K_\mathcal{X}$ can therefore be conceived of as a *classifying set* for the vector fields.

3.7. Retrieving the morphologies

$K_\mathcal{X}$ is intrinsically and canonically defined. Let $\sigma : W \to \mathcal{X}$ be the field of dynamics describing all the possible behaviors of our system S. The main modeling hypothesis is that the empirically observed morphology K_W can be retrieved (via the criterion I) from the inverse image

$$K'_W = \sigma^{-1}\left(K_\mathcal{X} \cap \sigma\left(W\right)\right) \tag{6}$$

of $K_\mathcal{X}$ relative to σ. To explain the morphologies K_W we therefore need good mathematical theories of structural stability and of the geometry of the bifurcation sets $K_\mathcal{X}$. These theories are very complex (see for instance the works of Poincaré, Birkhoff, Kolmogorov, Thom, Smale, Peixoto, Guckenheimer, Arnold, Ruelle, Sinaï, Herman, Yoccoz). In particular, for a general dynamical system X on M, there can be an infinite number of attractors, their basins can be inextricably intertwined, and their topology can be infinitely complex

3 Dynamical systems and singularities of differentiable maps constitute very technical and difficult topics. For mathematical details, see for instance Thom [379], Zeeman [418], Golubitsky-Guillemin [127], Arnold et al. [20], Chenciner [56], [57], Petitot [279].

("strange" attractors). On a stable strange attractor the dynamics is at the same time deterministic, structurally stable and chaotic.

3.8. Fast/slow dynamics

To explain the temporal evolution of S, we must consider *temporal paths* in the control space W. These paths are in general trajectories of dynamics in W. Such *external* dynamics must be carefully distinguished from the internal ones X_w. For neural networks, a few examples of external dynamics are well known: learning dynamics (e.g., the back-propagation algorithm), dynamics driving bifurcations (see, e.g., Amit's work [11] about cycles of attractors), cascades of dynamics (see, e.g., Hirsch [154]). Relative to the internal temporal scale of X_w, which is "fast", the external temporal scale is "slow". We can therefore suppose that the system S is always in an internal non-transient state (hypothesis of "adiabaticity").[4] As was strongly emphasized by Thom ([383]), this opposition between fast and slow dynamics is essential for the model.

> The main philosophical idea (...) is that every phenomenon, every spatio-temporal morphology owes its origin to a qualitative distinction between different acting modes of *time*. Any qualitative distinction in a space W (the substrate) can be attributed to two acting modes of time: a "fast" mode which generates in an internal space "attractors" which specify the local phenomenological *quality* of the substrate; and a "slow" mode acting in the substrate space W itself.

3.9. Lyapunov functions

Let A be an (asymptotically stable) attractor of X and let $B(A)$ be its basin. It can be shown that X is *dispersive* on $B(A) - A$ and that there exists a *Lyapunov function* on $B(A)$. X is called dispersive on N if for every $x, y \in N$, there exist neighborhoods U of x and V of y and $T > 0$ such that U and V become asymptotically disconnected, that is, such that, for every $t \geq T$ and $t \leq -T, U \cap \Gamma_t(V) = \emptyset$. X is dispersive on N iff X is trivial on N, that is if it is equivalent to a constant field.

A Lyapunov function f on $B(A)$ is a real continuous function $f : B(A) \to \mathbb{R}$ which is strictly > 0 on $B(A) - A$, $= 0$ on A and which decreases strictly along the trajectories of X. It is like a generalized "energy" which is minimized during the evolution of the system. There exist therefore essentially *two* types of dynamical behaviors: the *dissipative* behaviors which minimize a Lyapunov function f and contract the basins $B(A)$ on the attractors A, and the non-dissipative (asymptotic) behaviors which are in general chaotic, ergodic and conservative relative to an invariant measure (the Ruelle-Bowen-Sinaï measure).

---

4 Relatively to a slow dynamics, a fast dynamics is virtually "instantaneous". It loses its dynamical character and becomes in some sense "static". The "fast/slow" opposition is therefore intimately related to the "static/dynamic" opposition pointed out in the Introduction.

3.10. The reduction to gradient systems

We can therefore distinguish in the model the gradient-type dynamics on the basins $B(A) - A$ and the (chaotic) dynamics on the attractor A. As Thom [383] claimed:

> Personally, I think that it is not the too fine notion of attractor which plays the main role, but an equivalence class of attractors which are equivalent because they are encapsulated in the level variety of a Lyapunov function (a quasi-potential), provided that the attractor escapes implosions of an exceptional character. According to me, this is the way to find a mathematically satisfactory definition of the asymptotic stationary regime for a dynamics.

In this perspective we try to approximate a dynamical system by a gradient system and we look for gradient systems which have the same bifurcations (cf. Zeeman's strategy presented in Section 2). This reduction of the bifurcations to those of the Lyapunov functions is identifiable with a change of the observation level. It is like a "thermodynamical" mean field theory. It is a path from the micro level (small-scale) to the macro level (large-scale).

3.11. Contents and complex attractors

In brain modeling, we can suppose, due to the oscillatory nature of the brain, that the attractors come from the coupling of limit cycles. But it is a well known fact that quasi-periodic motions on tori (that is, products of limit cycles) become structurally unstable when the dimension of the tori is sufficiently large. In that case they bifurcate spontaneously towards strange attractors.[5] The (complex) topology of such a strange brain attractor can be identified with the content of the correlated mental state. In reducing the attractors to points in a quasi-gradient model we therefore *reduce* these mental contents to unanalyzable units. This reduction is equivalent in the morphodynamical approach to the classical reduction of semantic units to formal symbols. The main difference is that the relations between these units are no longer of a symbolic nature: they are dynamically generated.

3.12. Critical points, jets and Morse theory

When the generating dynamics X_w are gradient dynamics, that is when $X_w = -\operatorname{grad} f_w$, with $f_w : M \to \mathbb{R}$ a smooth real function, the theory becomes

5 This property was used by David Ruelle and Floris Takens in the early 1970's for their new theory of turbulence. The idea that the attractors of a general dissipative dynamical system are complex has been furthered by the recognition that these attractors can be self-organized critical states characterized by critical exponents and scaling laws. Self-organization provides emergent structures. For an introduction to critical phenomena see Petitot [279] and its bibliography. For an introduction to the various theories of emergence see EMC [98]. For self-organized critical states see Bak et al. [28] and Zurek [423].

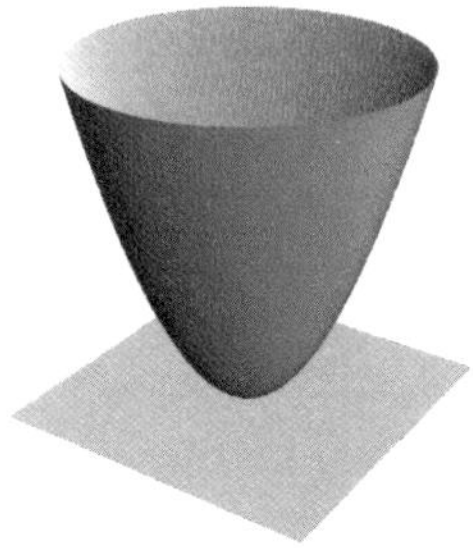

FIGURE 5. A critical point of $f(x,y)$ is a point where the tangent plane of the graph $z = f(x,y)$ is horizontal, that is, where the partial derivatives $\partial f/\partial x$ and $\partial f/\partial y$ vanish.

much simpler. Let $f \in \mathcal{F} = C^\infty(M,\mathbb{R})$ be such a potential on M. One of the highest achievements of modern differential geometry is to have shown that the qualitative *global* structure of such a geometrical entity is essentially encoded in its *local singularities.* Let G_f be the graph of f, that is the subset of $M \times \mathbb{R}$, $G_f = \{(x, f(x)) \mid x \in M\}$, constitued by the values of f over M. Let $a \in M$ be a point of M and $(x_1, ..., x_n)$ a system of local coordinates at a. The point a is called a *critical point* of f if the tangent space of G_f at the point $(a, f(a))$ is "horizontal", that is, parallel to the tangent space of M at a (see Figure 5). The technical condition is that the Jacobian of f at a, $J_f(a)$—that is, its gradient: the n-vector of its first partial derivatives $(\partial f/\partial x_1, ..., \partial f/\partial x_n)$—is 0 at a. This is an intrinsic geometric property, independent from the chosen coordinate system. Moreover, a is called a non-degenerate critical point if it is not the coalescence of several simpler critical points, that is, if it is as simple as possible, non-composite. The technical condition is that the Hessian of f at a—that is, the $n \times n$ symmetric matrix of its second partial derivatives $\left(\frac{\partial^2 f}{\partial x_i \partial x_j}\right)$—is of maximal rank ($= n$) at a. This is also an intrinsic geometric property of a critical point.

Non-degenerate critical points are minima, maxima or (generalized) saddles. Flex points are examples of degenerate critical points. *Generically,* the critical points of a potential are non-degenerate and their values are pairwise different. Technically, this fundamental result is proved in the following manner.

It can be shown (easily) that $a \in M$ is a non-degenerate critical point of f iff the 1-jet $j^1 f$ of f is *transverse* at a to the 0-section Σ^1 of the 1-jet space $J^1(M,\mathbb{R})$. The 1-jet of f at x is by definition the $(2n+1)$-uple:

$$(7) \quad j^1 f(x) = (x_1, ..., x_n; f(x); \partial f/\partial x_1, ..., \partial f/\partial x_n) = (x; f(x); J_f(x)) .$$

$j^1 f(x)$ belongs to a vector bundle $J^1(M, \mathbb{R})$ over $M \times \mathbb{R}$ which is locally of the form $U \times \mathbb{R} \times J$, where U is an open set of M and J the space of $(n \times n)$-matrices. The 0-section Σ^1 of $J^1(M, \mathbb{R})$ is the submanifold which is given locally as $U \times \mathbb{R} \times \{0\}$.

Now, one of the main theorems of the theory is *Thom's transversality theorem.* Let $f : M \to N$ be a differentiable map between two manifolds and let W be a submanifold of N. f is transversal on W if, for every $x \in M$ such that $f(x) \in W$, one has the equality of tangent spaces:

$$T_{f(x)}N = T_{f(x)}W + D_x f\,(T_x M) \tag{8}$$

where $D_x f$ is the linear tangent map of f at x. The theorem proves that if W is a submanifold of the jet space $J^1(M, \mathbb{R})$, then the set T_W of the $f \in \mathcal{F} = C^\infty(M, \mathbb{R})$ whose 1-jet is transversal to W is dense, and more precisely *residual*, that is, a countable intersection of dense open sets.[6] According to this result, transversality in jet spaces is a *generic property*, and therefore to be a non-degenerate critical point is also a generic property.[7]

If a is a critical point of f, $f(a)$ is called a *critical value.* In the same vein it can be shown that being a potential all of whose critical values are distinct is also a generic property.

A potential whose critical points are all non-degenerate with distinct critical values is called an *excellent Morse function.* Excellent Morse functions are generic in $\mathcal{F} = C^\infty(M, \mathbb{R})$: they can approximate every potential. Moreover (if M is compact) they are structurally stable. In fact, one of the main theorems of the theory, *Morse's theorem*, says that, if M is compact, $f \in \mathcal{F}$ is structurally stable iff it is an excellent Morse function. Morse's theorem is clearly crucial since it gives a simple geometrical characterization of structural stability and therefore of the causes of instability (the presence of degenerate critical points and equal critical values). To find such a characterization for general dynamical systems is one of the most difficult problems of global analysis.

3.13. Normal forms and residual singularities

Another fundamental theorem of Morse gives a *normal algebraic form* for f near a non-degenerate critical point a: there exists always a local coordinate system at a such that f becomes locally a quadratic form:

$$f(x) = f(0) - (x_1^2 + \dots + x_k^2) + x_{k+1}^2 + \dots + x_n^2\,. \tag{9}$$

---

6 The theorem is in fact true for the jet spaces $J^k(M, N)$ which generalize the $J^1(M, \mathbb{R})$. With the Whitney topology, the functional spaces $C^\infty(M, N)$ satisfy the Baire property: a countable intersection of dense open sets is dense. If M is compact these spaces are even Frechet manifolds.

7 In a Baire space a property is called *generic* if it is satisfied on a residual subset, that is, if it is equivalent to a countable conjunction of properties which are all stable (open) and dense.

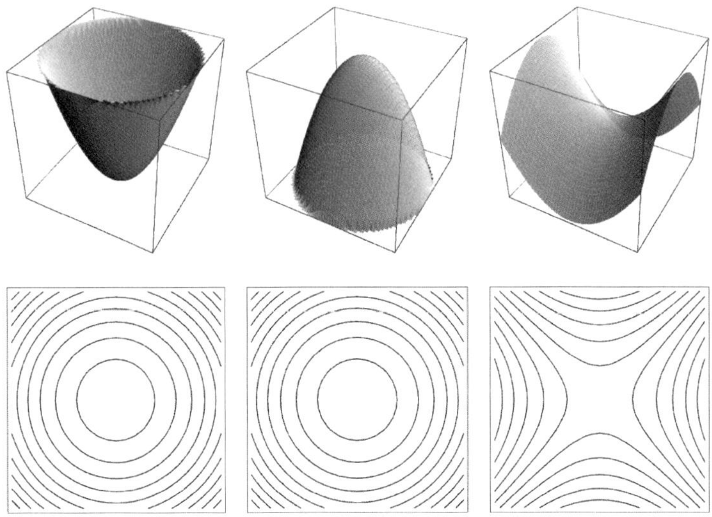

FIGURE 6. The three types of non-degenerate critical points in dimension 2. From left to right: a minimum, a maximum and a saddle point.

The number k possesses an intrinsic geometric meaning. It is called the *index* of the critical point a. $k = 0$ for a minimum, and $k = n$ for a maximum (see Figure 6).

Qualitatively, the structure of f near a non-degenerate critical point is therefore completely known. If a is degenerate, then another deep theorem, the *residual singularity theorem*, says that if the corank of f at a (corank $= n-$ the rank of the Hessian) is s, then there exists a local coordinate system $(x_1, ..., x_{n-s}; y_1, ..., y_s)$ such that, locally, $f = H(x)+g(y)$ where $H(x)$ is a non-degenerate quadratic form (the Hessian) and $g(y)$ is a function whose critical point a is totally degenerate (with a zero Hessian). This means that we can locally decompose M into complementary spaces, one along which f is non-degenerate, that is quadratic, and another along which f is totally degenerate.

3.14. The local ring of a singularity

The problem is therefore to study the local structure of potentials near a totally degenerate critical point. For this, the algebraic tools developed by John Mather in the late 1960's (see [225]) are essential. The main result is that if a is a totally degenerate critical point of f and if we localize the situation in the neighborhood of a in M and of f in $\mathcal{F} = C^\infty(M, \mathbb{R})$, then the orbit (the equivalence class) $\widetilde{f} = G(f)$ of f is a subspace of $\mathcal{F}$ admitting supplementary

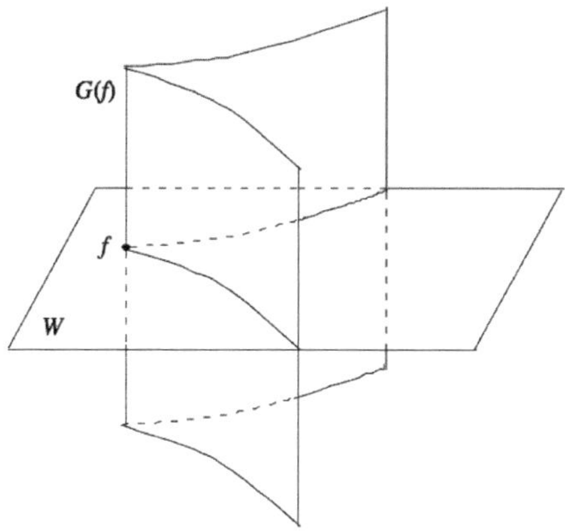

FIGURE 7. A local supplementary subspace $\mathcal{W}$ in $\mathcal{F}$ to the orbit $G(f)$ of a function f having a totally degenerate critical point.

subspaces $\mathcal{W}$ which are all equivalent (see Figure 7). The dimension c of these $\mathcal{W}$ is called the *codimension* of f at a.

More precisely, as long as we work locally, we can identify (M, a) with $(\mathbb{R}^n, 0)$. Let $\mathcal{E}_n$ be the space of germs[8] at 0 of potentials $f : \mathbb{R}^n \to \mathbb{R}$.[9] Since $\mathbb{R}$ is a ring, $\mathcal{E}_n$ is also a (commutative) ring. It can be easily shown that $\mathcal{E}_n$ is in fact a *local ring* whose unique maximal ideal $\mathfrak{m}$ is the ideal $(x_1, ..., x_n)$ of the germs f such that $f(0) = 0$. More generally, the quotient $\mathcal{E}_n/\mathfrak{m}^k$ is the ring of Taylor series of the f truncated at order k, that is, the ring of the germs of the $(k-1)$-jets $j^{k-1}f$. 0 is therefore a critical point of f iff $f \in \mathfrak{m}^2$ and it is totally degenerate iff $f \in \mathfrak{m}^3$.

Let $f \in \mathfrak{m}^3$. As f is structurally unstable, its orbit $\widetilde{f}$ for the C^∞-equivalence is not locally open at f. It can be shown that the "tangent space" of $\widetilde{f}$ at f is the product $\mathfrak{m}\Delta$ where Δ is the *Jacobian ideal* $(\partial f/\partial x_i)$ generated by the first partial derivatives of f. As $f \in \mathfrak{m}^3$, we have $\Delta \subset \mathfrak{m}$ and therefore $\mathfrak{m}\Delta \subset \mathfrak{m}^2$. The number $\dim_{\mathbb{R}}(\mathfrak{m}^2/\mathfrak{m}\Delta)$ is equal to the codimension c of f at 0. It is the dimension of every supplementary subspace of $\mathfrak{m}\Delta$ in $\mathfrak{m}^2$, that is, of a slice $\mathcal{W}$ in $\mathcal{F}$ which is transverse to $\widetilde{f}$ at f. It can also be defined as the dimension of the space $\mathfrak{m}/\Delta$.

This leads us to a concept of utmost importance, the concept of *universal unfolding.*

8 Intuitively, the germ of f at a is the restriction of f to an "infinitesimal" neighborhood of a.

9 We let f also denote the germ of f at 0.

3.15. Universal unfoldings and classification theorems

We have seen that when an external control $w \in W$ drives an internal dynamics f_w (or more generally X_w), bifurcations can naturally occur. They occur when $f_w \in K_{\mathcal{F}}$. Now the main fact is that, conversely, every unstable dynamics naturally generates an entire system of bifurcations. As far as the concept of bifurcation is the key concept allowing to work out a dynamical theory of constituent structures we need a clear understanding of this important fact.

Let $f \in \mathfrak{m}^3$ be a totally degenerate singularity of finite codimension c. In general, a small perturbation of f will yield a function g *less* degenerate than f: small perturbations have a stabilizing effect. But in general there will be many ways to stabilize f by small perturbations. The crucial fact is that it is possible to group all these possibilities together in a *single* structure: the *universal unfolding* of f.

For instance the 2-dimensional family of potentials of Figure 9 below is the universal unfolding $x^4 + ux^2 + vx$ of the codimension 2 singularity x^4 (normal form of a degenerate minimum). Such a degenerate critical point derives from the coalescence of two minima and a maximum, and it can therefore "explode" and split into 3 non-degenerate critical points. Its universal unfolding gathers the semi-stabilized small perturbations possessing only codimension 1 unstabilities of type "flex point" or "equality of two critical values", and the completely stabilized small perturbations possessing one simple minima or two simple minima separated by a simple maximum.

More precisely, let $\mathcal{W}$ be a slice of dimension c transverse to the orbit $\widetilde{f}$ at f. $\mathcal{W}$ being isomorphic to a neighborhood W of 0 in $\mathbb{R}^c$, the $f \in \mathcal{W}$ can be conceived of as potentials f_w parametrized by $w \in W$. $\mathcal{W}$ is therefore identifiable with a deformation f_w of $f = f_0$ controlled by c parameters. Such a deformation is called an *unfolding* of f. One can define a natural equivalence relation for unfoldings and a natural concept of structural stability. When the field of dynamics $\sigma : w \to f_w$ is transverse to $\widetilde{f}$ at $f = f_0$, the unfolding is called *transversal*. It is called *universal* if it is possible to reconstruct *all* the other unfoldings f_t of $f = f_0$, $t \in T$, using pull-backs of f_w, that is C^∞-maps $\varphi : (T, 0) \to (W, 0)$ such that $f_t = \varphi^*(f_w)$.

The main theorem concerning unfoldings says that the universal unfoldings of $f = f_0$ are the transversal unfoldings constructed from a basis $(h_1, ..., h_c)$ of $\mathfrak{m}/\Delta$ via the formula

$$f_w = f + \sum_{i=1}^{i=c} w_i h_i \,. \tag{10}$$

They are all C^∞-equivalent.

From this theorem, It is easy to derive *normal algebraic forms* for universal unfoldings. Consider for example the *double cusp* singularity $f(x, y) = x^4 + y^4$, which is the sum of two cusps defined on two *independent* variables x and y.[10] As $\partial f/\partial x = 4x^3$ and $\partial f/\partial y = 4y^3$, the space $\mathfrak{m}/\Delta$ is generated by x, y, x^2, xy, y^2, x^2y, xy^2, x^2y^2. We have therefore $\text{codim}(f) = 8$. The normal form for the universal unfolding is:

$$f_w = x^4 + y^4 + ax^2y^2 + bx^2y + cxy^2 + dx^2 + exy + fy^2 + gx + hy\ . \tag{11}$$

In the 10-dimensional space $\mathbb{R}^2_{x=(x,y)} \times \mathbb{R}^8_{w=(a,b,c,d,e,f,g,h)}$ which is the direct product of the internal and external spaces,[11] we consider the subspace

$$\Sigma\ = \{(x,w) \mid x \text{ is a critical point of } f_w\}\ . \tag{12}$$

The bifurcation set K_W is the subset of the control values $w \in W$ such that f_w presents at least one degenerate critical point. K_W is the apparent contour of the canonical projection $\chi : \Sigma \to W$ (see above, Section 2 on Zeeman's initial move). Σ is called the *catastrophe map* associated with the unfolding. Its geometry is very complex. K_W *classifies and categorizes* the different qualitative types (of germs) of potentials which can be derived from $f_0 = f$ by small deformations.

The *classification theorems* (Thom, Zeeman, Arnold, etc.) give such explicit algebraic normal forms for the singularities and their universal unfoldings up to codimensions which are not too large (around 12). They constitute one of the most important and beautiful achievements of modern differential geometry.

4. A few examples: cusp, swallowtail, butterfly

To be more concrete, we give the examples of the simplest singularities.

4.1. The cusp

The cusp is the universal unfolding

$$f_{u,v}(x) = x^4 + ux^2 + vx \tag{13}$$

or, better, after normalization,

$$f_{u,v}(x) = \frac{x^4}{4} + u\frac{x^2}{2} + vx \tag{14}$$

10 This singularity is already notably complex. It "contains" all the elementary catastrophes. We used it for modeling Lévi-Strauss' "canonical formula" for semionarrative structures (see [297]).

11 We indicate in subscript the coordinates (renaming x the bivariable (x,y)).

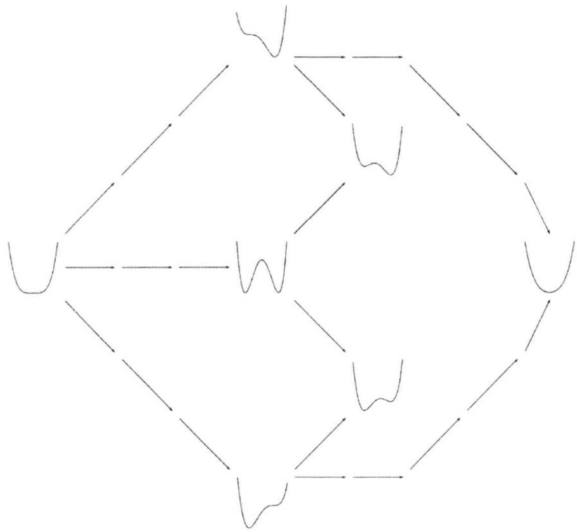

FIGURE 8. The 3 classes of stable fonctions resulting from the stabilization of the singularity $x^4/4$ and the 3 intermediary classes of codimension 1 partial stabilizations.

of the unstable codimension 2 singularity $f_0(x) = x^4$, where (u, v) varies in a neighborhood W of the origin 0. Zeeman called v the normal factor and u the splitting factor. The equation

$$f'_{u,v}(x) = x^3 + ux + v = 0 \tag{15}$$

is cubic and has therefore either a single real root (and two complex conjugates roots) or 3 real roots. Figure 8 shows the 3 classes of stable fonctions coming from the stabilization of f_0 and the 3 intermediary classes of codimension 1 partial stabilizations (they correspond to the values of (u, v) for which $f'_{u,v}(x) = 0$ has a double root).

To represent the total graph V of the unfolding $y = f_{u,v}(x)$, we would need a 4D space with coordinates (x, u, v, y). We will consider only the critical locus Σ and its apparent contour on W along the catastrophe map $\chi : \Sigma \to W$. In $\mathbb{R}^3(x, u, v)$, Σ is the set of critical points of $f_{u,v}$ (see Figure 9).

A first part K_b of the catastrophe set K in W is the set of (u, v) for which the cubic equation $f'_{u,v}(x) = x^3 + ux + v = 0$ has a double root (i.e., for which we also have $f''_{u,v}(x) = 0$). It is the *discriminant* of the equation and its equation in W is $4u^3 + 27v^2 = 0$. It is a semi-cubic parabola with a cusp at the origin $(u = 0, v = 0)$. But K contains another stratum K_c corresponding to the equality of the two minima (see Figure 10).

Let γ be a path in W parallel to the v-axis with $u < 0$ (Figure 11). Along this path, the graph of critical values is a "swallowtail". Outside the cusp, there

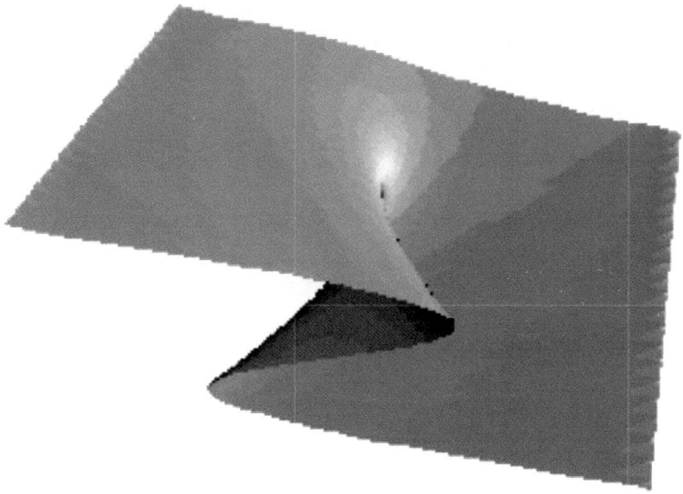

FIGURE 9. The critical surface Σ which is the set of critical points of the potentials $f_{u,v} = x^4/4 + ux^2/2 + vx$.

exists a single branch γ_1 (value of the single minimum). At the first crossing of K_b, two new branches γ_2 and γ_3 emerge. At the crossing of K_c, γ_1 and γ_2 intersect transversally. At the second crossing of K_b, γ_1 and γ_3 disappear together, the branch γ_2 being the only to survive.

4.2. The swallowtail

The swallowtail is the universal unfolding of the codimension 3 singularity x^5

$$f_{u,v,w}(x) = x^5 + ux^3 + vx^2 + wx \tag{16}$$

or, in normalized form

$$f_{u,v,w}(x) = x^5/5 + ux^3/3 + vx^2/2 + wx\ . \tag{17}$$

Its potential functions can present 0 or 2 (a minimum and a maximum) or 4 (2 minima and 2 maxima) quadratic critical points. Figure 12 shows the relations between the different classes of stable potentials and the codimension 1 and 2 intermediary potentials. Figure 13 shows the bifurcation part K_b of the catastrophe set K in the external space W with coordinates (u, v, w). And Figure 14 shows a 2D section of K for $u < 0$ which organizes the potentials of the previous figure.

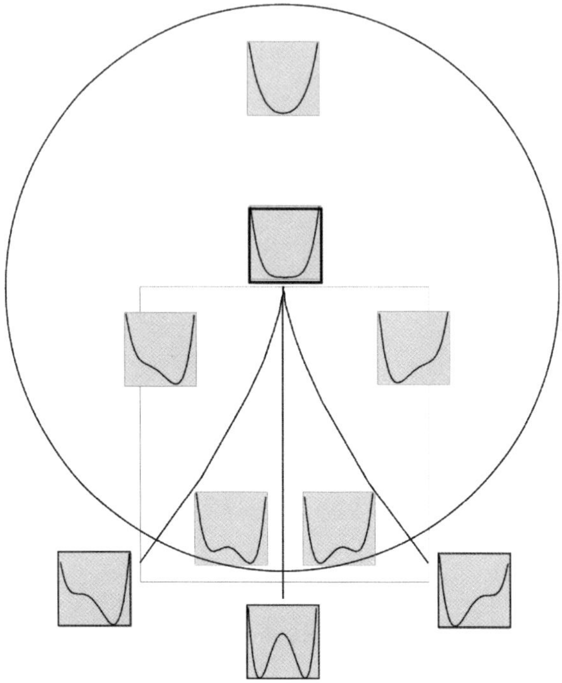

FIGURE 10. The catastrophe set of the cusp. It is composed of two bifurcation strata (where the maximum merges into a flex point with one of the minima) and a conflict stratum (where the minima have the same critical value).

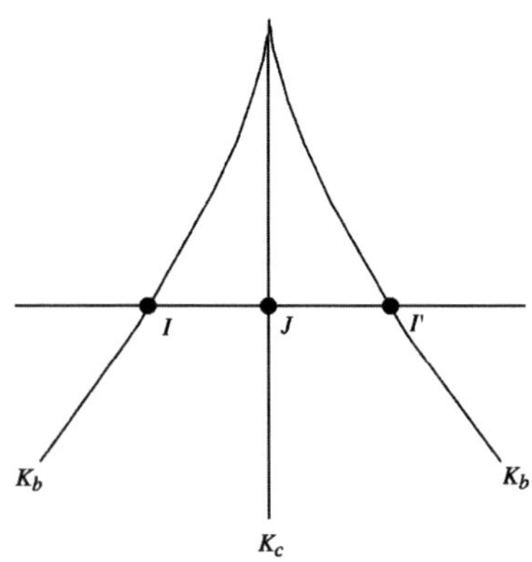

FIGURE 11. A path in the external space W of the cusp singularity.

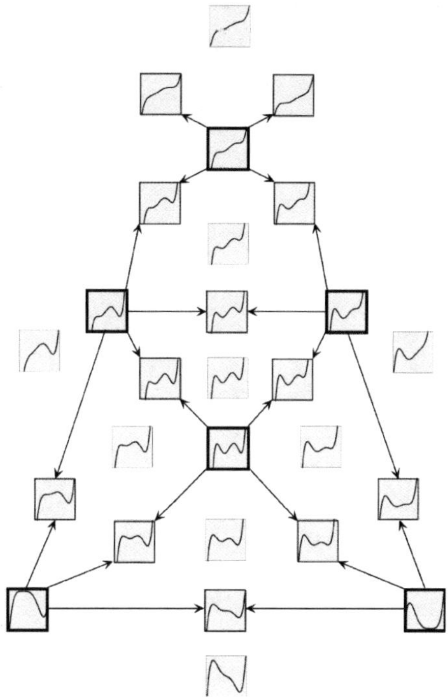

FIGURE 12. The relations between the different classes of stable potentials of the swallowtail and the codimension 1 and 2 intermediary potentials.

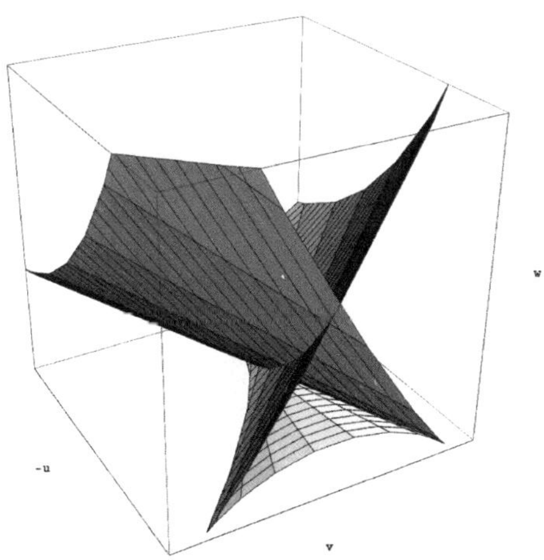

FIGURE 13. The bifurcation part K_b of the catastrophe set K of the swallowtail in the external space W with coordinates (u, v, w).

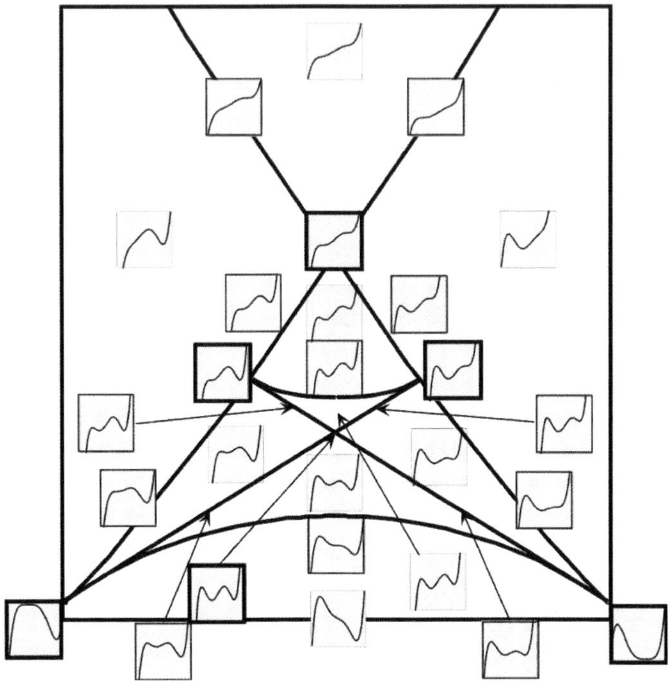

FIGURE 14. A 2D section of the catastrophe set K of the swallowtail for $u < 0$ with the associated potentials.

4.3. The butterfly

The butterfly is the universal unfolding of the singularity $f = x^6$, which presents at the origin a degenerate minimum where 5 critical quadratic points, 3 minima and 2 maxima, collapse. Its normal form is:

$$f_\tau = x^6 + tx^4 + ux^3 + vx^2 + wx \tag{18}$$

or better:

$$f_\tau = \frac{x^6}{6} + t\frac{x^4}{4} + u\frac{x^3}{3} + v\frac{x^2}{2} + wx \tag{19}$$

where $\tau = (t, u, v, w)$ varies in a neighborhood W of the origin of $\mathbb{R}^4$. The bifurcation part K_b of the catastrophe set is the apparent contour on W of the catastrophe map $\chi : \Sigma \to W$.

The strata of K_b correspond to the possible multiplicities of the roots of the equation $f'_\tau = 0$, that is to the different decompositions of 5 in sums of integers. These decompositions α are $(1,1,1,1,1)$, $(1,1,1,2)$, $(1,1,3)$ and $(1,2,2)$, $(2,3)$ and $(1,4)$, (5), and therefore the 3D hypersurface K_b has for singular locus a

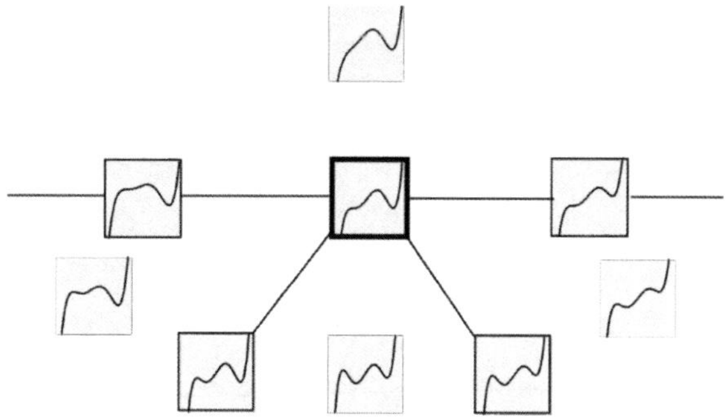

FIGURE 15. A "beak" point where a flex point (fold) has the same critical value as a quadratic point.

2D surface in $\mathbb{R}^4$ which glues two strata of codimension 2, a stratum of cusp points ($\alpha = (1,1,3)$), and a stratum of self-intersection ($\alpha = (1,2,2)$), along a skew curve presenting a cusp at the origin ($\alpha = (5)$) and composed of a stratum of swallowtail points ($\alpha = (1,4)$) and a stratum of "beak" points where a fold point has the same critical value as a quadratic point ($\alpha = (2,3)$) (see Figure 15).

To display the hypersurface $\Delta = K_b$, we look at 3D sections Δ_t for $t < 0$ in the (u, v, w) space, and to display Δ_t we show 2D sections $\Delta_{t,u}$ in the plane (v, w). Figure 16 shows a table of such sections.

For $-\sqrt{2/5} < u < \sqrt{2/5}$, $\Delta_{t,u}$ has 3 cusps and only one outside this interval, and the table of Figure 16 describes a process of *transfer* of cusps.

(i) For $u < -\sqrt{2/5}$, the process begins with a single cusp γ_1.
(ii) Then a swallowtail Γ_1 appears for $u = -\sqrt{2/5}$ and unfolds two new cusps γ_2 and γ_3. γ_3 begins to compete with γ_1.
(iii) For $u = 0$ the situation is symmetric.
(iv) Then γ_2 and γ_1 progressively collapse together and merge in a second swallowtail Γ_2 for $u = \sqrt{2/5}$.
(v) For $u > \sqrt{2/5}$, the process ends with a single cusp γ_3.

The bifurcation set of the butterfly is rather complex geometrically, see Figure 17.

In what concerns its internal potentials, the sections of the butterfly presented above correspond to a process of transfer. Figure 18 shows potentials along a typical path in the external space. Figure 19 shows the structure of the transfer: emission of an intermediary minimum by a first minimum, and transfer of the intermediary minimum to a second minimum that captures it.

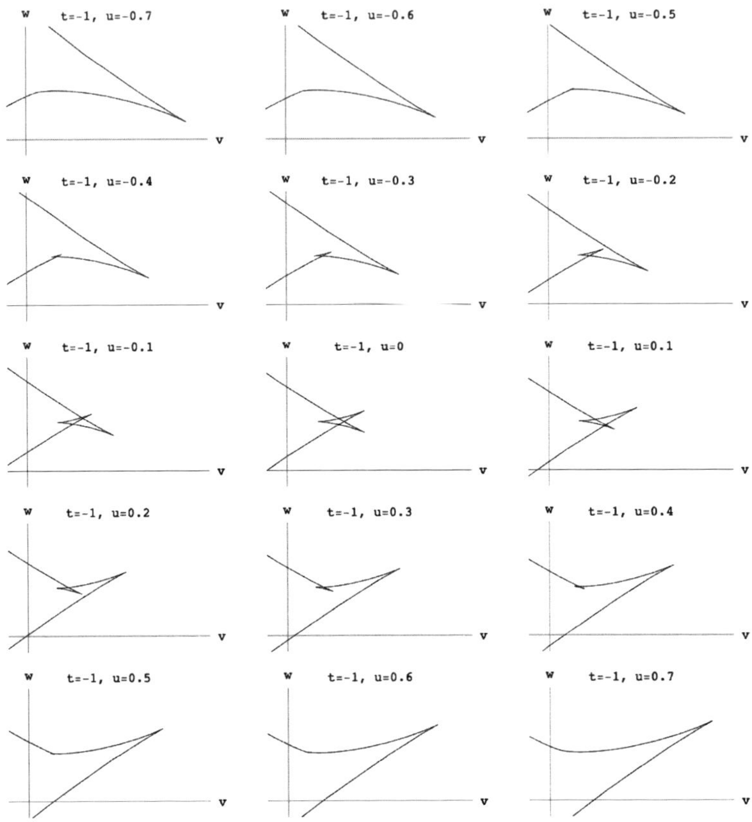

FIGURE 16. A table of 2D sections of the bifurcation set of the butterfly in the plane (v, w). t is fixed to -1 and u varies from -0.7 to 0.7.

5. Applications of Morphodynamics

5.1. Dynamical functionalism

Using these fundamental results of global analysis, bifurcation theory and singularity theory, René Thom designed a large research program leading from physics to cognitive sciences, including linguistics. His main idea was to use these tools to develop a mathematical theory of natural morphologies and cognitive structures.

He first showed that, as far as it concerns the system of connections that "organically" links parts within a whole in a structurally stable way, every structure can be reduced to a (self)-organized and (self)-regulated morphology. But, as we have seen, every morphology is itself reducible to a system

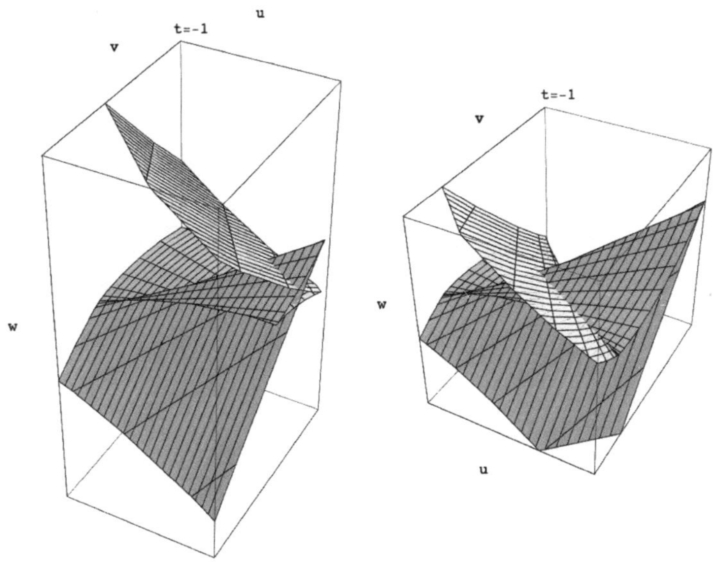

FIGURE 17. The 3D section of the bifurcation set of the butterfly for $t = -1$ in the (u, v, w) external space. Two 2D sections in the plane (v, w) are shown.

of qualitative discontinuities emerging from an appropriate underlying substrate (the substrate can be physical, purely geometrical, or even "semantic"). The theoretical problem was therefore to build dynamical mechanisms able to generate, in a structurally stable way, these discontinuities both at the local level (what was called by Waddington the theory of "morphogenetic fields" or "chreods") and at the global one (aggregation, combination, and integration of chreods).

The classification theorems allowed a revolutionary strategy which can be called *dynamical functionalism* (see Chapter 4, Section 9). Instead of first defining the generating dynamics explicitly and then deriving from it the observable discontinuities, one first describes the observable discontinuities geometrically and then derives from them a minimally complex generating dynamics. This minimal explicit dynamics must construed as a simplification of the real implicit generating dynamics.

> The essential idea proposed here is that the (processes of morphogenesis) are in fact determined by an underlying dynamics, which in general it is impossible to make explicit. (...) To a certain extent, one will be able to classify and predict the singularities of the morphogenesis of the system, even if one does not know, either the underlying dynamics of it or the macroscopic dynamics of its evolution. (...) In fact, in most cases one will proceed in the reverse direction: from the macroscopic

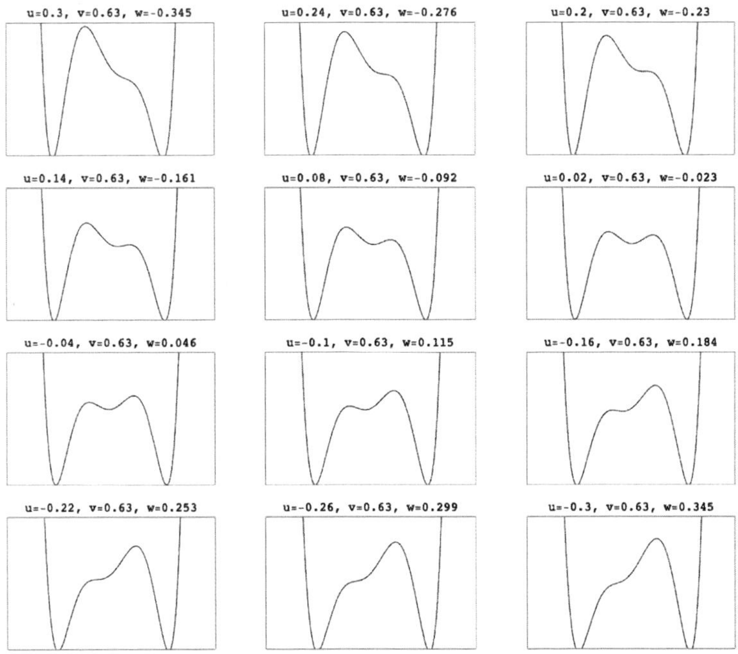

FIGURE 18. Twelve potentials along a typical path in the external space of the butterfly.

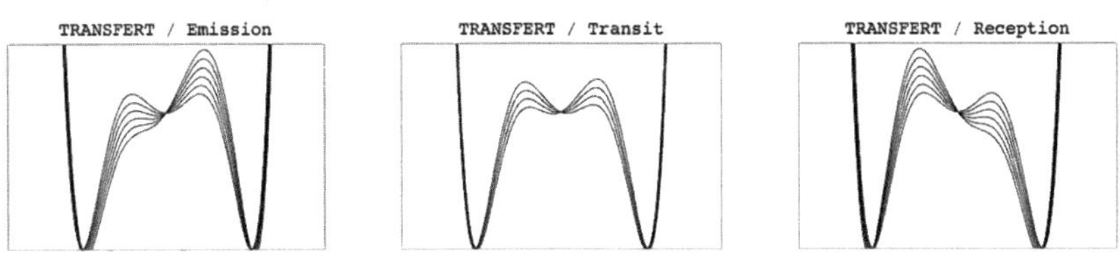

FIGURE 19. The structure of the transfer in the butterfly: emission of an intermediary minimum by a first minimum, transfer of this intermediary minimum to a second minimum that captures it.

> examination of the morphogenesis of the process, from the local or global study of its singularities, one will try to trace back to the generating dynamics. (Thom [382], p. 101).

As it is stressed in Andler, Petitot, Visetti [18], this dynamical functionalism is not of a classical (e.g., Fodorian) type. Indeed, classical functionalism entails a strict separation between the cognitive and physical levels, the relation between the two being a matter of mere compilation and implementation. This is no longer the case in an emergentist approach. But dynamical functionalism is nevertheless a "true" functionalism in the sense that classification theorems show that emergent structures share properties of universality which are to a large extent independent from the specific physical properties of the underlying substrate (see Section 6.2 of Chapter 1).

Such an explanatory paradigm has been extensively developed during the 1970's and the early 1980's. We now give briefly a few indications about these precursory trends.[12] As we already mentionned the applications in qualitative pheno-physics, we will now focus only on the cognitive and linguistic domains.

5.2. Actantial interactions and verbal nodes

For topological models of syntax, Thom's main idea was the following. We start with a general morphodynamical model of gradient type. Let f be a (germ of) potential on an internal manifold M. We suppose that f presents at a a singularity of finite codimension c. Let (f_w, W, K) be the universal unfolding of f and $\chi : \Sigma \subset M \times W \to W$ the catastrophe map associated with it. In other words, we consider the product $M \times W$ of the internal space M (on which the f_w are defined) by the external space W. We see it as a fibration $\pi : M \times W \to W$ over the external space W, and we consider the "vertical" potentials $f_w(x)$. We have seen in Section 4.2.2 of Chapter 4 that connectionist models have also used this idea, but only for the theory of learning (W is then the space of synaptic weights which vary slowly, adiabatically, along the trajectories of the back-propagation dynamics). Here, it is used for a completely different purpose: to model the *categorial difference between actant and verb.*

We use then the universal unfolding (f_w, W, K) as a geometrical generator for *events of interaction between attractors.* We introduce temporal paths $\gamma = w(t)$ in the external space W and consider that they are driven by slow external dynamics. When γ crosses K, bifurcation events occur. They are events of interaction of quadratic critical points. Thom's idea is then to interpret the minima of the f_w—the attractors of the internal dynamics—as "actants", the generating potential f_w as a generator of *relations* between them [13], a temporal path $f_{w(t)}$ as a *process* of transformation of these relations, and the interaction of actants at the crossing of K as a *verbal node.*

> If one interprets the stable local regimes (of the fast internal dynamics) as actants, it becomes possible to give the qualitative appearance of catastrophes a semantic interpretation, expressed in natural language. If (...) one introduces time (i.e. a slow external dynamics), (the bifurcations) are interpreted as verbs. (...) One gets that way what I think is *the universal structural table*, which contains all types of elementary sentences. (Thom [382], p. 188, re-issue of a 1972 paper).

12 For a general panorama, see the Proceedings of Thom's Cerisy Conference *Logos et Théorie des Catastrophes* [214]. See also Petitot [271].

13 It is this interpretation of relations between entities via a generating potential that is of the utmost technical and philosophical importance. It constitues the hard core of Thom's "morphodynamical turn".

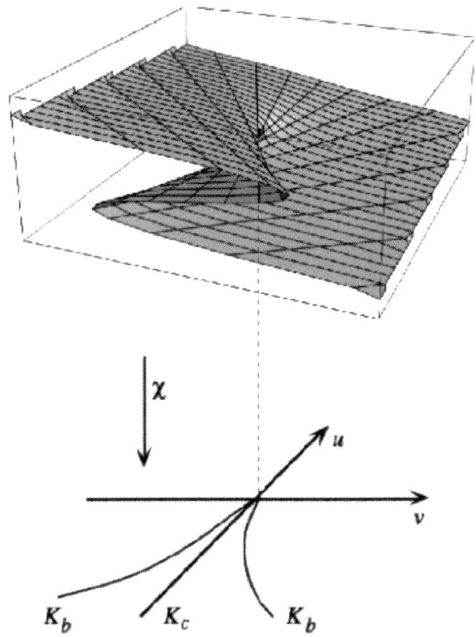

FIGURE 20. The cusp map.

We therefore get a complete and coherent dynamical interpretation of the Langackerian hierarchy listed in Section 3.2.5 of Chapter 1:

- terms $\equiv$ minima of generating potentials;
- relations $\equiv$ relations between the minima defined by the potentials;
- processes $\equiv$ temporal deformations of relations;
- events $\equiv$ interactions between minima;
- agentivity $\equiv$ causal control of actions and interactions;
- semantic roles $\equiv$ types of transformations and interactions (configurational definition).

Let us take a simple example using the cusp catastrophe. We look a the cusp map (Figure 20) and select in the universal unfolding the path of Figure 21. The temporal evolution of the potential is shown in Figure 22. It corresponds to an event of "capture" of an actant X by an actant Y.

It must be strongly emphasized that in such interactional events, the "actants" are reduced to pure abstract places—*locations*—which must be filled by "true" participants (even when these participants are concrete places as in sentences like "John sends an e-mail to Bloomington"). They play for cognitive grammars almost the same role as symbols do when one symbolizes a

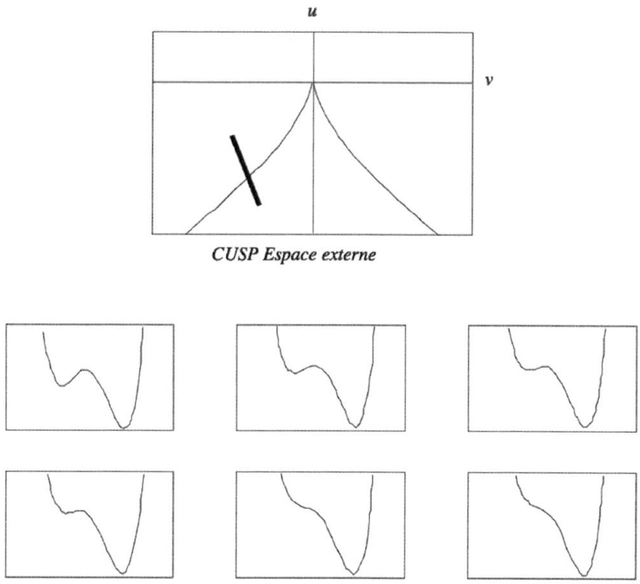

FIGURE 21. A path of "capture" in the universal unfolding of the cusp.

sentence such as "John takes the book" by a symbolic formal expression such as "$X\ R\ Y$". The main difference is that, in the topological-dynamical paradigm, a basic "local" content ($X =$ "source", $Y =$ "object", etc.) can be retrieved from the morphology of the event itself. Such a configurational definition of purely local semantic roles is a major consequence of iconicity in syntax. It has no equivalent in the symbolic classical paradigm.

It must also be emphasized that an archetypal dynamical event has nothing to do with "objective" outer space-time and can be semantically interpreted in several ways. It is only a dynamical invariant. If for instance Y is an agent and X an object it will yield sentences such as "Y catches X", etc. If Y is a place and X an agent or an object, it will yield sentences such as "X enters in Y", etc.

5.3. Actantial paradigms and their temporal syntagmation

It must be emphasized that, when so interpreted, an unfolding (f_w, W, K) constitutes what can be called an *actantial paradigm*. It is a system of actantial relations and interactions with no temporal structure. It is only when one introduces temporal paths $\gamma = w(t)$ in the external space W that the paradigmatic system is converted into a syntagmatic process.

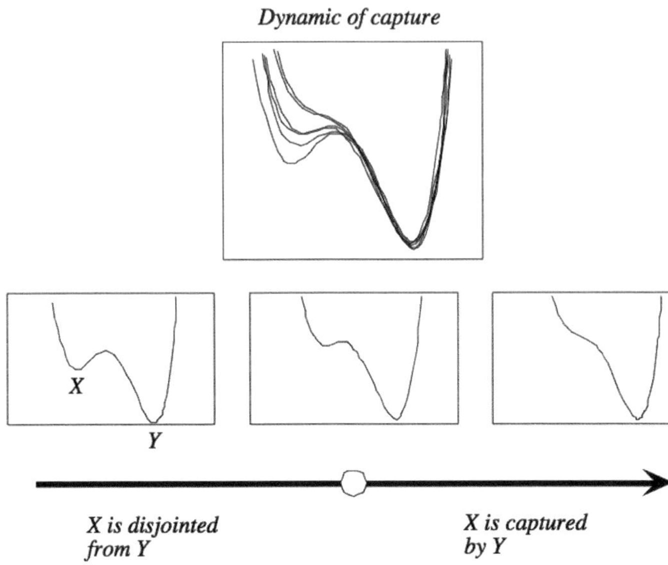

FIGURE 22. Interpreted actantially, a typical path in the cusp catastrophe corresponds to an event of "capture" of an actant X by an actant Y.

5.4. Actantial graphs and their combinatorics

To such a process derived from a temporal path in a universal unfolding we can associate a combinatorial structure which is called its *actantial graph* and belongs to a sort of *algebraic topology of syntactic structures.* The idea is to reduce the attractors to points (the minima of the generating potentials), and to look at their temporal trajectories (their world-lines) and at the nodes (the vertex) where these world-lines interact. Actantial graphs are extremely interesting structures. Indeed, they are image-schemata deeply rooted in perception which are still of an (abstract) perceptual nature but already of a (proto) linguistic nature. They provide *types* of actantial interactions and therefore a *categorization* of the possible interactional events.

This is really the key point. As was pointed out by Benjamin Bergen and Nancy Chang in their paper "Spatial Schematicity of Prepositions in Neural Grammar" [31], image-schematic roles are perceptually rooted:

> Although image schemata have often been likewise characterized in terms of such roles (also called components or elements), it is crucial to note that these roles are abstractions over individual perceptual experiences, and that a full representation of image schemata must at some level involve representations based on the perceptual system. That is, although these roles can be represented in symbolic terms,

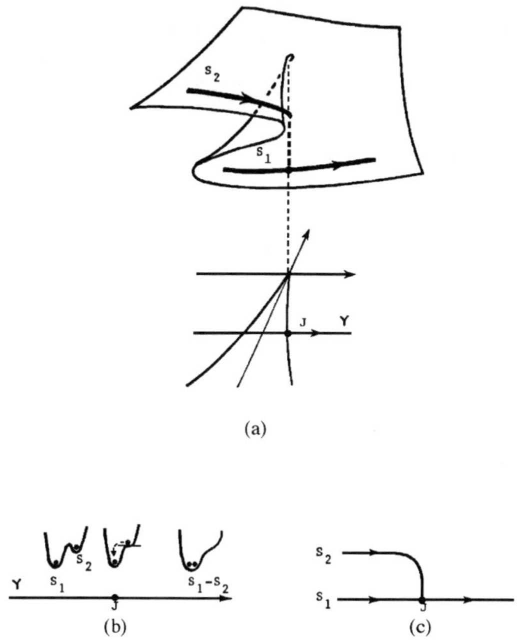

FIGURE 23. The actantial "capture" graph deduced from the cusp. (a) The universal unfolding of the cusp and a path γ. (b) The capture of the minimum S_2 by the minimum S_1 along γ. (c) The actantial graph.

> this symbolic representation serves only to parameterize, and not to replace, the perceptual properties of the schema in question.

Let us be a bit more precise. Consider a spatio-temporal scene, e.g., a scene describable by a sentence such as "X takes Y". The schematization leads to an actantial graph which can be deduced from the cusp catastrophe using paths such as those described in previous sections (see Figure 23).

Look now at a scene describable by a sentence such as "X gives Z to Y". We split the semanticism of the scene in two completely different parts.

(i) One part concerning the purely positional (local) content of "give" as an image-schema of "transfer" type in the sense of Section 4.3. This local content is like a frame, or script:
- X, Y, Z are places (locations);
- in the initial state, Z is linked with the "source" X which "emits" it;
- in the final state, Z is linked with the "target" Y wich "receives" (or "captures") it;

- between these two states there is a "movement" of "transfer" type;
- the "movement" is controlled by X and this intentional control of the action makes X an Agent.

(ii) Another part concerning the semantic lexical content of X, Z, Y and "give" as a gift action.

After having separated these two types of semanticism, we have to model the local type. For this, we must retrieve the local information from perceptual data (in a bottom-up and data-driven manner). We must therefore extract syntactic invariants ("syntactic" in the configurational sense) from the scenes. This is a very difficult task because the local information is coarser than the topological information. It belongs to an algebraic topology of events of interaction between actants. It is here that the actantial graphs become essential. They belong to the relevant level of representation. Moreover, they can be explicitely generated by generating potential functions which dynamically define the relations that they consist of. Their main function is to define in a purely configurational manner the local semanticism. Look, for example, at the "capture graph" of Figure 23. Its morphology and its generating potential characterize the local content of the actants (Agent and Object).

Now, we observe that actantial graphs share all the properties of symbolic structures. Many combinatorial operations and transformations can be performed on them. Let us again take the very simple exemple of "capture".

The graph "Y captures X" is constructed by *gluing 3 components*: an event E and two actants X and Y (see Figure 24(a)).

(i) The difference between "Y captures X" and "Y emits X" is schematized[14] by the reversing of the time arrow (see Figure 24(b)).

(ii) The thematization of Y[15] in the formulation "Y captures X" is schematized by the fact that Y and E are glued together before X is glued to the resulting complex (see Figure 24(c)).

(iii) The thematization of X (as in the passive diathesis "X is captured by Y") is schematized by the fact that X and E are first glued together (see Figure 24(d)).

(iv) The difference between "Y captures X" and "X captures Y" is schematized by the exchange of the two lines of the graph (see Figure 24(e)).

To be more complete we must introduce (at least) two supplementary elements. First, we must take into account the different "ontological" categories to which the actants belong. An actant can be, for example, a localized material thing, a pure locus or a diffuse field (light, heat, etc.). As a thing, it can

14 "Schematization" of a linguistic procedure here denotes the way in which this procedure is reflected in the actantial graph (which is not a linguistic structure but a geometrical one).

15 "Thematization" consists in focalizing attention on an actant in such a way that it is taken in charge at the grammatical level by the grammatical subject.

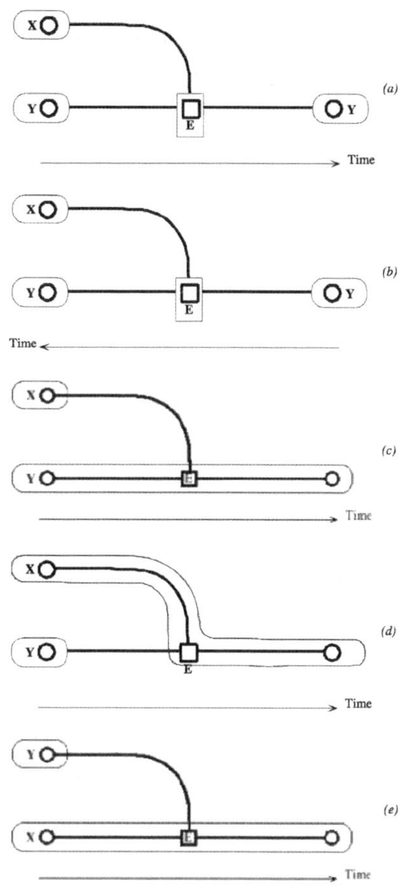

FIGURE 24. Combinatorial operations and transformations performed on the graph "Y captures X".

be animate or inanimate, etc. Secondly, we must also take into account which actant possesses the control and the intentionality of the process (see below Section 8.3.3).

5.5. Summary of the principles

Let us summarize the main principles of this topological and dynamical conception of syntax and verbal valence:

1. we introduce an underlying implicit fast dynamics X_w;
2. we use X_w to model the relations between the attractors of X_w: relations are no longer logical entities but dynamical ones;

3. we use the (complex) topology of these attractors in order to model the semantic content of the corresponding "actants": we obtain this way a geometrical functional semantics;
4. we introduce Lyapunov functions and we reduce the dynamics to the associated quasi-gradient dynamics: such a reduction expresses the shift of level from semantics to deep iconic structural syntax;
5. we introduce temporal paths in universal unfoldings, that is slow dynamics, so as to get actantial processes and actantial interactions;
6. we interpret such actantial processes and interactions as verbs;
7. we use the classification theorems about universal unfoldings to establish a universal table of syntactic archetypes;
8. we associate with the syntactic archetypes actantial graphs which support combinatorial operations analog to those that operate on symbolic structures;
9. we interpret agentivity in terms of control dynamics.

The linguistic specificity of these theoretical steps was very precisely pointed out by Wolfgang Wildgen ([405], pp. 264-265). Let us read a long quotation:

> The structure of the elementary interactions, which are derived from paths in the bifurcation space of elementary catastrophes, defines different roles which can be roughly compared to the "schémas actantiels" proposed by Tesnière and to the "case frames" classified by Fillmore. The basic difference between these structures and the semantic archetypes consists:
>
> (1) In the *preverbal* character of archetypes. The structures proposed by Tesnière, Fillmore and others are only generalizations of linguistic structures found in natural languages.
>
> (2) The foundation of the classification of archetypes in a formalism which is supposed to be basic for many biological systems. It is therefore universal in a very deep sense and it is of interdisciplinary relevance.
>
> (3) The semantic archetypes are *irreducible Gestalts.* They are not composed in a single combinatorial way. This fact constitutes a major difference in Thom's theory against all theories proposed up to now. Some of these have tried to describe field-like structures, but as no tool for consequently doing so was available they all drove away, irresistibly attracted by the static-logical paradigm.

The main advantages of such a point of view are the following.

(i) Since elementary interactions between local spatio-temporal actants are mathematically characterizable and classifiable, they provide a theoretical mean for deducing case universals on the basis of prior basic principles.

(ii) Since the actantial content of case roles is purely local and positional, one avoids the well known vicious circle of a semantic interpretation of deep structures that plagues classical case grammars.

(iii) Since the local content of the same case can change according to the topological complexity of the relational scheme where it is located, one can understand in this way why it is idle to search for a small list of universal case labels.

It is essential to understand that in order to elaborate a correct theory of interaction between attractors we really need a theory of universal unfoldings. To get interactions we need temporal processes, that is, temporal deformations f_t of systems f of actantial relations. But these temporal paths "live" in spaces that must be generated by the relations themselves. They must contain potentially the possibility of changes of the relations. The most relevant way to do this is to use the key concept of universal unfolding.

We must also stress the fact that the use of universal unfoldings allows to group temporal processes in equivalence classes and therefore to work out a theory of transformation of these processes themselves. Indeed, let $F = (f_w, W, K)$ be a universal unfolding. We consider paths $\gamma_t = f_{w(t)}$ in F. If they are structurally stable (it is a generic property), they can intersect only the strata of codimension 1 of K and only transversally.[16] Consider now an *homotopy* in W between two such paths γ_t and η_t, that is, a deformation $H(t,i)$ ($i \in I = [0,1]$) such that $H(t,0) = \gamma_t$ and $H(t,1) = \eta_t$. If H is structurally stable (it is a generic property), it can intersect only the strata of codimension 2 of K and only transversally. Let $K^{(2)}$ be the union of the strata of K which are of codimension ≥ 2. The homotopy classes of $W - K^{(2)}$ correspond to equivalence classes of paths, and the crossing of a singularity of codimension 2 by a homotopy of paths in W corresponds to a transformation of the corresponding processes.

6. Import and limits of Thom's paradigm

Thom's works introduced revolutionary mathematical methods into the linguistic and cognitivist establishment which, at that time, was quite exclusively dominated by the formalist symbolic paradigm. They gave an extraordinary new impulse to traditions such as Gestalt theory, phenomenology and structuralism. It was the first time that, in cognitive and linguistic matters, differential topology substituted for formal logic as the main mathematical tool.

Roman Jakobson—who, as far as we know, was with Waddington the first to support the advances of this outstanding mathematical genius outside the domain of hard sciences—said that he acknowledged three "great structuralists": Prince Trubetskoy, Claude Lévi-Strauss, and René Thom.[17]

---

16 In the case of low codimension, K is naturally stratified in submanifolds (called strata) K_i^l which are the different equivalence classes of the f_w of codimension $l = 0,..., c = \text{codim}(f_0)$. The codimension of a stratum K_i^l in W (that is, $\dim(W) - \dim\left(K_i^l\right)$) is l. Moreover the K_i^l possess "good" properties of incidence (Whitney conditions for a stratification).

17 See Holenstein [157].

Personally, we spent many years trying to make more explicit the historical and philosophical background of this dynamical structuralism. It was indeed necessary to be very cautious with the foundational issues to avert and ward off any misunderstanding. We remember that in 1975, when we explained these new ideas concerning the topological and dynamical schematicity of deep linguistic structures at the Chomsky-Piaget Royaumont meeting organized by Massimo Piatelli-Palmarini (see [258]), the breach was complete between the dominating symbolic paradigm and the emerging dynamical one.

But in their breakthrough, Thom and Zeeman proceeded as mathematicians, not in a "bottom-up" manner, from empirical data first to ad hoc models and then, at the end, to theoretical principles, but rather in a "top-down" manner, from fundamental principles and mathematical structures to empirical data. The advantage of such a strategy was that their perspective was theoretically very well grounded and mathematically very strong.

But the limits of their dynamical functionalism, and even its partial failure, were the lack of an effective computational theory to undergird it. Indeed, what can the *cognitive origin* of the generating dynamics be? In the following chapter we will address this issue.

On the other hand, connectionist models have proceeded in a "bottom-up" manner, elaborating models that are computationally effective, but lack grounding in theoretical principles and mathematical strength. For instance, the key concept of universal unfolding is not used, the stratified geometry of the bifurcation sets K_W in the control spaces of synaptic weights is unknown, the back-propagation algorithms provide external dynamics in the W spaces whose behavior at the crossing of the K_W is not analyzed, the complex topology of strange attractors coming from a coupling between neural oscillators is not used for semantic purposes, nor are their bifurcations used for modeling syntactic structures, etc.

7. Morphodynamics and Attractor Syntax

We recall now briefly how in *Morphogenèse du Sens* and other works ([261], [266], [267]), then in further works by Per Aage Brandt [44], [45], Thom's dynamical conception of syntactic iconicity was linked with some basic linguistic trends, from Fillmore's and Anderson's case grammars to Talmy's and Sweetser's "Force dynamics".

7.1. The mathematization of Fillmore's scenes

Thom's perspective was the first schematic and iconic theory of syntactic structures. It succeeded in defining the actantial roles not conceptually, using semantic labels, but configurationally. This is the only way to resolve the tension between two equally important requirements. On the one hand, if they are really universal, case universals must constitute a very limited set. On the other hand, in order to perform their syntactic discriminative function, they must be sufficiently numerous.[18] A configurational definition of the semantic roles solves the problem because it makes case universals depend on *relational* configurations. It therefore provides an answer to what Charles Fillmore has called the "truly worrisome criticism of case theory", that is, the fact that

> nobody working within the various versions of grammars with "cases" has come up with *a principled way* of defining cases, or *principled* procedures for determining how many cases there are. (Fillmore [106], p. 70, our emphasis)

In fact, regarding case grammars, Thom's morphodynamics yields a natural mathematization of the theory of cognitive *scenes* developed by Charles Fillmore in "The case for case reopened" [106].

Fillmore's slogan "meanings are relativized to scenes" leads to a dividing of the verbal semanticism between two components: on one hand, the contextual component of the semantic fields associated with the scenes, and, on the other hand, the pure actantial component of the case frames and of the semantic roles. Such a definition of case values is *conceptual* in the cognitive sense of "conceptual structure":

> such descriptions (are) in some sense intuitively relatable to the way people thought about the experiences and events that they (are) able to express in the sentences of their language. (Ibid., p. 62)

When case grammars are revisited by topological syntax and cognitive grammars, the actants (the semantic roles) are reinterpreted in a localist, topological-dynamical key. The main consequence is that their content is no longer construed as a semantic one (that is, as a meaning) but as an *iconic-schematic* one (that is, as an abstract image). And, as we have seen, the main problem (which we will address in the following chapter) is to retrieve such an abstract iconic-schematic content from perceptual data.

In this new interpretation, a scene Σ consists of the following components :

(i) A *semantic "isotopy"* (e.g., the "commercial" context in the prototypical commercial scene); its semanticism is not reducible to a localist one.

(ii) A *global scheme* G of interaction between purely positional actants P_i: these positional actants (source, agent, object, goal, instrumental, etc.) are actantial positions in an abstract external space Λ underlying the scene. It must be strongly emphasized that in this morphodynamical approach to syntax, the positional actants P_i play the same role as symbols do in the symbolic classical aproaches and that the global actantial graph G is an image-schema in the sense of cognitive grammars. G is not a linguistic entity but a Gestalt-like entity and it defines in a schematic (topological-dynamical) way the purely *local* content of case roles.

18 For the question of language universals, see Section 8 of Chapter 1. For an introduction to case grammars, see Petitot [261].

(iii) Specializations of the positional actants P_i either into effective actants (human beings, objects, forces, etc.) or into places.

The scene Σ defines conceptually and semantically (because of (i) and (iii)) the semantic roles involved. But it also defines them configurationally (because of (ii)). In general, Σ is spatio-temporally localized in space-time $\mathbb{R}^4$ by means of an embedding $j : \Lambda \hookrightarrow \mathbb{R}^4$ of its underlying external space Λ in $\mathbb{R}^4$. Through j, the positional actants which are specialized in places become pure spatio-temporal actants while the positional actants which are specialized in persons, objects, etc. become concretely localized. Localization is linguistically expressed by adverbial complements.

There exists a (restricted) number of *local archetypic schemes* $\Gamma_1, ..., \Gamma_n$ that determine case universals. What Fillmore calls the "orientational or perspectival structuring" of a scene consists in covering the global particular scheme G by gluing together such local archetypes. In general, there will be many different possible coverings. The gluing operators are *anaphoric.*

The choice of an archetype Γ_i is linguistically expressed by the choice of a verb ('sell', 'buy', 'pay', 'cost', etc. for the "commercial" scene). Through its semanticism, the verb excites the whole scene Σ. But through its valence and case schema it selects an archetype of type Γ_i. What Fillmore calls the "saliency hierarchy" determines what is the minimal part of G that must be covered if we want to describe the scene adequately. A case hierarchy then determines the manner in which the actants of the Γ_i selected for covering G are taken over at the surface level by grammatical relations. The part of G that is not covered by the selected Γ_i can be described by other sentences (using anaphoric gluing) or by adverbs, subordinate sentences, etc. After their grammaticization, the nuclear sentences coming from the Γ_i covering G become inputs for transformational cycles. At this stage, the approach becomes akin to classical symbolic conceptions of grammar.

7.2. The localist hypothesis (LH)

The morphodynamical schematicity of deep actantiality gives a rigourous status to one of the main hypothesis of the linguistic traditions, namely, the *localist hypothesis* (LH). In a nutshell, LH claims that the grammatical relations standing for abstract actantial relations grammaticize the *spatio-temporal* interactions between spatio-temporal actants, i.e., actants whose identity is reducible to their localization.

Historically, LH goes back to the Byzantine grammarians Theodorus Gaza and Maximus Planudes.[19] Its key thesis is that cases are grammatical *and* local determinations. As was claimed by Hjelmslev ([155], p. 15)—who, after the Kantian linguist Wüllner, was the leading modern linguist to support LH—, it

19 See Hjelmslev [155], Anderson [13] and Petitot [261].

> recognizes as equivalent the concrete or local and the abstract or grammatical manifestations

of the case dimensions (for instance direction as in "source ≡ nominative" and "goal ≡ dative"). In fact, Hjelmslev strongly emphasized the necessity of substituting the logical and formal conception of syntactic relations with a schematic and iconic spatial conception. The idea of a "space grammar" in the sense of Talmy or of Langacker is very much like Hjelmslev's structuralist perspective.

> The spatial conception is necessary if one wants to give a tangible and plastic interpretation of the relations *in abstracto*. To limit oneself to abstract relations without giving them an intuitive stand by means of which it becomes possible to represent them is to forbid a clear and evident explanation of the facts. (Ibid., p. 45)

Classically, LH seems to be a hypothesis about the nature of language. In fact, it concerns the topological dynamics of actantial relations and has to do with the underlying structural syntax of grammars, which is of a *cognitive* nature.

Along with Jean-Pierre Desclés, we shall distinguish four forms of localism:

(i) a weak localism limited to stating a compatibility between language and perception;
(ii) a cognitive localism stating that certain cognitive representations coming from language are in some way equivalent to certain cognitive representations coming from perception;
(iii) a strong localism—of a metalinguistic nature—stating that the deep syntactic schemata that encode perceptual situations give rise to syntactic schemata that encode non-perceptual situations;
(iv) a hyper-strong localism—of a linguistic nature—making the same assertion about surface structures.

In agreement with Desclés, we think that the localist hypothesis deploys its true meaning in the foundation of a moderate localism, i.e., as a cognitive hypothesis. In this respect, it constitutes the great hypothesis in the history of linguistics that anticipated the thesis of cognitive grammars (in particular Jackendoff's idea of an ontology of the projected world).

7.3. The uses of external dynamics

The other contributions of morphodynamics that we want to point out in the linguistic domain concern the different uses of the *external spaces* (W, K).

7.3.1. *Aspectuality.* Firstly, one can interpret *aspectuality*, i.e., the temporal structure of the processes, using such devices (Petitot [278]).

The role of aspectuality is, by definition, to convert actantial relations into *processes*. Spatialization and temporalization correlate the actions of the subjects with the temporal succession of the cut out places that they fill. It concerns the linkage of partial spaces (passages, boundary crossings, movements, etc.). Aspectuality is grafted on temporality, and represents it by means of categories such as inchoative, terminative, durative, perfective, imperfective, etc. Thus actantial statements are converted into spatio-temporal processual statements that can refer to states of affairs. In such a conversion, the statements undergo a mutation of their status. Indeed, actantial statements are of an abstract nature, while processual statements are of a spatio-temporal nature.

Consequently, the main features of aspectualization are:

(i) spatio-temporalization of statements produced by actantial syntax;
(ii) aspectual qualification of this spatio-temporalization;
(iii) logical→topological mutation of the status of the statements;
(iv) constitutive role of perception.

Aspectuality is, therefore, one of the linguistic mechanisms that transform an abstract syntax of actants into a spatio-temporal and dynamical processual syntax. Mediating between the symbolic pole of formal syntax and the morphological pole of dynamics, it constitutes a pivotal feature. Its status is bimodal, governed by a principle of symbolic/topological complementarity. For reasons pertaining to the history of ideas, the symbolic pole has always been privileged to the detriment of the topological pole. Traditionally, the logical-symbolic level is projected into the depths of linguistic structure, then one "climbs" up in successive stages to a topological surface level. In a very general way, it is assumed that there exists a deep level of primitive, relational and abstract structures, of a formal nature, and that through a successive series of levels of representation and construction operations (involving predication, modality, aspect, focus, theme, categorization, quantification, qualification, etc.) one reaches the superficial levels of manifestation. In such approaches, the aspectual operations only appear to "embed" formal relations into space-time: they allow to spatio-temporally locate the predicative representation constructed by a statement, and refer the internal cognitive states to external states of affairs.[20] Thus the topological and dynamical levels function only superficially here. Deep structures remain essentially discontinuous and discrete.[21]

We have seen, however, that the topological-dynamical level is as deep as the logical-symbolic level, and as constitutive for the actantial structures. This gives aspectuality a new status linked to the topological-dynamical schematization of actantiality that we have developed. If we consider aspectuality going from the logical domain towards the topological domain (i.e., from the subject toward the world, from the interoceptive to the exteroceptive) then this allows logical-symbolic structures to become compatible with the continuum.

20 See for example Desclés [85].
21 Bernard Pottier has much criticized this point. See Pottier [313], [314], [315].

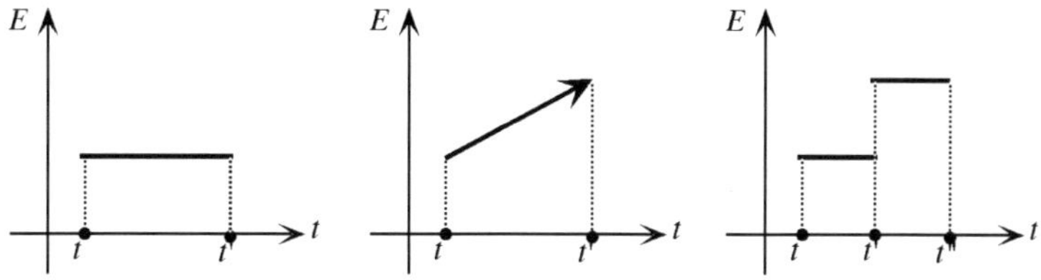

FIGURE 25. Aspectuality as temporal structure of processes. From left to right: a stable state, a process, an event.

If we consider it, instead, from the topological domain toward the logical domain, then this allows, on the contrary, the topological-dynamical structures to become compatible with the symbolic stance.

Therefore, aspectuality concerns the grammaticization of "becoming" (movement, processes, crossings of obstacles, etc.), i.e., action of time on substances, qualities and states of affairs constituting the qualitative ontology of the phenomenological world.

If we consider the generic morphological accidents generated by the action of time on substances and states of affairs, we note that aspectuality, which specifies these accidents grammatically, is largely deducible from them. Indeed, we immediately find again the main features identified long ago by linguists:

(i) stable states reversibly occupying a temporal interval;
(i) processes irreversibly occupying a temporal interval;
(iii) the limits of intervals as beginnings and ends of processes (inchoative, terminative);
(iv) events corresponding to qualitative discontinuities, etc.

We therefore have to consider internal dynamics on which a slow time acts as a control, i.e., as an external dynamics. In a nutshell, the idea is the following. Consider a temporal path $\gamma = w(t)$ (driven by a slow dynamics) in a universal unfolding (f_w, W, K). The stable states reversibly fill open temporal intervals. The processes irreversibly fill temporal intervals. The boundaries of such intervals correspond to the beginning (inchoativity) and the end (terminativity) of the processes. The events correspond to the actantial interaction points where γ crosses the bifurcation set K. Aspectuality is grounded in the topological structure of the temporal line intervals. Its well known interval models here become models for the topology of the embedded paths $\gamma : I \to (W, K)$ (where I is an interval) (see Figure 25).

7.3.2. *Agentivity.* Another problem concerns "agentivity", i.e., how intentional agents control actions. It is clear that a purely topological (positional) definition of the actantial case roles cannot be sufficient. To explain, for instance, the difference between Agent, Beneficiary, Patient, Instrumental, etc., we also need *causal* relations between them. But one of the main benefits of the morphodynamical models is that actantial configurations are derived from internal generative dynamics. Agentivity can thus be modeled by a *feedback* from the internal dynamics to the *external dynamics.* Indeed, insofar as it is driven by one of the actants, this feedback expresses precisely that this actant *controls* the action (the evolution in control space) and is therefore an agent.

7.3.3. *Modal external dynamics.* There exists a deep connection between agentivity and *modality.* In his fundamental 1986 work *La Charpente modale du Sens* (The Modal Scaffolding of Meaning), Per Aage Brandt [44] has shown, using Talmy's theory of *Force Dynamics* [371], that external dynamics in W can be interpreted as *modal dynamics.* Talmy has shown that modal systems grammatically specify a dynamical conceptual content concerning the notions of force, obstacle, resistance, clamping, overtaking, cooperation, competition, interaction, etc. The associated schemata are agonistic and schematize the possible force relations between actants. Talmy's thesis is that this "force dynamics" constitutes

> the semantic category that the modal system as a whole is dedicated to expressing. (Talmy [371], p. 1)

As we will see in the following section, it is easy to model such schemata using morphodynamical models. We consider a canonical model (f_w, W, K) and assume that

(i) the actants possess an *internal energy,* and
(ii) each of them can control some external dynamics on W (many external dynamics being therefore in competition).

Their internal energy allows actants to jump over potential barriers and their control of external dynamics allows them to act on each other's dynamical situation.[22]

This point is of such importance that we must develop it in a new section.

8. Force dynamics from Talmy to Brandt

8.1. The key idea

The leading idea developed by Talmy in his seminal article of 1985, *Force Dynamics in Language and Thought* [371], is that

> the force dynamic system is a major conceptual organizing system (p. 1).

Talmy begins with the consideration of a stationary conflict between two forces. Language specifically represents a difference of actantial roles between two entities exerting forces. It selects one of them as the Agonist and the other one as the Antagonist. Furthermore, language attributes a force to each one of these actants, *Ago* and *Ant,* i.e., an intrinsic dynamic tendency toward either

22 For details, see Petitot [270].

rest or movement. Applying the respective forces of the actants, the resultant of the interaction is either rest or action. It is easy to formulate the combinatorics of the dynamical states of interaction of forces and assign linguistic representations to them. For example, if *Ago* possesses an intrinsic tendency to rest, *Ant* will possess an intrinsic tendency toward movement, and if *Ant*'s force is greater than *Ago*'s, then the resultant of the interaction will be *Ago*'s movement. Conjunctions such as 'because of', 'owing to', 'despite', 'though', 'against', etc., or quasi-auxiliaries such as 'keep', 'force to', 'manage to', etc. are all instances that grammatically specify the category of Force. Thus Talmy analyzes in detail the changes of dynamical states due to the action of *Ant* or its removal, or to the inversion of the direction of force relative to *Ago*. He thereby rediscovers (not too surprisingly) some of the results well known to the actantial theories.

As far as the linguistic expression of these "force dynamics" patterns are concerned, several possibilities arise. On the one hand, the actantial roles *Ago* and *Ant* can be taken over by cases (semantic roles having a syntactic function), themselves taken over by grammatical functions such as Subject (*S*), Direct Object (*DO*), Oblique Object (*OO*), etc., as in Relational Grammars.[23] On the other hand, concerning the verbalization of action, we could privilege either the intrinsic tendency of *Ago* or the resultant of the *Ago/Ant* conflict. Thus we obtain a rich combinatorics that provides a dynamical analysis of expressions such as (*VP* is a verb phrase):

- *S* (*Ago*)—'overcoming'—*VP*—"owing to"—*OO* (*Ant*);
- *S* (*Ant*)—'making'— *DO* (*Ago*)—*VP* (focus on the resultant);
- *S* (*Ant*)—'resisting'—*DO* (*Ago*)—*VP* (focus on the intrinsic tendency of Ago);
- *S* (*Ant*)—'blocking'—*DO* (*Ago*)—*VP*;
- *S* (*Ant*)—'letting'—*DO* (*Ago*)—*VP*;
- *S* (*Ant*)—'helping'—*DO* (*Ago*)—*VP*; etc.

Talmy remarks that we obtain an analysis of the four English verbs 'make', 'let', 'have', and 'help', which, like the auxiliaries and the modals, possess verbal complements without 'to'. This grammatically definable closed class therefore possesses a dynamical content.

The force dynamics category is more primitive than the category of "cause". A dynamical analysis of causality shows that they are part of a larger system of notions containing in addition to 'causing' or 'forcing to do', notions such as 'letting', 'helping', 'resisting', 'trying', etc.

Talmy then shows that this somewhat "physical" dynamical analysis naturally extends to psychology and opens onto a grammatical specification of "psychodynamical" contents. He explains phenomena, well known in semiotic actantial theory, as the "actorial" syncretism or non-syncretism of the actants.

23 For a brief summary of relational Grammars, see Section 8 of Chapter 1.

If *Ago* and *Ant* are in syncretism in an acting subject endowed with intentionality, wish, or desire, then we obtain conflicts, tensions, etc., that belong to a "divided self", divided in a Freudian manner between a desiring ego (*Ago*) and a superego (*Ant*). If on the contrary *Ago* and *Ant* are actorialized in two different intentional actors, then we obtain "sociodynamical" figures of intersubjectivity. All this is well known[24], but the main originality of Talmy is to use dynamical analysis to identify the content specified by the modal category. His thesis, let us repeat it, is that the notion of Force is "the semantic category that the modal system as a whole is dedicated to express" ([371], p. 1) and, therefore, as far as the modal auxiliaries are concerned, that it yields "the core of their meanings" (ibid. p. 27). Upon such a basis, we can analyze 'Can', 'May', 'Must', etc., in their physical, psychodynamic, sociodynamic and epistemic uses ('Can': *Ant* is an obstacle, an opponent; 'May', 'Must': deontic modalities, *Ant* is a Receiver; etc.).

The conclusion is that

> the semantic category of force dynamics (...) must be recognized as one of preeminent conceptual organizing categories in language. (ibid. p.41).

Force dynamics is grammatically specified in language in a rather subtle and diversified way. In fact, the force can be (ibid. p. 42):

1. present/absent (a state of a thing can be dynamically neutral);
2. internal/external (*Ago*/*Ant* opposition);
3. oriented towards action/rest;
4. more or less big (i.e., virtual, actual, realized, overcome, cancelled, etc.);
5. extended-continuous/punctual (in space or in time: aspectuality, etc.);
6. permanent-determinist/random;
7. physical/psychological;
8. supported by a conflict *Ago*/*Ant* inserted in the same actor (syncretism) or projected onto two different actors (non-syncretism);
9. localized/distributed (diffuse);
10. of uniform application/of gradient type;
11. repulsive/attractive;
12. of opposition/cooperation.

This shows that

> conceptual models of certain physical and psychological aspects of the world are built into the semantic structure of language. (Ibid. p. 37)

24 We cannot say that these considerations are without precedent. When Len Talmy declares: "to my knowledge systematic applications of force concepts to the organizing of meaning in language remained neglected (until now)" (Talmy [371], p. 2), he is taking into account only the recent North American traditions. There are, however, as we have seen, important European forerunners.

In its very form, language specifies a "naïve" physics, a "naïve" biology, and a folk psychology pertaining to common sense[25] and establishing for example an asymmetry between *Ago* and *Ant* as well as between action and reaction.

8.2. Eve Sweetser's systematization

Eve Sweetser has elaborated upon this fundamental idea of Len Talmy—which, as all fundamental ideas, is simple and powerful—and systematically applied it in modal analysis. Her hypothesis was that modalities grammatically specify the socio-physical scenes that depend on a general dynamics of forces and resistances extended to the framework of intentional causality (p. 35). The basic schema is the following: an entity E (*Ago*) follows a path C and encounters resistance R (*Ant*). Its internal force F enters into conflict (ago-antagonism) with the force F' of R.

Variations on this basic schema allow us to develop in a very economical way a *unified* semantics—simultaneously deontic and epistemic—of the modals (Sweetser [365] p. 484).

Examples:

(i) Permission ('may'): the resistance R is potential.
(ii) Ability ('can'): F is sufficiently big for E to be able to overcome R.
(iii) Obligation ('must'): F' (*Ant*) dominates E and directs it toward a goal.
(iv) In English, 'ought' indicates a moral obligation, 'have to' indicates that the necessity is internal to the Subject E (*Ago*). In other words, the forces and barriers may be surmountable/insurmountable, social/moral, internal/external.

8.3. Modal dynamics according to P. Aa. Brandt

8.3.1. *Modality as dynamical principle.* In *La Charpente modale du Sens* [44], Per Aage Brandt has brought fundamental extensions to the morphodynamical approach of the actantial structures.

As the title suggests, the work is devoted to the reexamination of the concept of modality and its promotion as a basic concept for semiolinguistic theory. The central idea is that the modal level does not constitute a relatively superficial level that just operates on the already constituted deep structures, but rather expresses the shift from a formalist conception to a dynamical conception of deep structures. Consequently, the relationship with the morphodynamical schematism becomes intrinsic.

Brandt has developed three main orientations.

1. Show that, as much in the actantial modalities as in the linguistic analysis of modal auxiliaries, modals refer to a dynamism of actantial relations

25 We know that studies on common sense cognition and naïve physics—not to be confused with qualitative physics—are currently well under way. See, e.g., the works of Gentner, Stevens, Lakoff, Hayes, Hobbes-More, etc.

and, particularly, to what in the narrative theory are called the relations of junction between subjects and objects, the relations of conflict between subjects and anti-subjects, as well as the relations of contract between subjects and Addressers [26].

2. Apply in a suitable way the tools of morphodynamical schematism.
3. Draw consequences for a reformulation of the classical actantial theories.

We have seen in Section 5.2 of Chapter 5 how morphodynamical models become actantial models. We interpret the minima of the elementary catastrophes (EC) as actantial determinations. We thus obtain paradigms of actantial relations. By introducing temporal paths in the external spaces W, such actantial paradigms are transformed into scenes, frames and scripts. Events of conjunction/disjunction between subjects and objects (S/O), or of conflict between subjects and anti-subjects ($S/\overline{S}$) occur at the crossing of the strata of the catastrophe set K in the external spaces W (hence the aspectual dimension, see previous section). The paths are regrouped into classes of equivalence (classes of homotopy in $W - K^{(2)}$ where $K^{(2)}$ is the set of strata of K of codimension ≥ 2). Hence a theory of variants and transformations (see Section 5.5).

Three things remained unquestioned in this conception—which may be called the "standard" morphodynamical theory—of actantial structures.

(i) We assumed that the internal dynamics (the potentials) confined actants in their minima (their attractors). In other words, we considered only the stable "asymptotic" states without taking into account the fact that the trajectories of the actants can be determined not only by the potential functions, but also by additional dynamics. This is usual in physics.

(ii) We did not take into account the "energetic" content of the thresholds separating the actants.

(iii) Finally, and most importantly, we did not consider external dynamics proper. We certainly considered paths in the external spaces W (which constituted, as we have seen, one of the main achievements of the model), but did not treat them as trajectories of specific slow dynamical systems.

In other words, the dynamism of the models was restricted to internal dynamics generating actantial paradigms and to the "syntagmatization" of such paradigms along temporal paths. It included *neither* "subjective" intra-actantial dynamics *nor* external dynamics. It is precisely these new resources that Brandt used for dynamically schematizing the linguistic description of the modal category.

26 In narrative semiotics, "addressers" ("destinateurs" in French) are actants making subjects ("addressees", "destinataires" in French) act in specified ways.

8.3.2. *Modalities in narrative semiotics.* To understand correctly the contribution of Brandt, we must return for a moment to the semionarrative conceptions of actantiality. For Brandt, the modal is coextensive with narrativity, as

> a modal statement is the minimal unit of the subject's narrative course. (p. 13)

In this perspective, there exists a complementarity between the modal structure and the deep semic structure (the Greimasian fundamental semantics that defines the "values" invested in the objects with which the subjects seek to be conjoined). There exists a "continuity" between the semiotic object and the narrative route of the semiotic subject. When the object is the focus, it is the semic structure that comes first. When it is, on the contrary, the subject, it is the modal structure that comes first (p. 13). Let us explain this point.

Generally, one considers that modalities (for example deontic modalities) modalize the activity of the subjects. In other words, one assumes three levels (p. 13):

(i) the level of being: the junctions between subjects and objects $J(S, O)$ (where $S =$ Subject of being);
(ii) the level of doing: transformations of junctions $T(S, J)$ (where $S =$ Subject of doing);
(iii) the level of modality: modalization of transformations $M(S, T)$ (where $S =$ Addresser).

A typical (deontic) modal statement will be for example::

(20) $$M\,(/\,\text{obligation}\,/)\,(S_3, T\,(S_2, J\,(S_1, O)))$$

where the Addresser S_3 modalizes deontically the subject of doing S_2, whose obligation of doing/not doing bears on the junction (conjunction/disjunction) between the subject of being S_1 and the object O. In other words, S_3 "programs" S_2 (positively in the form of a prescription or negatively in the form of a prohibition). It confers a competence to S_2. All the actorial syncretisms between S_1, S_2, and S_3 are possible. Hence an obvious combinatorics, which is explained in great detail by the standard theory.

Beside the deontic modalities, three other types of modalities can be defined:

(i) First, the modalities described by Brandt as *ontic*. They have to do with being (and not with doing) and are thus directly concerned (without the mediation of a subject of doing) with the junctions $J(S, O)$. In an ontic modal statement $M\,(S_3, J\,(S_1, O))$, the Addresser S_3 acts in some ways as a "fatal cause" (p. 17). It effectively "destines" the subject of being S_1. The idea therefore is that a junction is always modalized.

> The junction *contains*, for its subject S, a modal force that gives value to the object. (p. 16)

In this perspective, the doing becomes "a deontic operation applied to an ontic state" (p. 17). Consequently, the modalization of doing becomes a remodalization of being. It transforms not the state of the subject, but the "ontic determination of the state". Hence the general formula of a "deontic-ontic grammar" (p. 18):

(21) $$M \text{ deontic } (S_3, T\,(S_2, M \text{ ontic } S'_3\,(J\,(S_1, O))))$$

where S_3 = deontic Addresser and S'_3 = ontic Addresser.

(ii) Second, the alethic modalities. They are concerned with the situations where the object O is itself the representation of a junction (for example, perceptual) J' of a subject with an object. The having-to-be corresponds here to necessity, the having-not-to-be to impossibility, the not-having-to-be to contingency, and the not-having-not-to-be to possibility (see Greimas-Courtès [135]).

(iii) Finally, the epistemic modalities (also called "veridictory") concerning truth and falsehood of junctions. Hence the general formula of an "epistemic-alethic grammar" (p. 19):

(22) $$M \text{ epistemic } (S_3, T\,(S_2, M \text{ alethic } (S'_3\,(J\,(S_1, J'\,(S, O))))))$$

where S_3 = epistemic Addresser, S'_3 = alethic Addresser and S_1 = veridictory Subject.

Therefore, according to Brandt, there would exist an asymmetry of veridiction to the extent that there is

> the superimposition of an epistemic modalization on an alethic modalization in relation to the same veridictory statement of junction (p. 21).

We must insist on the importance of the ontic modalities that modalize junctions. They impose constraints on doing and allow us to grasp the subtle difference that exists between a subject S related to an object O and a subject of being whose junction with O is already an effect of "destination". In every intentional relation linking a subject with an object, a fundamental modality (a common root of having-to and being-able-to) would intervene and belong to

> a modal framework of syntactically structurable meaning

without being, however,

> a syntactic fact that a functional grammar might (...) be able to capture." (p. 26)

Now, underlying modalities, there is a "history of forces" that must be described, explained and modeled (p. 229).

8.3.3. *The basic hypotheses.* Brandt's initial hypothesis is that the actant *Ago* E is armed with a force, its "own internal energy" F. The second hypothesis is that E is submitted to the action of a potential f_w as in the EC model. The internal dynamics f_w [27] constrains E to occupy (passively) a minimum of f_w. But its own internal energy F allows it to step over the potential barriers R that are the maxima of f_w. We meet again Sweetser's basic schema with

27 The dynamics is the vector field $-\text{gradient}\,(f_w)$.

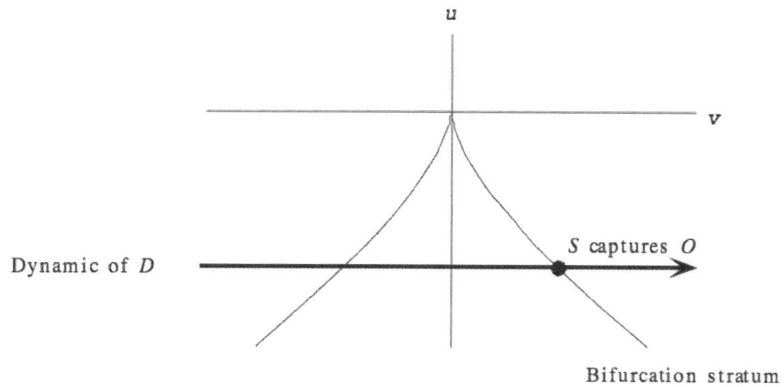

FIGURE 26. A conjonction between a subject S and an object O induced by an Addresser.

the F/R dialectic involving paths in the external space. Let us suppose that the goal of E as "subject" is, for example, a change of state (passing from a minimum of f_w into another) or the capture of an object O actualized in another minimum. Several solutions are then possible. For example:

(i) storing internal energy to be able to cross the threshold R;
(ii) lowering the threshold R, i.e., adequately deforming f_w by varying w in the external space W.

The paths in the external spaces transform the relations between F and R and can thus be *modally* interpreted on the basis of the dynamical conception of Talmy-Sweetser. Consequently, the external dynamics can be considered as modal dynamics.

Let us note that a very simple way to generate the external dynamics is to *polarize* positively or negatively the *strata* of the catastrophe set K. Let us for example consider the model of the cusp for the conjunction $S \cap O$ between a subject and an object. To say that the conjunction is "desirable" amounts to saying that the bifurcation stratum of the minimum of O is polarized positively on the $S \cap O$ side and negatively on the $S \cup O$ side.

If this external dynamic is induced by the subject E, it models the modality of "wanting". If it is induced by an Addresser, it models the modality of "having-to" (see Figure 26).

But the conjunction may be impossible to attain, the polarization (+) projected by E being weaker than a polarization (−) projected by an anti-Addresser or an anti-subject. Thus there will be a conflict between two external dynamics, an *Ago* dynamics and an *Ant* dynamics.

We see how the junctions $S - O$ (i.e., the bifurcation strata) can be modalized from the outset. It is the same thing for the conflicts opposing intentional subjects and giving rise to *mimetic* relations in René Girard's sense. Figure

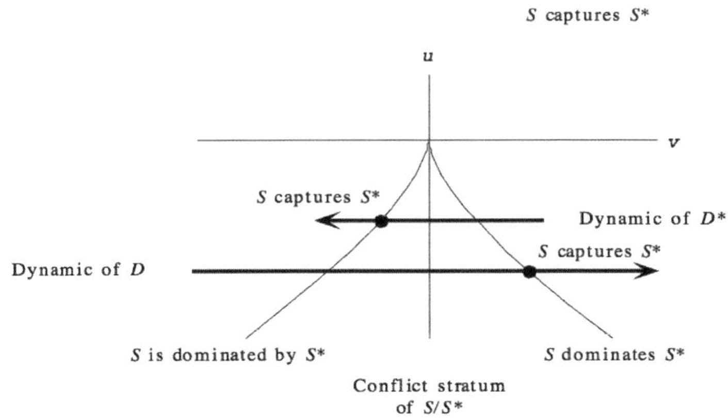

FIGURE 27. A conflict between a Subject S and an Anti-subject S^* can be controlled by Addressers and Anti-addressers.

27 shows how a conflict between a Subject S and an Anti-subject S^- can be controlled by Addressers and Anti-addressers.

Figure 28 shows how the butterfly catastrophe organizes the possible relations between two subjects S and S^* and an object O. The two symmetric bifurcation strata of "capture" $S \cup O \to S \cap O$ and $S^* \cup O \to S^* \cap O$ are linked through a *triple point* singularity with a stratum of pure conflict S/S^*. This is a model of René Girard's mimetic conflict (see e.g., Girard [124]).

8.3.4. *The dialectic of the subjective and the objective.* Its internal energy F endows the actant *Ago* E with an "interiority". But this interiority is not yet a "subjectivity". For E to become an *Ego*, it is still necessary that

(i) E can represent itself, memorize and anticipate ("subjectivize") the way in which it is situated in the dynamic landscape f_w, and that
(ii) E can react to the effects induced by the control of f_w by the external space W.

There is a subtle interplay here between the "subjective" side and the "objective" side, which Brandt proposes to explain in the following manner.

We assume that the landscape $\mathrm{Gr}(f_w)$ (Gr for "graph") defines an "objective environment" for the *Ago* E, some sort of "geography" constraining its behavior. Temporal paths in the external space induce "objective" deformations of this environment. We further assume (reflexivity hypothesis) that E can represent its own environment and "react" to it as a function of its own energy F. Thus E is not subjected, let us repeat, to passively occupying a minimum of f_w. It can evolve in $\mathrm{Gr}(f_w)$ by following an internal trajectory and it is the composition of this internal movement with the external movement that Brandt calls the dynamical route (or course) of E. The external

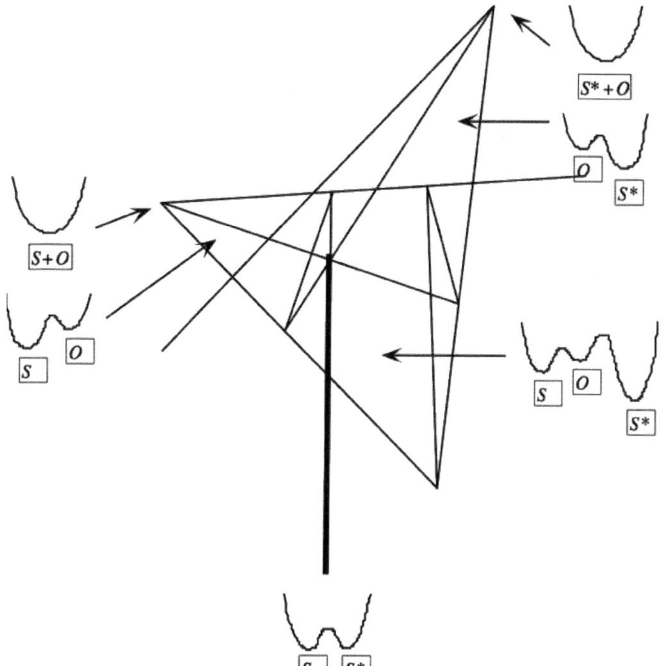

FIGURE 28. The mimetic conflict model provided by the butterfly catastrophe. It involves two subjects S and S^* and an object O. The two symmetric strata of "capture" $S \cup O \to S \cap O$ and $S^* \cup O \to S^* \cap O$ are linked through a triple point singularity.

paths deform the "objective" potentials, but the dynamical routes are, on the contrary, "subjective". They are liftings of external paths γ in W into the complete space $M \times W$ (where M is the internal space).

But how to determine the dynamical routes along which E reacts to its "objective" environment? It is here that, in an elegant and economical way, Brandt reintroduces a *semantic* concept. He supposes that we can invest the minima of f_w by *semes* X, Y, Z, etc., which act on E by attraction/repulsion. This somewhat "internal" semic polarization is expressed by the "external" polarization of the strata of K that we mentioned above. Indeed, if the place X is polarized positively relative to E, this means that the conjonction $E \cap X$ is an intentional target of E and "therefore" the modal dynamics pushes E along a path γ "attracted" by the bifurcation stratum of the minimum X. This stratum is consequently polarized positively. The dialectic between the modal and the semic dimensions thus becomes an aspect of the general dialectic between the external and the internal dimensions that is constitutive of the morphodynamical schemata.

8.3.5. *Semiotic subjectivity and modalization.* Brandt's modal theory thus appears to be a hierarchized theory of actantial "subjectivity".

(a) First, there is the basic actantial level. Actants are without "interiority". They provide only an actantial interpretation of the minima of generating potentials. The definition of their content as semantic roles is purely configurational. Involving neither modality nor intentionality, this level constitutes the interface between actantial morphodynamics and the linguistic theories of case.

(b) Then, there is the basic dynamical modal level described by Brandt, following Talmy and Sweetser. Actants remain without subjectivity but acquire a proto-interiority and a proto-intentionality. They have their own internal energy (they are *Ago* actants and not just syntactic actants) and they can follow dynamical routes driven by external modal dynamics.

(c) Then, there is the "subjective" level that may well be called *actorial.* An *actor* subject—an *Ego*—endowed with perceptual and linguistic representations possesses the capacity of

(i) representing to itself an actantial paradigm and the set of scenes which are derivable from it by temporal syntagmation, and
(ii) positioning itself, acting, and reacting according to these representations.

It positions itself by identifying itself with an actant of some schema (even with several actants in case of syncretism). It acts and reacts projectively by identifying other external actors with other actantial positions in the schema. By such imaginary identifications, it "narrativizes" its "objective" environment according to actantial structures. We must therefore distinguish between:

(i) the constitution of subject positions, and
(ii) the intersubjective plays expressing the imaginary interactions between different subject positions: when an actor subject S identifies another actor S' with another actantial position, the latter may possess its own subject position and there is obviously no reason for the position of S' to conform to what S attributes to it and vice versa.

(d) It is essentially to the modal constitution of such subjective positions that Brandt has devoted the rest of his efforts.

(i) The basic modalization is "ontic" and means that the *Ago* actant is localized in an EC (see (b)). The force F and resistance R (i.e., f_w and the internal energy of E) here characterize "the imaginary representation of the dynamism" (p. 43).
(ii) F and R are then treated as external "controls" of this ontic internal dynamism. This means that we consider deformations of f_w controlled by W. Hence, as we have seen, the occurrence of external modal dynamics, which are called "deontic". The deontic modality modalizing the ontic modality thus appears as an over-modalization.

Such a manipulation of variables in the already modalized space of E may be conceived as (deontic) over-modalization. (p. 43)

(iii) Now, E can "subjectivize" the internal dynamics f_w and the events induced by the crossings of the catastrophe set K in the external space W (p. 60). E becomes then an *Ego* in the sense of (c), endowed with an "epistemic synthesis" and a "reflexive" representation allowing a "global vision" and "a synthetic panorama" (p. 235) of the EC where it is ontically and deontically embedded. It is now able to "see himself acting" by deforming f_w (p. 60). Brandt calls "perspectives" the internal dynamics thus "subjectivized". The subject E is capable of perspectives and has the capacity for varying the latter imaginarily, that is, of "mentally" simulating its actions and, further, of epistemically evaluating them in veridictory terms. It has a "modal perception" (p. 236).

(iv) We must then take into account the co-presence (eventually conflictual) between, on the one hand, an "internalized" external dynamics that corresponds to the wanting-to-be of E, its expectatives and anticipations, and, on the other hand, deontic modal external dynamics driven by the Addressers. It is in this sense that the fundamental modality is indistinguishably being-able-to and having-to: when the subject is submitted to the reasons of the Addresser and contracts with it an external modal dynamics, this is because it believes this perspective to be "true".

The epistemic modality anchors the deontic in truth.

Through the contract,

the doing is seen to be anchored in the being. (p. 61)

(e) At the level of intersubjectivity, the basic actantial graphs (a), become considerably enriched (p. 240). Indeed, the actants are now the *Ego* actors, each endowed with "perspectives", "modal perception", Addressers, modal dynamics, dynamic routes and semanticization. Their interactions can be very complex. They raise a fascinating theoretical problem. In an intersubjective interaction, each actant A_i possesses an "internal simulation" which is an EC M_i. The graph of interaction serves as a framework of interaction for the M_i. *But what is an interaction of EC*? Certainly, in an interaction between M_i and M_j some actants of M_i will be identified with some actants of M_j, but obviously the interaction cannot be reduced to this trivial procedure. In fact, the *couplings* between EC are dynamical processes of great subtlety which result in a considerable growth of the geometrical complexity of the initial ECs. For example, the interaction of two cusps results in the non-elementary catastrophe called the *double cusp*, catastrophe of codimension 8 of extreme complexity.[28] Thus it is for intrinsic syntactic reasons that the plays of intersubjectivity cannot be easily formally mastered.

---

28 See Section 3.15 of Chapter 5: we used the double cusp for schematizing the canonical formula of the myth proposed by Lévi-Strauss.

By founding the actantial theory on a modal basis inside the framework of the morphodynamical schematism so as

> to explain narrativity in general as a semiotic universal characterizing the human imaginary and preceding all semanticization (p. 226),

by developing the central thesis of the dynamical nature of the modal "constitutive secret of meaning" (p. 278), by locating meaning

> between natural causality and cultural conventionality (p. 280),

and thus by showing that modality could indeed embody the subjectivity in the organic body and represent the

> missing link between biology and semiotics (p. 278),

Per Aage Brandt has remarkably broadened the research program of a "physics of meaning".

CHAPTER 6

Attractor Syntax and Perceptual Constituency

1. "From pixels to predicates": the seven pillars of cognition

In this last chapter we want to unify singularity theory with attractor syntax and draw a link between three things:

(i) the perceptual morphological models developed in Chapter 3;
(ii) the attractor syntax in the sense of Chapters 4 and 5;
(iii) the concept of actantiality in structural syntax, case grammars, and cognitive grammars.

The problem is the following. If we want to complete the research program eloquently qualified by Pentland [255] by the slogan "From pixels to predicates", we have to articulate at least seven different levels of representation. Moving up from perception to language we meet (at least) four levels:

P1. perception first provides static visual scenes (3D-models: objects with relations);

P2. time provides a temporal evolution of these configurations;

P3. their schematization provides image-schemata;

P4. their further categorization (in the sense of a morphological "algebraic topology") provides actantial graphs.

Moving down from language to perception, we meet (at least) three levels:

L1. linguistic surface structures;

L2. deep formal predicative structures;

L3. AI symbolic structures such as frames or scripts.

We will first show how it is possible to define an equivalence—a reciprocal coding, a translation, a reformatting, a *representational redescription* "iconic $\leftrightarrow$ symbolic"—between **P4** and **L3**.[1] We will then address the other question of this chapter, namely the nature of the link between **P3** and **P4**, i.e., between image-schemata and attractor syntax.

2. Apparent motion and the perception of intentionality

To begin with, let us consider experimental data concerning the remarkable faculty of *perceiving* intentionality.

1 The importance of reformatting or redescription has been emphasized by specialists of cognitive development such as Annette Karmiloff-Smith [175].

The spontaneous linguistic interpretation of spatial, inanimate object motion (even mere dots) in terms of verbs of movement reveals *a direct perception of intentional actions.* More precisely, perceptual inferences lead to spontaneously ascribing to objects the *semantic roles* of animate and intentional subjects. This constitutes one of the great discoveries in cognitive psychology during the second half of the 20th century.

The first careful experiments in this area were conducted in 1944—following Michotte's [234] famous work on the direct perception of reality—by F. Heider and M. Simmel in their landmark article "An experimental study of apparent behavior" (Heider-Simmel [148]). They studied how the movement (including speed, acceleration, deceleration, change of direction, etc.) of simple shapes (e.g., two triangles, one circle and one bigger rectangle) were spontaneously interpreted as finalized actions that are *intentionally caused* by the objects. They examined how those actions were verbalized by verbal lexemes such as 'enter', 'exit', 'hide', 'escape', 'run away', 'chase', 'attack', 'give', 'let go', 'force inside', etc. (which showed that high-level semantic features such as factitives can also be detected). A purely kinematical movement can thus be described by an elaborate verbal proposition such as "J walks out of the house and goes to B and R, fights with B about R; then R hides in the house; J comes in, closes the door and assaults him".

Many works have been dedicated to these problems, e.g., J. N. Bassili showing that actantial descriptions are induced by correlations among movements; S. Weir [400] on the perception of motion; A. M. Leslie on the perception of causality in children; J. Scholl and P. D. Tremoulet [336] on the perception of causality and animacy of objects; D. Premack (expert in the language of primates) on the theory of self-propelled objects in children (self-propelling is able to trigger a discrimination between agent and patient), R. Gelman, F. Durgin and L. Kaufman on the animate/inanimate opposition; S. J. Blakemore and J. Decety [37] on the perception of action and the understanding of intention. They present fine experimental analyses concerning the interpretation of the topological-dynamical (i.e., morphological) information contained in visual scenes by a spontaneous application of intentional verbs of action and interaction. Maria Elisabetta Zibetti's PhD thesis, *Contextual categorization and understanding of events visually perceived and interpreted as actions* [420], offers a thorough review of these phenomena (see also [421]).

The "agentive" interpretation of purely kinematical configurations—i.e., the spontaneous attribution of intentionality to apparent motion—constitutes a broad generalization of Michotte's perception of causality. As was emphasized by Johan Wagemans et al. ([399], p. 3):

> Michotte became convinced that we can perceive actions performed by objects or animate beings ("agents") on one another in the same way as we can see simple kinetic movements.

The main point is that, contrary to common wisdom, this perception is low-level, automatic, modular, and hardwired, "rooted in automatic visual processing" (ibid., p. 12). It is non-conceptual and seems to originate precisely from a *categorization* of spatio-temporal interaction scenarios corresponding to the generic interactions described by image-schemata, frames or scripts. Such a categorization is of *second order* relatively to shape categorization. We studied this question in several works because it is at the core of the analysis of relations between perception and language and shows that an important part of verbal semantics is grounded in perceptual Gestalts.

It is no longer the place here to delve into the cognitive problems that these categorizations raise, but they are significant nevertheless. The perceived scenes can vary continuously, while categorization is on the contrary discrete. We point again to the reference work of M. E. Zibetti [420]. First, one must segment the temporal flow of objects into action segments, then extract relational universals from the spatio-temporal interactions, then apply interpretative inferences in order to select lexemes whose verbal semantics is compatible with the topological-dynamical morphology of the perceived scene.

3. From actantial graphs to cognitive archetypes

Between the descriptive and predicative levels of structure, there are *intermediate* levels, such as scripts, frames, mental schemata, mental imagery in the sense of Kosslyn, and so on.

To show the equivalence **P4** $\leftrightarrow$ **L3**, we use Desclés' theory of *cognitive archetypes* [86].

3.1. Cognitive archetypes

Cognitive archetypes (CAs) are data structures analog to those proposed by linguists and AI theorists such as Fillmore, Schank, Minsky or Winograd. They are *intermediate* structures between image-schemata and symbolic predicative structures. As elements of knowledge representation, they are symbolic *non-predicative* structures that schematize generic and stereotypical situations, allowing natural or artificial minds to make anticipations and inferences on specific, real situations. They are systems of slots and relations with default settings that are filled with specific fillers or possibly other slots.

Let us take a most elementary example, a verbal image-schema such as [ENTER] that expresses a temporal transformation of spatial relations. How can this *topological and dynamical* information be converted into a *symbolic predicative* information? The archetype is the following (see Figure 1):

(i) SIT_1 and SIT_2 are stative situations (initial and final states).

(ii) $SIT_1[y]$ is described by the following symbolic descriptor of positional relations: $y \in_0 \mathbf{ex}(Loc)$, where $\in_0$ is a localization operator of an object y relatively to a locus Loc. The relation $y \in_0 \mathbf{ex}(Loc)$ is equivalent

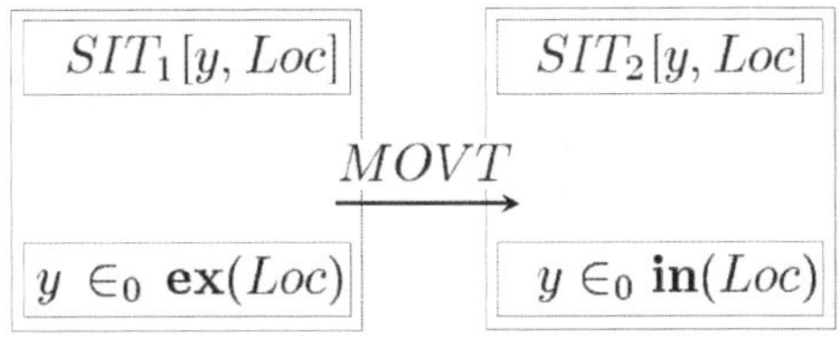

FIGURE 1. The cognitive archetype for a semantic of [ENTER] type. *Loc* is the location where the object y enters. The process evolves from an "out" relation to an "in" relation.

to Langacker's *association* relation $[y$ ASSOC $Loc]$ in Section 3.2.3 of Chapter 1.

(iii) $SIT_2[y]$ is described by $y \in_0 \mathbf{in}(Loc)$.

(iv) $MOVT$ is an operator of movement that modifies the stative states.

Now, using combinatorial logic and applicative grammar, Desclés showed how such a cognitive archetype can be automatically converted into a predicative structure. He first used the following symbolic expression for the cognitive archetype:

$$ENTER \equiv MOVT(\in_0 (\mathbf{ex}Loc)y)(\in_0 (\mathbf{in}Loc)y)$$

Then, he associated it with a *predicative* structure of the form $E(Loc, y)$ ("y enters in *Loc*") where $E(\bullet, \bullet)$ is a binary predicate. To do that he posited:

$$E = \Psi(B\Phi\Phi MOVT)(B \in {}_0)\mathbf{ex}\ \mathbf{in}$$

where Ψ, B and Φ are the combinators:

- $\Psi XYZU \to X(YZ)(YU)$,
- $BXYZ \to X(YZ)$ (composition),
- $\Phi(XYZ)U \to X(YU)(ZU)$ (intrication).[2]

It is easy to verify that, starting from the definition $E(Loc, y)$ and applying these rules sequentially,[3] we arrive at the symbolic expression of the cognitive archetype [ENTER]. The derivation is interesting because it shows that the semantic meaning of an item like [ENTER] is *twofold*. First, it contains the *local* (positional and dynamical) content expressed by the primitives $\in_0$, **ex-in**, $MOVT$. Second, it also contains the *formal* content expressed by the combinatorial operations on predicates. As was stressed by J.P. Desclés, the lexical law

$$E(Loc, y) \equiv MOVT(\in_0 (\mathbf{ex}Loc)y)(\in_0 (\mathbf{in}Loc)y)$$

2 X, Y, Z, U are any combinatorial expressions.

3 We first apply Ψ to $X = B\Phi\Phi MOVT, Y = B \in {}_0, Z = \mathbf{ex}, U = \mathbf{in}$. We get $B\Phi\Phi MOVT(B \in {}_0\mathbf{ex})(B \in {}_0\mathbf{in})$. Then, we apply $B\Phi\Phi MOVT = \Phi(\Phi MOVT)$ and Φ with $U = (Loc)(y)$, where *Loc* and y are the two terms filling the two slots of the archetype.

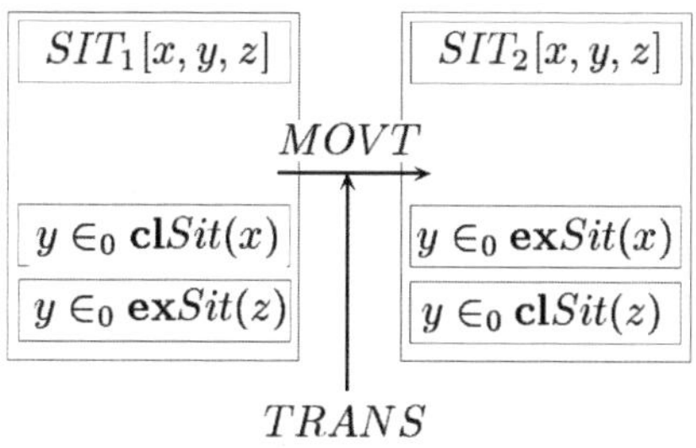

FIGURE 2. The [GIVE] cognitive archetype according to Desclés [86].

is

> a "compilation" of the linguistic expression, encoded with the grammatical constraints of language, in a system of semantic representations organized by means of cognitive archetypes. (Desclés [86], p. 307)

Consider another example, such as [GIVE] (see Figure 2).

(i) SIT_1 and SIT_2 are stative situations (initial and final states).
(ii) $SIT_1(x, y, z)$ is described by the two following symbolic descriptors of positional relations:
- $y \in_0 \mathbf{cl}Sit(x)$ where $Sit(x)$ is the locus of x and **cl** the topological closure;
- $y \notin_0 \mathbf{cl}Sit(z)$ i.e. $y \in_0 \mathbf{ex}Sit(z)$.

(iii) $SIT_2(x, y, z)$ is described by:
- $y \notin_0 \mathbf{cl}Sit(x)$ i.e. $y \in_0 \mathbf{ex}Sit(x)$,
- $y \in_0 \mathbf{cl}Sit(z)$.

(iv) $MOVT$ is an operator of movement which modifies the stative states.
(v) $TRANS$ is an operator which expresses the fact that x is an "agent" who *controls* the movement ($TRANS \equiv DO + CONTROL$).

3.2. Reformatting actantial graphs

It is easy to show that the information encoded in such data structures, which can be converted in predicative structures, is essentially the same as the information encoded in *actantial graphs* (AGs). Figure 3 shows the case of [ENTER] and Figure 4 the case of [GIVE].

We therefore get a *reformatting* (a redescription) of the information CA $\leftrightarrow$ AG. So if we return to the tower of levels

$$\mathbf{P1} \rightarrow \mathbf{P2} \rightarrow \mathbf{P3} \rightarrow \mathbf{P4} \leftrightarrow \mathbf{L3} \leftarrow \mathbf{L2} \leftarrow \mathbf{L1}$$

we see that we have already modeled the **P1** $\rightarrow$ **P2** (from objects to relations) and **P2** $\rightarrow$ **P3** (from relations to image-schemata) shifts, that we get an equivalence iconic $\leftrightarrow$ symbolic for **P4** $\leftrightarrow$ **L3**, and that the linguistic part

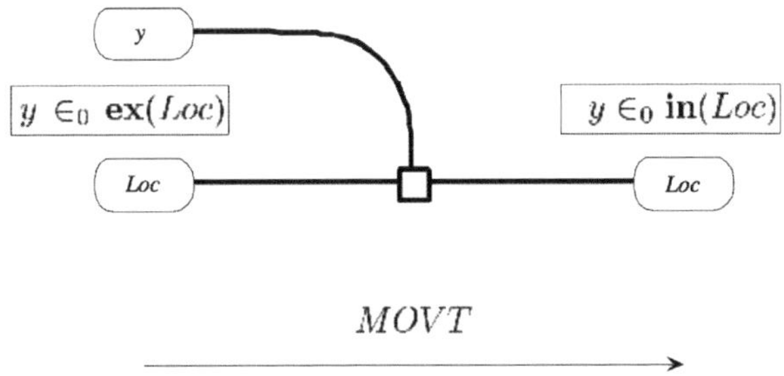

FIGURE 3. The equivalence of the symbolic information of the cognitive archetype [ENTER] with the actantial graph of "capture".

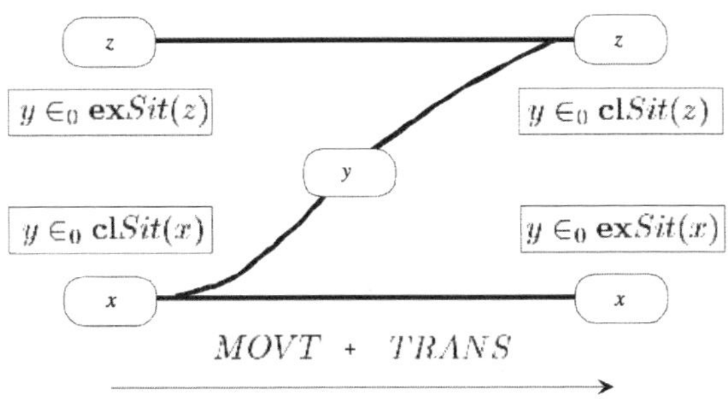

FIGURE 4. The equivalence of the symbolic information of the cognitive archetype [GIVE] with the actantial graph of "transfer".

L3 ←**L2** ←**L1** is also already modeled. The main remaining point is therefore to model the **P3** →**P4** shift from syntactic image-schemata to attractor syntax. The last challenge we have to tackle is to go from the morphological analysis of relations given in Chapter 3 to the attractor syntax of Chapters 4 and 5.

4. Contour diffusion and singular encoding of relations

We reduce the problem to the simpler case of 2D objects, identify positional actants with topological domains in 2D ambiant space E, and consider *configurations* $\mathcal{A} = \{A_1, ..., A_n\}$ of such domains. These configurations can evolve in time. The problem is to scan their relational and temporal profiles in view of an actantial syntax.

4.1. The general strategy for solving the main problem

We use contour diffusion/propagation as general cognitive algorithms. Indeed, these are algorithms which perform the transition from the local to the global level starting from initial conditions provided by the scanning of qualitative discontinuities (boundaries).

In general, these algorithms are used in computational vision according to a coarse-to-fine strategy. But to extract the positional information contained in syntactic image-schemata we shall use them according to a *fine-to-coarse* strategy.

In a nutshell, the strategy is the following.

(i) to assume the general cognitive validity of these algorithms;
(ii) to apply them to gestaltic configurations of locations similar to Langacker's syntactic schemata;
(iii) to use them according to a fine-to-coarse strategy.

The problem is now to know if such local algorithms are able to extract the gestaltic information that we want.

Let $\mathcal{A} = \{A_1, ..., A_n\}$ be a configuration of objects. Looking at the common background of the A_i, we treat it *as a pattern* and apply a morphological analysis. According to the strategy already defined in Section 2.2 of Chapter 3, we analyze not the objects themselves (they are morphologically trivial) but their *complementary set* $W - \{A_i\}$. There are two possibilities:

1. using a propagation routine and looking at the interfaces between the influence zones of the A_i. It is the SKIZ routine presented in Chapter 3;
2. using a diffusion routine and following the level curves of the contour diffusion of the A_i's boundaries $B_i = \partial A_i$.

Next, we explain this second routine.

4.2. Contour diffusion, cobordism, and Morse theorem

Using *Morse theory* we can characterize the static gestaltic configurations of locations by *local* and *informationally finite* necessary and sufficient conditions. The idea is the following:

(i) starting from the boundaries of the locations (objects, terms): these initial boundaries are composed of as many connected components (topological circles) as there are objects,
(ii) triggering a contour difusion, and
(iii) following the diffusion until the boundaries become trivial (i.e., topologically a circle).

As the initial contour (which has many connected components) and the final contour (which has only one connected component) are not of the same topological type, the contour diffusion must cross *critical points* where the topological type of the contour changes catastrophically. Morse theory proves that the spatial distribution and the temporal order of these critical points characterize the configuration.

The *singularities* are local and finitely characterizable entities that can be generically detected and addressed using point processors. According to

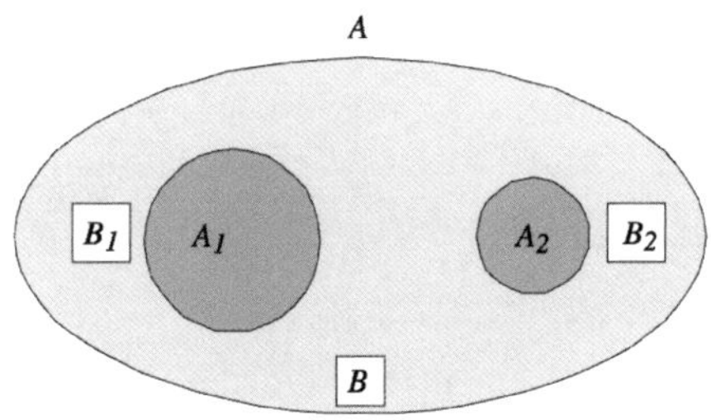

FIGURE 5. The relation of *association*: two domains A_1 and A_2 with respective boundaries B_1 and B_2 are included in a superordinate domain A with boundary B.

Morse theorem *they locally encode global configurations* and provide local and finite necessary and sufficient conditions. They therefore allow us to escape the "global Gestalt/local computation" dilemma.

4.3. The example of the Association relation

Let us consider for instance the simplest, but also the most important, case of the Langackerian basic relations, namely the relation of *association* already presented in Section 3.2.3 of Chapter 1. We start with two domains A_1 and A_2 (with respective boundaries B_1 and B_2) included in a superordinate domain A (with boundary B) (see Figure 5).

Thus, initially, the distribution of activity $I(x, y)$ is equal to 1 inside A_1 and A_2 and to 0 outside. We trigger a diffusion process I_s, the value of I_s being clamped to 1 inside the A_i (the A_i are identified with constant sources). We consider then the diffusion fronts B^s. There are many ways to define them. For instance, we can consider the level curves L_z of I_s for $z = h_s$, where h_s is some indicator of the figure/ground (profile/base) separation. We can also take the curves where the gradient ∇I_s of I_s has maximal norm $\|\nabla I_s\|$. We can further modify the diffusion equation. During the diffusion, the virtual contours B^s *propagate*. For some value of s (which we can normalize to $s = 1$), B^s can play the role of the outer contour B.

The diffusion process allows to construct a *continuous deformation* between the initial boundary $B^0 = B_1 + B_2$ and the final boundary $B^1 = B$—what is called in differential topology a *cobordism*.[4] As the initial and final contours B^0 and B^1 are not of the same topological type, there must exist a *critical value* c of s for which the diffusion front B^c is critical. B^c makes the transition

4 Let N and N' be two smooth, compact, connected, orientable n-manifolds. A cobordism between N and N' is a smooth compact, oriented $(n+1)$-manifold with boundary M such that (i) the oriented boundary ∂M of M is the algebraic sum $\partial M = N - N'$ of N and N', (ii) the inclusions $N \hookrightarrow M$ and $N' \hookrightarrow M$ are homotopy equivalences.

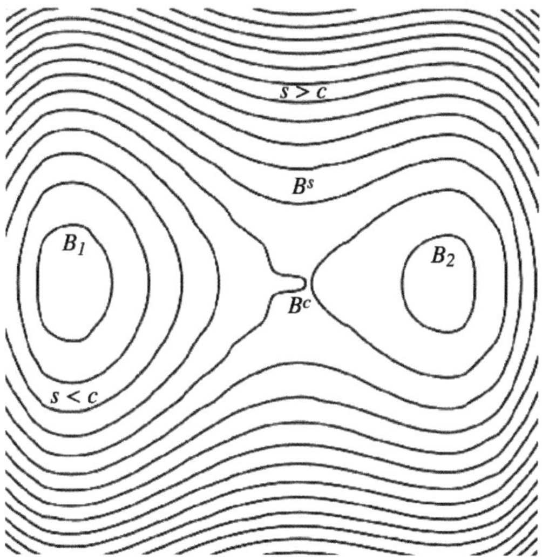

FIGURE 6. A diffusion process induces a cobordism B^s between an initial boundary $B_1 + B_2$ and a final boundary B. There exists a critical value c of s for which B^c makes the transition between the fronts B^s for $s < c$ and the fronts B^s for $s > c$, which are not of the same topological type.

between the fronts B^s for $s < c$, which have two components and the fronts B^s for $s > c$, which have only one component. It presents a saddle-type singularity (see Section 6).

But in Morse theory one can prove the following result.

Theorem. *A configuration $\mathcal{A}$ is an association relation iff the contour diffusion process presents only one singularity of saddle type.*

We have therefore succeeded in solving the main problem in this elementary but fundamental case. In fact, we can generalize the solution to the following construct.

4.4. Generating potentials

So far, we have considered only the external (outward) contour diffusion. But from the initial boundaries B_1 and B_2 we can also trigger an internal (inward) contour diffusion.[5] The critical points of this secondary diffusion are the centers of the initial blobs. If we suppose that these inward propagating contours B^s correspond to decreasing s, we construct in this way a *potential function* $f_{\mathcal{A}} : \mathbb{R}^2 \to \mathbb{R}$. The B^s are its level-curves. The initial boundary $B_1 + B_2 = B^0$

5 "External" and "internal" are used here in their naive, not dynamical, sense.

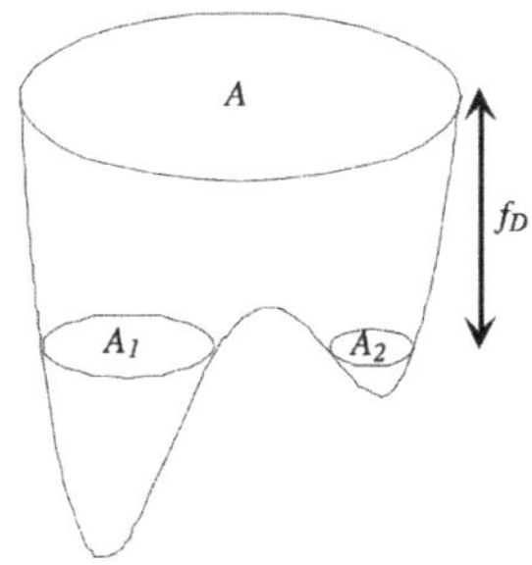

FIGURE 7. The generating potential $f_{\mathcal{A}}$ of a configuration $\mathcal{A} = \{A, A_1, A_2\}$. f_D corresponds to the outward diffusion.

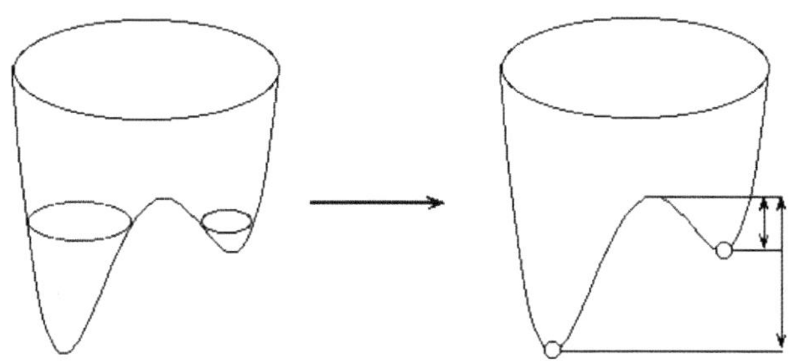

FIGURE 8. Dominance relations between proto-actants.

corresponds to a particular level-curve $f_{\mathcal{A}} = a_0$ and the outward (resp. inward) propagating contours B^s to level-curves $f_{\mathcal{A}} = a$ with $a > a_0$ (resp. $a < a_0$).[6] The graph of $f_{\mathcal{A}}$ has the qualitative global shape of a potential pit A with A_1 and A_2 as subpits. We call $f_{\mathcal{A}}$ the *generating potential* of the configuration $\mathcal{A} = \{A, A_1, A_2\}$. The respective size of regions A_i, i.e., their dominance relations, can be encoded in the *depth* of the associated minima: the deepest minimum is associated with the largest region (see Figure 7). The positional relations between proto-actants can therefore be schematized by the dominance relations between associated pits of potential (see Figure 8).

One can also *reduce* the dimension of the M space, on which the generating potential f is defined, to a *minimal* value by using the concept of codimension of a singularity (see Section 3 of Chapter 5). In the example of the association relation, the codimension is equal to 1, and potentials can therefore be reduced to 1D spaces (see Figure 9).

6 We can normalize by taking $a = s$ and $a_0 = c$.

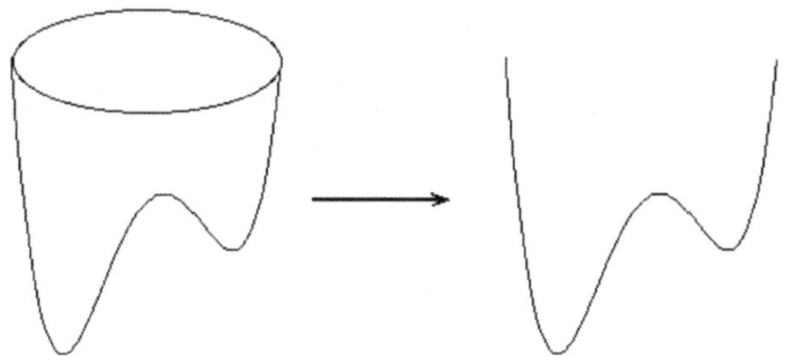

FIGURE 9. Dimensional reduction of the internal space of generating potentials.

Thus, through the contour diffusion routine, a cognitive-grammar schematization becomes equivalent to a morphodynamical schematization using generating potentials. In other words, the contour diffusion routine applied to a configuration $\mathcal{A}$ provides a generating potential $f_{\mathcal{A}}$ for $\mathcal{A}$.

4.5. Processes and potential deformations

In this new representation, the contour diffusion processes constitute fast internal dynamics and their temporal evolution, slow external dynamics. Processes become temporal deformations f_t of generating potentials, and events of actantial interaction become bifurcations of generating potentials. The events of interaction are therefore essentially events of *fusion* or *splitting* of locations.

We thus come to our main claim: *a morphodynamical syntax can be elaborated using bifurcation theory.* In Figure 10 we display the dynamic of a "capture" event as a stack of stages. In Figure 11 we display the dynamic of a "transfer" process.

4.6. Morse theory

In this appendix, we give a few elementary details concerning Morse theory.

Morse theory uses the theorems concerning Morse functions (Section 3.12 of Chapter 5) for analyzing the topology of manifolds. Let $f : M \to \mathbb{R}$ be an excellent Morse function defined on a compact smooth n-manifold M. Since M is compact, its image $f(M)$ is also compact. $f(M)$ is therefore a bounded closed set of $\mathbb{R}$. This implies that f has an absolute minimum $\underline{m}$ (with critical value $f(\underline{m}) = \underline{e}$) and an absolute maximum $\overline{m}$ (with critical value $f(\overline{m}) = \overline{e}$). Since f is an excellent Morse function:

(i) $\underline{m}$ and $\overline{m}$ are non-degenerate isolated critical points,
(ii) their critical values $\underline{e}$ and $\overline{e}$ are distinct, and
(iii) $f^{-1}(\underline{e}) = \{\underline{m}\}$ and $f^{-1}(\overline{e}) = \{\overline{m}\}$.

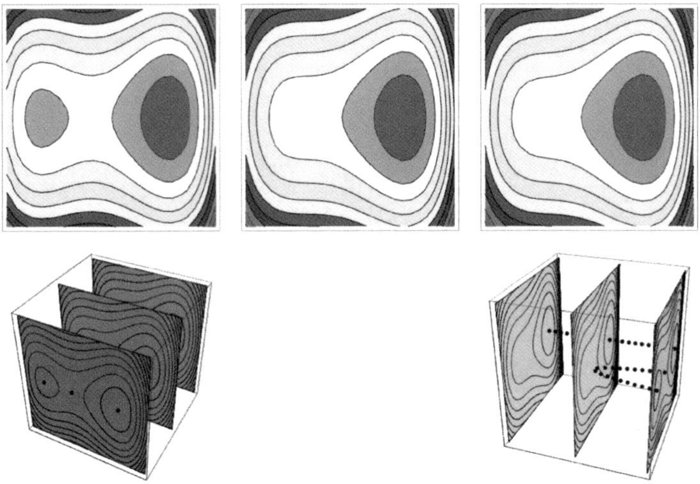

FIGURE 10. The dynamic of a "capture" event. Top: the temporal evolution of the generating potentials. Bottom-left: a stack representation. Bottom-right: the trajectories of the three critical points; the merging of the saddle point in which one of the minima corresponds to the event of capture.

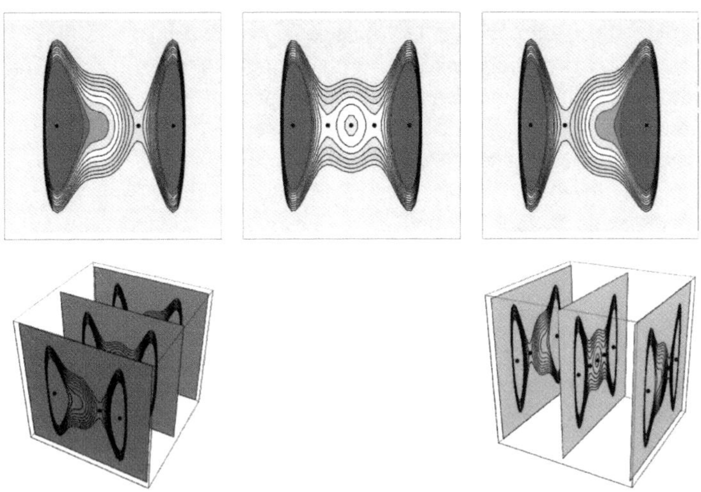

FIGURE 11. The dynamic of a "transfer" process with the temporal evolution of the five critical points. For simplicity, we did not represent their complete trajectory.

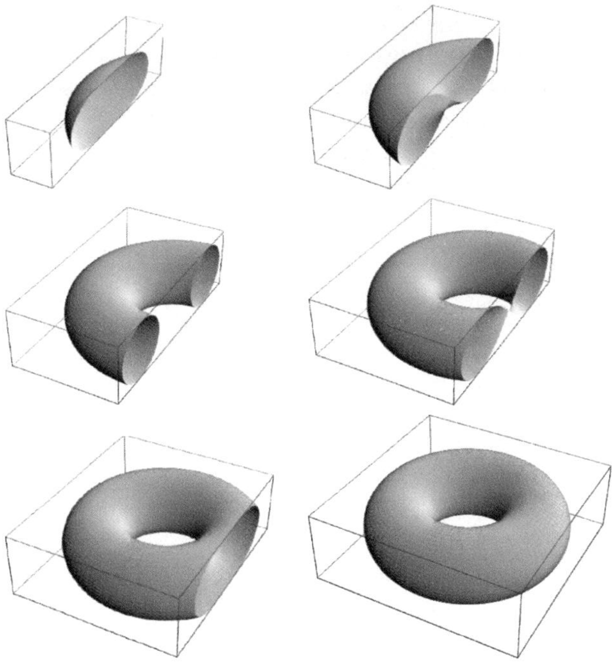

FIGURE 12. Reconstruction of a torus from a Morse function.

Let c be an intermediary value $c \in]\underline{e}, \overline{e}[$. One can prove that if the inverse image $M_c = f^{-1}(c)$ is regular, that is, contains no critical point, then:

(i) M_c is a smooth submanifold of M of codimension 1 (hence of dimension $n-1$),
(ii) f is locally trivial in the neighborhood of M_c, that is, there exists an open neighborhood U of c such that the open strip $M_U = f^{-1}(U)$ is diffeomorphic with the direct product $M_c \times U$.

According to this result, the slices M_c of M can change their topological type only when c crosses a critical value of f. When M is a surface (case $n = 2$), the regular level curves M_c are finite sets of smooth topological circles. Consequently only three types of topological accidents can occur when c increases from $-\infty$ to $+\infty$:

(i) the emergence of a new circle: c crosses the critical value of a minimum of f (critical point of index 0);
(ii) the vanishing of a circle: c crosses the critical value of a maximum of f (critical point of index 2);
(iii) the fusion of two circles or the splitting of one circle: c crosses the critical value of a saddle (critical point of index 1) (see Figure 12).

The fundamental result is that this decomposition of M by means of its critical points—what is called a *handle presentation* of M—is in some sense reversible: the topological structure of M can be retrieved from the critical structure of any of its Morse functions.

Theorem. *Let M be a compact surface. Then M is topologically determined by any of its Morse functions.*[7]

4.7. Representing positional information

We can see that, for the [ENTER] archetype, there is an equivalence between four modes of representation of positional information:

1. the *profiling* of the verb [ENTER] in the sense of cognitive grammar;
2. the *contour diffusion* process and its temporal deformation;
3. the temporal path that parametrizes the deformation of the generating potential f_t and generates the actantial graph transforming the disjunction $A_1 \cup A_2$ into the conjunction $A_1 \cap A_2$;
4. the *cognitive archetype* [ENTER].

However, although these four representations are equivalent, they do not belong to the same *mode* of representation.

1. The profiling provides an intermediate (figurative) schematic representation that can neither be directly derived from elementary cognitive routines (scanning, propagation), nor inserted as such into a symbolic predicative computation.
2. The deformation by contour diffusion provides a dynamical representation endowed with psychological content, which is grounded on elementary cognitive routines (scanning, propagation) and compatible with the perceptual anchoring of natural language.
3. The deformation of the generating potential and the associated actantial graph provide a morphodynamical representation that inserts contour diffusion into mathematical theories of qualitative dynamics, in particular into dynamical systems theory, including attractors and bifurcations, and algebraic topological theories.
4. Finally, the cognitive archetype provides a symbolic representation that can be directly inserted into a formal predicative calculus.

---

7 It is in using a strong generalization of these elementary results to higher dimensions $n \geq 5$ that Stephen Smale was able to demonstrate in 1962 his celebrated h-cobordism theorem. Consider a cobordism M between N and N'. Let $n \geq 5$ and N and N' be simply connected. Then M is trivial, that is, diffeomorphic to $N \times [0, 1]$. It is the same to say that every Morse function on M is trivial, all its critical points being removable using a deformation. The h-cobordism theorem implies Poincaré's conjecture for the dimension $n \geq 6$: if an n-manifold M is homotopically equivalent to the n-sphere S^n, then it is homeomorphic to S^n (it is for this extraordinary result that Smale won the Fields medal).

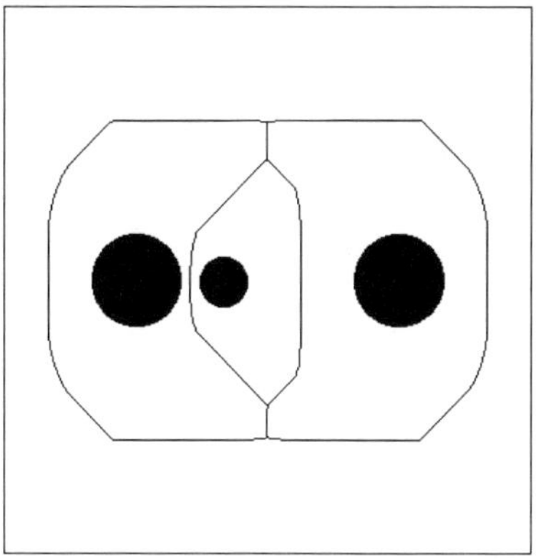

FIGURE 13. The external cut locus (SKIZ) of a configuration of three objects in a bounded window. The most external component reflects the boundary.

5. Contour propagation and the cut locus theory

There is another way to use the figure/ground complementarity and look at the background of a configuration as the spatial medium unifying the objects and expressing via its form the relations between them.

Instead of a contour diffusion routine, we can consider the *cut locus* (CL) of the background. This means that we use a contour propagation routine and look at its singular locus. As the objects are trivial, the CL is the same thing as their SKIZ (see Section 5.4.2 of Chapter 3). The singularities of the SKIZ, essentially its triple points, encode the geometrical information of the configuration (see Figure 13).

If we then consider a temporally evolving configuration leading to interactions between actants, we have to consider a temporally evolving background cut locus. The dynamics generating the cut locus is a fast dynamics, but the temporal evolution of the cut locus itself is a slow dynamics, according to which the background's cut locus evolves and may present bifurcations: emergence and vanishing of branches, or splitting of branches. These bifurcations encode events of interaction between actants. We can in this way develop a program analogous to contour diffusion. Figure 14 gives an example for the "transfer" type.

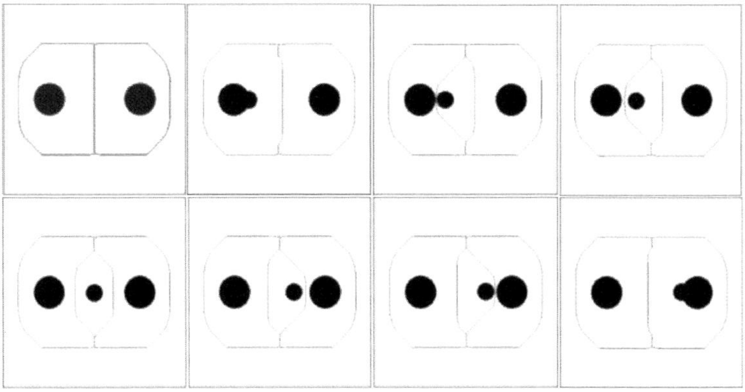

FIGURE 14. The temporal evolution of the SKIZ during a scenario of "transfer" type.

Conclusion

In his paper *A Suggestion for a Linguistics with Connectionist Foundation* [199], George Lakoff generalized Regier's idea (see Section 5.3 of Chapter 3) to the conjecture that

> Ullman-style visual routines (...) are sufficient to characterize all known structures in cognitive topology.

We think that we have demonstrated Lakoff-Regier's conjecture for actantial relations and interactions.

Starting from the Smolensky/Fodor-Pylyshyn debate concerning connectionist modeling of constituency, we first stressed that the main problem was to achieve a connectionist configurational definition of semantic (actantial) roles, that is, according to Jerry Fodor and Zenon Pylyshyn, of that "geometrical whole, where the geometrical relations are themselves semantically significant", and which constitutes the geometrical basis of constituent-structures.

We have then emphasized that, in order to solve this non-trivial problem, we first need a "good" linguistic theory. We selected cognitive grammars in Langacker's, Talmy's, Jackendoff's and Lakoff's sense and we stressed the central role of the localist hypothesis. Using this general perceptual, iconic and schematic grounding of basic elementary syntactic structures, we reduced the main problem to "perceptual" constituency.

We then introduced contour diffusion/propagation routines that generalize, to higher-order representational levels, well-known routines of computational vision. We treated two cases of spreading activation triggered by boundaries: contour diffusion (heat equation) and contour propagation (wave equation). In these two cases we showed, according to deep theorems such as Morse's theorem, that the singularities of the diffusion/propagation processes are singular structures that can be locally and finitely processed, and encode in a local and finite way the global holistic structure of the input configurations. This was the first key idea: constituency is retrievable from the detection and identification of *singularities*.

The contour diffusion/propagation routines solve the "global Gestalt/local computation" dilemma. They allow the scanning of profiled positional relations. With such a result, we can then explain how to scan actantial processes and interaction schemata. It is therefore possible to elaborate a connectionist theory of a configurational conception of semantic (actantial) roles.

The second key idea was that a syntactic interaction between actants that are modeled by attractors of some underlying dynamics can be modeled by a bifurcation of these attractors.

Thanks to such effective models, we can easily construct actantial graphs and therefore combinatorial structures which share the combinatorial properties and the systematicity requirements characteristic of symbolic structures. We can also, in a second step, implement external control dynamics to achieve a causal theory of agentivity and modality in Brandt's sense.

In conclusion, we have shown that this kind of purely morphodynamical response to the main challenge makes computationally effective the pioneering topological and dynamical syntax created by René Thom in the late 1960's.

In a nutshell, we have shown that by adding to higher cognitive levels of computational vision new modules performing contour diffusion/propagation and singularity extraction, it was possible to build a bottom-up and data-driven theory of *constituency.* Relations are encoded in virtual singular structures that underlie image-schemata, which are themselves linguistically grammaticized into symbolic predicative structures.

It will perhaps seem difficult to accept such a cognitive relevance of virtual singular structures. But since Gestalt theory, it has become a widely confirmed experimental fact that virtual boundaries are essential for perceptual structuring. Moreover, as we have seen with Talmy, a detailed analysis of the relationships between language and perception shows that many virtual structures are linguistically encoded and act as "*organizing Gestalts*".

In summary, we have shown that it is possible to work out a dynamical conception of constituent-structures using virtual constructs that share the properties of a formal "syntacticity". These geometrical constructs do possess an internal structure. Moreover their generating physical mechanisms are "structure-sensitive".

An attractor syntax "works". It relies on a "morphodynamical" functionalism that shares the characteristic properties of classical functionalism.

The fundamental difference between the classical symbolic paradigm and the morphodynamical paradigm is to be found in their conception of instantiation and implementation. To do syntax we have generalized bottom-up, data-driven and self-organizing perceptual algorithms of profiling (e.g., contour detection) and categorization. The crucial epistemological point is the following: mathematically, physical models are in general of a geometric-dynamical nature. Every physics is a geometrodynamics. Therefore, if we are able to extract syntactic structures by abstracting invariants from such a geometrodynamics we become able to understand the link between an ideal formal "syntacticity" and the underlying (neuro)physics. It is in that sense that geometry and dynamics are key to formal syntax.

We have applied this strategy here. All the formal tools we have used (contour detection, wavelets, diffusion, propagation, cobordism, singularities, Morse-Whitney-Thom-Smale-Mather's theories, algebraic topology, etc.) belong to the geometrodynamics which governs the physics of neural networks. They can therefore explain how symbolic constituent-structures can emerge from their (neuro)physical underlying processes, and how "syntax is to perception what algebraic topology is to differentiable manifolds".

Bibliography

[1] AAAI (1995). *Fall Symposium on Computational Models for Integrating Language and Vision.* Cambridge, MA: American Association for Artificial Intelligence, 1995.

[2] Abeles, M. (1982). *Local Cortical Circuits.* Berlin: Springer-Verlag, 1982.

[3] Abeles, M. (1991). *Corticonics.* Cambridge, UK: Cambridge University Press, 1991.

[4] Abeles, M., Hayon, G., Lehmann, D. (2004). "Modeling compositionality by dynamic binding of synfire chains." *Journal of Computational Neuroscience*, 17 (2004): 179–201.

[5] Adamatzky, A., ed. (2002). *Collision-Based Computing.* Berlin: Springer-Verlag, 2002.

[6] Afraimovich, V. S., Snor, S. N., Hale, J. K. (1997). "Synchronization in lattices of coupled oscillators." *Physica D*, 103 (1997): 442–451.

[7] Albertazzi, L., ed. (1999). *Shapes of Forms.* Dordrecht: Kluwer, 1999.

[8] Alvarez, L., Lions, P. L., Morel, J.M. (1992). "Image selective smoothing and edge detection by non-linear diffusion." *SIAM J. Numer. Anal.*, 29 (1992): 845–866.

[9] Ambrosio, L. (1995). "A new proof of the SBV compactness theorem." *Cal. Var. Partial Differential Equations*, 3 (1995): 127–137.

[10] Ambrosio, L., Tortorelli, V. (1990). "Approximation of Functionals Depending on Jump by Elliptic Functionals via Γ-convergence." *Communications on Pure and Applied Mathematics*, 43 (1990): 999–1036.

[11] Amit, D. (1989). *Modeling Brain Function.* Cambridge, UK: Cambridge University Press, 1989.

[12] ———(1995). "The Hebbian Paradigm Reintegrated: Local Reverberations as Internal Representations." *Behavioral and Brain Science*, 18 (1995): 617–626.

[13] Anderson, J. M. (1971). *The Grammar of Case. Towards a Localistic Theory.* Cambridge, UK: Cambridge University Press, 1971.

[14] ——— (1975). "La grammaire casuelle" and "Maximi Planudis in Memoriam", *Langages*, 38 (1975): 18–64 and 81–103. Paris: Didier-Larousse.

[15] Andler, D., (1987). "Progrès en situation d'incertitude." *Le Débat*, 47 (1987): 5–25.

[16] ——— (1990). "Connexionnisme et Cognition: à la recherche des bonnes questions." In *RS (1990)*: 95–127.

[17] ———, ed. (1992). *Introduction aux Sciences Cognitives.* Paris: Gallimard, 1992.

[18] Andler, D., Petitot, J., Visetti, Y.-M. (1991). "Dynamical Systems, Connectionism and Linguistics." In *COMPCOG I (1991).*

[19] Anttila, R. (1977). "Dynamic fields and linguistic structure: A proposal for a Gestalt linguistics." *Die Sprache*, 23, 1 (1990): 1–10.

[20] Arnold, V., Varchenko, A. N., Gusein-Zade, S. M. (1985). *Singularities of differentiable maps.* Boston: Birkhäuser, 1985.

[21] Ashby, R. (1952). *Design for a brain.* London: Chapman and Hall, 1952.

[22] Asperti, A., Longo, G. (1991). *Categories, Types, and Stuctures.* Cambridge, MA: MIT Press, 1991.

[23] Atiya, A., Baldi, P. (1989). "Oscillations and Synchronisation in Neural Networks: an Exploration of the Labeling Hypothesis." *International Journal of Neural Systems*, 1, 2 (1989): 103–124.

[24] Azencott, R. (1988). "Simulated Annealing." *Séminaire Bourbaki, 697*, Paris: Astérisque, 1988.

[25] Bach, E., Harms, R. T., eds. (1968). *Universals in Linguistic Theory.* New York: Holt, Rinehart & Winston.

[26] Baciu, M., Koenig, O., Vernier, M.-P., Bedoin, N., Rubin, C., Segebarth, C. (1999). "Categorical and coordinate spatial relations: fMRI evidence for hemispheric specialization." *NeuroReport*, 10 (1999): 1373–1378.

[27] Bak, P. (1996). *How Nature Works: The Science of Self-Organized Criticality.* New York: Copernicus, 1996.

[28] Bak, P., Tang, C., Wiesenfeld, K. (1988). "Self-organized criticality." *Physical Review A*, 38 (1988): 364–374.

[29] Bell, J. (2005). "The Development of Categorical Logic." In *Handbook of Philosophical Logic*, Vol. 12. Berlin: Springer, 2005.

[30] Bennett, D. G. (1975). *Spatial and Temporal Uses of English Prepositions. An Essay in Stratificational Semantics.* London: Longman, 1975.

[31] Bergen, B. K., Chang, N. C. (2000). "Spatial Schematicity of Prepositions in Neural Grammar." In *Fifth International Conference on Conceptual Structure, Discourse and Language (CSDL-V)*, University of California at Santa Barbara.

[32] Berti, A., Rizzolatti, G. (2002). "Coding near and far space." In *The cognitive and neural bases of spatial neglect*, H.-O. Karnath, D. Milner, G. Vallar eds. Oxford, UK: Oxford University Press, 2002, 119–130.

[33] Bialek, W., Rieke, F., de Ruyter van Steveninck, R., Warland, D. (1991). "Reading a neural code." *Science*, 252 (1991): 1854–1857.

[34] Bienenstock, E. (1992). "Suggestions for a Neurobiological Approach to Syntax." In *COMPCOG II (1992).*

[35] ——— (1995). "A model of neocortex." *Network*, 6 (1995): 179–224.

[36] ——— (1996). "Composition." In *Brain Theory*, A. Aertsen, V. Braitenberg eds. Amsterdam: Elsevier, 1996, 269–300.

[37] Blakemore, S. J., Decety, J. (2001). "From the perception of the action to the understanding of intention." *Nature Reviews Neuroscience*, 2 (2001): 561–567.

[38] Blum, H. (1973). "Biological shape and visual science." *Journal of Theoretical Biology*, 38 (1973): 205–287.

[39] Bonhöffer T., Grinvald A. (1991). "Iso-orientation domains in cat visual cortex are arranged in pinwheel-like patterns." *Nature*, 353 (1991): 429–431.

[40] Bonnet A., David G. (2001). *Cracktip is a global Mumford-Shah minimizer.* Paris: Astérisque, vol. 274, Société Mathématique de France, 2001.

[41] Boroditsky, L. (2000). "Metaphoric structuring: understanding time through spatial metaphors." *Cognition*, 75 (2000): 1–28.

[42] Bose, A., Kopell, N., Terman, D. (2000). "Almost-synchronous solutions for mutually coupled excitatory neurons." *Physica D*, 140 (2000): 69–94.

[43] Bosking, W., Zhang, Y., Schofield, B., Fitzpatrick, D. (1997). "Orientation Selectivity and the Arrangement of Horizontal Connections in Tree Shrew Striate Cortex." *Journal of Neuroscience*, 17, 6 (1997): 2112–2127.

[44] Brandt, P. Aa. (1986). *La Charpente modale du Sens*, Doctoral Thesis, University of Paris III. Amsterdam: John Benjamins, 1992.

[45] ——— (1994). "Pour une phrastique intégrale." In *Linguistique cognitive et Modèles dynamiques*, J. Petitot ed. *Sémiotiques*, 6-7 (1994) 121–136.

[46] Brown, R. (1973). *A First Language.* Boston: Harvard University Press, 1973.

[47] Buades, A., Coll, B., Morel, J.-M. (2005). "A Review of image denoising algorithms, with a new one." *Multiscale Model. Simul.*, Society for Industrial and Applied Mathematics, 4, 2 (2005): 490–530.

[48] Buser, P., Imbert, M. (1987). *Vision.* Paris: Hermann, 1987.

[49] Campbell, S. R., Wang, D. L. (1996). "Synchronization and desynchronization in a network of locally coupled Wilson-Cowan oscillators." *IEEE Transactions on Neural Networks*, 7 (1996): 541–554.

[50] Campbell, S. R., Wang, D. L., Jayaprakash, C. (1999). "Synchrony and desynchrony in integrate-and-fire neurons." *Neural Computat.*, 7 (1999): 1595–1619.

[51] Carlson, L. A., Regier, T., Lopez, W., Corrigan, B. (2007). "Attention Unites Form and Function in Spatial Language." *Spatial Cognition and Computation*, 6, 4 (2007): 295–308.

[52] Carlson, L. A., West, R., Taylor, H. A., Herndon, R. W. (2002). "Neural correlates of spatial term use." *Journal of Experimental Psychology: Human Perception and Performance*, 28 (2002): 1391–1408.

[53] Chalmers, D. J. (1990a). "Syntactic transformations on distributed representations." *Connection Science*, 2 (1990): 53–62.

[54] ——— (1990b). "Why Fodor and Pylyshyn Were Wrong: The Simplest Refutation." In *Proceedings of the Twelfth Annual Conference of the Cognitive Science Society*, 340–347.

[55] Chenciner, A. (1973). "Travaux de Thom et Mather sur la stabilité topologique." *Séminaire Bourbaki*, 424.

[56] ——— (1980). "Singularités des Fonctions Différentiables." Paris: *Encyclopædia Universalis*, 1980.

[57] ——— (1985). "Systèmes dynamiques différentiables." Paris: *Encyclopaedia Universalis*, 1985.

[58] Chambon, R. (1974). *Le Monde comme Perception et Réalité.* Paris: Vrin, 1974.

[59] Chomsky, N. (1965). *Aspects of the Theory of Syntax.* Cambridge, MA: MIT Press, 1965.

[60] Chua, L., Roska, T. (1998). *Cellular Neural Networks and Visual Computing.* Cambridge, UK: Cambridge University Press, 1998.

[61] Churchland, P.M. (1984). *Matter and Consciousness.* Cambridge MA: MIT Press, 1984.

[62] Clark, A. (2004). "Feature-Placing and Proto-objects." *Philosophical Psychology*, 17, 4 (2004): 443–469.

[63] Cole, P., Sadock, J. M., eds. (1977). *Grammatical Relations. Syntax and Semantics*, 8. New York: Academic Press, 1977.

[64] COMPCOG I (1991). *Interdisciplinary Workshop on Compositionality in Cognition and Neural Networks I.* D. Andler, E. Bienenstock, B. Laks eds. Abbaye de Royaumont, 1991.

[65] COMPCOG II (1992). *Interdisciplinary Workshop on Compositionality in Cognition and Neural Networks II.* D. Andler, E. Bienenstock, B. Laks eds. Abbaye de Royaumont, 1992.

[66] Coseriu, E. (1974). "Les universaux linguistiques (et les autres)." In *Proceedings of the Eleventh International Congress of Linguists*, L. Heilmann ed. Bologna: Il Mulino, 1974, I, 47–73. Also in E. Coseriu. *L'homme et son langage.* Louvain-Paris: Peeters, 2001, 69–107.

[67] Coullet, P., Emilsson, K. (1992). "Strong resonances of spatially distributed oscillators: a laboratory to study patterns and defects." *Physica D*, 61 (1992): 119–131.

[68] Coventry, K. R., Garrod, S. C. (2004). *Saying, seeing, and acting: The psychological semantics of spatial prepositions.* New York: Psychology Press, 2004.

[69] Crawford, L. E., Regier, T., Huttenlocher, J. (2000). "Linguistic and non-linguistic spatial categorization." *Cognition*, 75 (2000): 209–235.

[70] CSR (2006). "New insights in categorical and coordinate processing of spatial relations (Editorial)." *Neuropsychologia*, 44 (2006): 1515–1518.

[71] Culioli, A. (1970). "La formalisation en linguistique. Considérations théoriques à propos du traitement formel du langage." *Documents de Linguistique quantitative*, 7, Paris: Dunod.

[72] ——— (1971). "A propos d'opérations intervenant dans le traitement formel des langues naturelles." *Quelques aspects de la formalisation en linguistique*, 1, 34 (1971): 7–15.

[73] Culioli, A., Desclés, J.P. (1981) *Systèmes de Représentations Linguistiques et Métalinguistiques*. Laboratoire de Linguistique Formelle, Université de Paris VII.

[74] Culioli (2006). *Antoine Culioli. Un homme dans le langage*, Colloque de Cerisy, D. Ducard, C. Normand eds. Paris: Éditions Ophrys, 2006.

[75] Daido, H. (1990). "Intrinsic Fluctuations and a Phase Transition in a Class of Large Populations of Interacting Oscillators." *Journal of Statistical Physics*, 60, 5–6 (1990): 753–800.

[76] ——— (1993). "A solvable model of coupled limit-cycle oscillators exhibiting partial perfect synchrony and novel frequency spectra." *Physica D*, 69 (1993): 394–403.

[77] Damasio, H., Grabowski, T. J., Tranel, D., Ponto, L. L. B., Hichwa, R. D., Damasio, A. R. (2001). "Neural correlates of naming actions and of naming spatial relations." *NeuroImage*, 13 (2001): 1053–1064.

[78] Damon, J. (1988). *Local Morse Theory for Solutions to the Heat Equation and Gaussian Blurring*. Technical Report, University of North Carolina.

[79] Das A., Gilbert C.D. (1995). "Long range horizontal connections and their role in cortical reorganization revealed by optical recording of cat primary visual cortex." *Nature*, 375 (1995): 780–784.

[80] David, G., Semmes, S. (1996). "On the singular sets of minimizers of the Mumford-Shah functional." *J. Math. Pures Appl.*, 75 (1996): 299–342.

[81] De Giorgi, E., Carriero, G., Leaci, A. (1989). "Existence theorem for a minimum problem with free discontinuity set." *Arch. Rational Mech. Anal.*, 108 (1989): 195–218.

[82] Deleuze, G. (1969). *Logique du Sens*. Paris: Editions de Minuit, 1969.

[83] ——— (1973). "A quoi reconnaît-on le structuralisme?" *Histoire de la Philosophie*, F. Châtelet ed. Paris: Hachette, 1973.

[84] Desclés, J.-P. (1981). *Opérateurs-Opérations*. Laboratoire de Linguistique Formelle, Paris: Université de Paris VII.

[85] ——— (1986). "Représentation des connaissances, Archétypes cognitifs, Schèmes conceptuels, Schémas grammaticaux." *Actes Sémiotiques*, VII, 69–70 (1986).

[86] ——— (1990). *Langages applicatifs, Langues naturelles et Cognition*. Paris: Hermès.

[87] Dougherty, E. R. (1992). *An introduction to morphological image processing*. Bellingham, WA: SPIE Optical Engineering Press.

[88] Diessel, H. (2006). "Demonstratives, joint attention, and the emergence of grammar." *Cognitive Linguistics*, 17 (2006): 463–490.

[89] Doursat, R., Petitot, J. (1997). "Modèles dynamiques et linguistique cognitive: vers une sémantique morphologique active." In *Actes de la 6ème École d'été de l'Association pour la Recherche Cognitive*. Centre National de la Recherche Scientifique, 1997.

[90] ——— (2005a). "Bridging the gap between vision and language: A morphodynamical model of spatial categories." In *Proceedings of the International Joint Conference on Neural Networks*. Montréal, 2005.

[91] ——— (2005b). "Dynamical systems and cognitive linguistics: toward an active morphodynamical semantics." *Neural Networks*, 18 (2005): 628–638.

[92] Doya, K., (1992). "Bifurcations in the learning of recurrent neural networks." In *Proceedings of 1992 IEEE International Symposium on Circuits and Systems*, IEEE, 6 (1992): 2777–2780.

[93] Doyon, B., Cessac, B., Quoy, M., Samuelides, M. (1993). "Control of the transition to chaos in neural networks with random connectivity." *Int. Jour. Bif. Chaos*, 3, 2 (1993): 279–291.

[94] Ducrot, O. (1968). *Le Structuralisme en Linguistique*. Paris: Le Seuil, 1968.

[95] Dupuy, J.-P. (2000). *The Mechanization of the Mind: On the Origins of Cognitive Science*, (transl. M. B. DeBevoise). Princeton: Princeton University Press, 2000.

[96] Edelman,G. M. (1987). *Neural Darwinism. The Theory of Neuronal Group Selection.* New York: Basic Books, 1987.

[97] Elman, J. (1989). "Representation and Structure in Connectionist Models." In *Cognitive Models of Speech Processing*, G. T. M. Altmann, ed. Cambridge, MA: MIT Press, 1989, 345–382.

[98] EMC, (1992). *Emergence dans les modèles de la cognition*, B. Amy, J-J. Ducret, A. Grumbach eds. Paris: Ecole Nationale Supérieure des Télécommunications.

[99] Engel, A., König, P., Gray, C., Singer, W. (1992). "Temporal Coding by Coherent Oscillations as a Potential Solution to the Binding Problem: Physiological Evidence." In *Non-Linear Dynamics and Neural Networks*, H. Schuster ed. Berlin: Springer, 1992.

[100] Ermentrout, B. (1998). "Neural networks as spatio-temporal pattern-forming systems." *Reports on Progress in Physics*, 61 (1998): 353–430.

[101] Ermentrout, B., Kopell, N. (1991). "Multiple pulse interactions and averaging in coupled neural oscillators." *J. Math. Biol.*, 29 (1991): 195–217.

[102] Evans, G. (1982). *The Varieties of Reference.* Oxford: Clarendon Press, 1982.

[103] Evans, L. C., Spruck, J. (1991). "Motion of Level Sets by Mean Curvature. I." *J. Differential Geometry*, 33 (1991): 635–681.

[104] Feldman, J., Lakoff, G., Bailey, D., Narayanan, S., Regier, T., Stolcke, A. (1996). "L_0: The First Five Years." *Artificial Intelligence Review*, 10 (1996): 103–129.

[105] Field, D. J., Hayes, A., Hess, R. F. (1993). "Contour integration by the human visual system: evidence for a local 'association field'." *Vision Research*, 33, 2 (1993): 173–193.

[106] Fillmore, C. F., (1977). "The Case for case reopened." *Grammatical Relations*, P. Cole, J.M. Sadock eds. *Syntax and Semantics*, 8 (1977): 59–81.

[107] ——— (1997). *Lectures on deixis.* Stanford: CSLI Publications.

[108] FitzHugh, R. A. (1961). "Impulses and physiological states in theoretical models of nerve membrane." *Biophysical Journal*, 1 (1961): 445–466.

[109] Florack, L. M. J. (1993). *The Syntactical Structure of Scalar Images.* Thesis. University of Utrecht.

[110] Florack, L. M. J., Ter Haar Romeny, B. M., Koenderink, J. J., Viergever, M. A. (1992). "Scale and the differential structure of images." *Image and Vision Computing*, 10, 6 (1992): 376–388.

[111] Fodor, J., McLaughlin, B. P. (1990). "Connectionism and the problem of systematicity: Why Smolensky's solution doesn't work." *Cognition*, 35 (1990): 183–204.

[112] Fodor, J., Pylyshyn, Z. (1981). "How direct is visual perception? Some reflections on Gibson's 'ecological approach'." *Cognition*, 9, (1981): 139–196.

[113] ——— (1988). "Connectionism and cognitive architecture: a critical analysis." *Cognition*, 28, 1–2 (1988): 3–71.

[114] French, R. M., Thomas, E. (2001). "The Dynamical Hypothesis in Cognitive Science: A review essay of *Mind as Motion*." *Minds and Machines*, 11, 1 (2001): 101–111.

[115] Fuchs, C., Victorri, B., eds. (1994). *Continuity in Linguistic Semantics.* Amsterdam: John Benjamins.

[116] Gage, M., Hamilton, R. S. (1986). "The Heat equation shrinking convex plane curves." *J. Differential Geometry*, 25 (1986): 69-96.

[117] Gärdenfors, P. (1994). "Frameworks for Properties: Possible worlds vs. Conceptuel Spaces." *Linguistique cognitive et Modèles dynamiques*, J. Petitot ed. *Sémiotiques*, 6–7 (1994): 99–120.

[118] ——— (2000). *Conceptual Spaces. The Geometry of Thought.* Cambridge, MA: MIT Press, 2000.

[119] Geman, S., Bienenstock, E., Doursat, R. (1992). "Neural networks and the bias/ variance dilemma." *Neural Computation*, 4 (1992): 1–58.

[120] Gentner, D. (2001). "Spatial metaphors in temporal reasoning". In *Spatial schemata and abstract thought*, M. Gattis ed. Cambridge, MA: MIT Press, 203–222.

[121] Gentner, D., Goldin-Meadow, S. (2003). *Language in Mind.* Cambridge, MA: MIT Press, 2003.

[122] Gibson, J. J., (1979). *The Ecological Approach to Visual Perception.* Boston: Houghton Mifflin.

[123] ——— (1982). *Reasons for Realism: Selected essays of James J. Gibson*, E. Reed & R. Jones eds. Hillsdale, NJ: Lawrence Erlbaum.

[124] Girard, R. (1972). *La Violence et le Sacré*, Paris: Grasset. (*Violence and the Sacred*, transl. P. Gregory. Baltimore: Johns Hopkins University Press, 1977).

[125] Gobbino M. (1998). "Finite Difference Approximation of the Mumford-Shah Functional." *Communications on Pure and Applied Mathematics*, LI (1998): 197–228.

[126] Goel, P., Ermentrout, B. (2002). "Synchrony, stability, and firing patterns in pulse-coupled oscillators." *Physica D*, 163 (2002): 191–216.

[127] Golubitsky, M., Guillemin, V. (1973). *Stable Mappings and their Singularities.* Graduate Texts in Mathematics, 14. New-York, Heidelberg, Berlin: Springer, 1973.

[128] Granger, G.-G. (1980). *La notion de contenu formel.* Also in *Forme, opération, objet.* Paris: Vrin, 1994.

[129] ——— 1988. *Pour la Connaissance philosophique.* Paris: Odile Jacob.

[130] Grassberger, P., Nadal, J.-P., eds. (1994) *From Statistical Physics To Statistical Inference and Back*, NATO ASI Series C: mathematical and physical sciences, 428. Dordrecht: Kluwer Acad. Pub., 1994.

[131] Gray, C. M., König, P., Engel, A. K., Singer, W. (1989). "Oscillatory responses in cat visual cortex exhibit intercolumnar synchronization which reflects global stimulus properties." *Nature*, 338 (1989): 334–337.

[132] Grayson, M. (1987). "The Heat Equation Shrinks Embedded Plane Curves to Round Points." *J. Differential Geometry*, 26 (1987): 285–314.

[133] Greenberg, J. H., ed. (1963). *Universals of Language.* Cambridge, MA: MIT Press, 1963.

[134] ———, ed. (1978) *Universals of Human Language*, Volume 1: *Method and Theory*, Volume 2: *Phonology*, Volume 3: *Word Structure*, Volume 4: *Syntax.* Stanford: Stanford University Press, 1978.

[135] Greimas, A.J., Courtès, J. (1979). *Dictionnaire raisonné de la théorie du langage.* Paris: Hachette, 1979.

[136] Grossberg, S. (1988). *Neural Networks and Natural Intelligence.* Cambridge, MA: MIT Press, 1988.

[137] Grossmann, A., Morlet, J. (1984). "Decomposition of Hardy functions into square integrable wavelets of constant shape." *SIAM J. Math. Anal.*, 15 (1984): 723–736.

[138] Grush, R. (1997a). "Yet another design for a brain? Review of Port and van Gelder (eds) *Mind as Motion.*" *Philosophical Psychology*, 10, 2 (1997): 233–242.

[139] ——— (1997b). "The architecture of representation." *Philosophical Psychology*, 10, 1 (1997): 5-23.

[140] Guckenheimer, J., Holmes, P. (1983). *Non-linear Oscillations, Dynamical Systems and Bifurcations of Vector Fields.* New-York, Heidelberg, Berlin: Springer, 1983.

[141] Haiman, J., ed. (1985). *Iconicity in Syntax.* Amsterdam: J. Benjamins, 1985.

[142] Hamy, H. (1997). *Méthodes géométriques multi-échelles en vision computationnelle.* PhD thesis. Paris: Ecole Polytechnique.

[143] Haralick, R.M. (1984). "Digital step edges from zero crossing second directional derivatives." *IEEE Trans. Pattern Anal. Mach. Intell. PAMI-6*, (1984): 58–68.

[144] Harris, C. (1990). "Connectionism and cognitive linguistics." *Connection Science*, 2, 1–2 (1990): 7–33.

[145] Harnad, S. (1987). *Categorical perception.* Cambridge, UK: Cambridge University Press, 1987.

[146] Haspelmath, M. (1997). *From space to time: temporal adverbials in the world's languages.* Newcastle, UK: Lincom Europa, 1997.

[147] Hayward, W. G., Tarr, M. J. (1995). "Spatial language and spatial representation." *Cognition*, 55 (1995): 39–84.

[148] Heider, F., Simmel, M. (1944). "An experimental study of apparent behavior." *American Journal of Psychology*, 57 (1944): 243-259.

[149] Herskovits, A. (1985). "Semantics and Pragmatics of Locative Expressions." *Cognitive Science*, 9, 3 (1985): 341–378.

[150] ——— (1986). *Language and Spatial Cognition. An Interdisciplinary Study of the Prepositions in English.* Cambridge, UK: Cambridge University Press, 1986.

[151] Herzog, G. (1995). "Coping with Static and Dynamic Spatial Relations." In *Proceedings of TSM'95, Time, Space, and Movement: Meaning and Knowledge in the Sensible World*, P. Amsili, M. Borillo, L. Vieu eds. Toulouse: Université Paul Sabatier, 1995, C 47–59.

[152] Herzog, G., Rohr, K. (1995). "Integrating Vision and Language: Towards Automatic Description of Human Movements." In *Advances in Artificial Intelligence*, C.-R. Rollinger, W. Brauer eds. Berlin, Heidelberg: Springer, 1995.

[153] Heyman, K. (2006). "The map in the brain: Grid cells may help us navigate." *Science*, 312 (2006): 680-681.

[154] Hirsch, M. (1989). "Convergent activation dynamics in continuous time networks." *Neural Networks*, 2 (1989): 331–349.

[155] Hjelmslev, L. (1935). *La catégorie des cas.* München: Wilhelm Fink Verlag, 1972.

[156] Hoffman, W. C. (1989). "The visual cortex is a contact bundle." *Applied Mathematics and Computation*, 32 (1989): 137–167.

[157] Holenstein, E. (1992). "Phenomenological Structuralism and Cognitive Semiotics." *Scripta Semiotica*, 1 (1992): 133–158.

[158] Hubel, D. H. (1988). *Eye, Brain and Vision.* Scientific American Library, 22. New York, Oxford: W.H. Freeman, 1988.

[159] Hummel, R., Moniot, R. (1989). "Reconstructions from zero crossings in scale space." *IEEE Transactions on Accoustics, Speach and Signal Processing*, 37, 12 (1989): 2111–2130.

[160] Husserl, E. (1900–1901). *Logische Untersuchungen.* Halle, Max Niemeyer, 1913. *Logical Investigations*, transl. J. N. Findlay. London: Routledge, 1973.

[161] ——— (1907). *Ding und Raum.* Vorlesungen 1907, Husserliana XVI. The Hague: Martinus Nijhoff, 1973. *Thing and Space: Lectures of 1907*, transl. R. Rojcewicz. Oxford: Basil Blackwell, 1997.

[162] ——— (1913). *Ideen zu einer reinen Phänomenologie und phänomenologischen Philosophie*, Husserliana III–IV. The Hague: Martinus Nijhoff, 1950.

[163] ——— (1939). *Erfahrung und Urteil. Untersuchungen zur Genealogie der Logik.* Hamburg: Claassen&Goverts, 1939. *Experience and Judgement*, transl. J. S. Churchill and K. Ameriks. London: Routledge, 1973.

[164] Huttenlocher, J., Hedges, L., Duncan, S. (1991). "Categories and particulars: prototype effects in estimating spatial location." *Psychological Review*, 98 (1991): 352–376.

[165] Ikegaya, Y., Aaron, G., Cossart, R., Aronov, D., Lampl, I., Ferster, D., Yuste, R. (2004). "Synfire chains and cortical songs: temporal modules of cortical activity." *Science*, 304 (2004): 559–564.

[166] INLV (1994). Integration of Natural Language and Vision Processing. *Artificial Intelligence Review Journal*, 8, Special Issue, 1994.

[167] Izhikevich, E. M. (2000). "Phase equations for relaxation oscillators." *SIAM J. Appl. Math.*, 60, 5 (2000): 1789–1805.

[168] Izhikevich, E. M., Gally, J. A., Edelman, G. M. (2004). "Spike-timing dynamics of neuronal groups." *Cerebral Cortex*, 14 (2004): 933–944.

[169] Jackendoff, R. (1983). *Semantics and Cognition.* Cambridge, MA: MIT Press, 1983.

[170] ——— (1987). *Consciousness and the Computational Mind.* Cambridge, MA: MIT Press, 1987.

[171] Jager, G., Postma, A. (2003). "On the hemispheric specialization for categorical and coordinate spatial relations: a review of the current evidence." *Neuropsychologia*, 41 (2003): 504–15.

[172] Johnson, J. L. (1994). "Pulse-coupled neural nets: translation, rotation, scale, distortion and intensity signal invariance for images." *Applied Optics*, 33, 26 (1994): 6239–6253.

[173] Johnson, M. (1987). *The Body in the Mind.* Chicago, London: The University of Chicago Press, 1987.

[174] Kanizsa, G. (1980). *La Grammatica del Vedere.* Bologna: Il Mulino. *Organization in Vision: Essays on Visual Perception* . New York: Praeger, 1979.

[175] Karmiloff-Smith, A. (1992). *Beyond Modularity: A Developmental Perspective on Cognitive Science.* Cambridge, MA: MIT Press.

[176] Kemmerer, D. (2005). "The spatial and temporal meanings of English prepositions can be independently impaired." *Neuropsychologia*, 43 (2005): 797–806.

[177] ——— (2006). "The semantics of space: Integrating linguistic typology and cognitive neuroscience." *Neuropsychologia*, 44 (2006): 1607–1621.

[178] ——— (2007). "A Neuroscientific Perspective on the Linguistic Encoding of Categorical Spatial Relations." In *Language, Cognition and Space*, V. Evans and P. Chilton eds, *Advances in Cognitive Linguistics.* London: Equinox Publishing Co, 2007.

[179] Kimia, B. B. (2003). "On the role of medial geometry in human vision." In *Neurogeometry and Visual Perception*, J. Petitot and J., Lorenceau eds. *Journal of Physiology Paris*, 97, 2–3 (2003): 155–190.

[180] Kitcher, P. (1988). "Marr's Computational Theory of Vision." *Philosophy of Science*, 55 (1988): 1–24.

[181] Koenderink, J. J. (1984). "The Stucture of Images." *Biological Cybernetics*, 50 (1984): 363-370.

[182] ——— (1990a). "The brain as a geometry engine." *Psychological Research*, 52 (1990): 122–127.

[183] ——— (1990b). *Solid shape.* Cambridge, MA: MIT Press, 1990.

[184] Koenderink, J. J., van Doorn, A. J. (1986). "Dynamic Shape." *Biological Cybernetics*, 53 (1986): 383–396.

[185] ——— (1987). "Representation of local geometry in the visual system." *Biological Cybernetics*, 55 (1987): 367–375.

[186] König, P., Schillen, T. B. (1991). "Stimulus-dependent assembly formation of oscillatory responses." *Neural Computation*, 3 (1991): 155–166.

[187] Kopell N., Ermentrout, G. B. (1986). "Symmetry and phaselocking in chains of weakly coupled oscillators." *Comm. Pure Appl. Math.*, 39 (1986): 623–660.

[188] ——— (1990). "Phase Transitions and Other Phenomena in Chains of Coupled Oscillators." *SIAM J. Appl. Math.*, 50, 4 (1990): 1014–1052.

[189] Kopell N., Somers, D. (1995). "Anti-phase solutions in relaxation oscillators coupled through excitatory interactions." *J. Math. Biol.*, 33 (1995): 261–280.

[190] Kosslyn, S. M. (1980). *Image and Mind.* Cambridge, MA: Harvard University Press, 1980.

[191] ——— (1987). "Seeing and imagining in the cerebral hemispheres: a computational approach." *Psychological Review*, 94 (1987): 148-75.

[192] ——— (1994). *Image and Brain.* Cambridge, MA: MIT Press, 1994.

[193] ——— (2006). "You can play 20 questions with nature and win: Categorical versus coordinate spatial relations as a case study." *Neuropsychologia*, 44 (2006): 1519–1523.

[194] Kuramoto, Y. (1984). *Chemical Oscillators, Waves and Turbulence.* Berlin: Springer-Verlag, 1984.

[195] Kuramoto, Y., Nishikawa, I. (1987). "Statistical Macrodynamics of Large Dynamical Systems. Case of a Phase Transition in Oscillator Communities." *Journal of Statistical Physics*, 49, 3–4 (1987): 569–605.

[196] Laeng, B., Chabris, C.F., Kosslyn, S.M. (2003). "Asymmetries in encoding spatial relations." In *The asymmetrical brain*, K. Hugdahl and R.J. Davidson eds. Cambridge, MA: MIT Press, 303–339.

[197] Lakoff, G. (1987). *Women, Fire, and Dangerous Things: What Categories Reveal about the Mind.* Chicago: The University of Chicago Press, 1987.

[198] ——— (1988a). "Cognitive semantics." In *Meaning and Mental Representations*, U. Eco, M. Santambrogio, P. Violi eds. *Advances in Semiotics.* Bloomington: Indiana University Press, 1988, 119–154.

[199] ——— (1988b). "A suggestion for a linguistics with connectionist foundations." In *Proceedings of the 1988 Connectionist Models Summer School.* San Francisco: Morgan Kaufmann.

[200] Lakoff, G., Johnson, M. (1980). *Metaphors We Live By.* Chicago: The University of Chicago Press, 1980.

[201] ——— (1999). *Philosophy In The Flesh: The Embodied Mind and Its Challenge to Western Thought.* New York: Basic Books, 1999.

[202] Landau, B., Jackendoff, R. (1993). " 'What' and 'where' in spatial language and spatial cognition." *Behavioral and Brain Sciences*, 16 (1993): 217–265.

[203] Langacker, R.W. (1987). *Foundations of Cognitive Grammar, Vol. I.* Stanford: Stanford University Press, 1987.

[204] ——— (1991). *Foundations of Cognitive Grammar, Vol. II.* Stanford: Stanford University Press, 1991.

[205] ——— (1994). "Structural Syntax: The View from Cognitive Grammar." *Linguistique cognitive et Modèles dynamiques*, J. Petitot ed. *Sémiotiques*, 6–7 (1994): 69–84.

[206] ——— (1999). *Grammar and Conceptualization.* Cognitive Linguistics Research, 14. Berlin, New York: Mouton de Gruyter, 1999.

[207] ——— (2010). "Reflections on the Functional Characterization of Spatial Prepositions." In *Espace, Préposition, Cognition. Hommage à Claude Vandeloise*, G. Col, C. Collin eds. Corela, http://edel.univ-poitiers.fr/corela/document.php?id=2657

[208] Lantuéjoul, C. (1978). *La squelettisation et son application aux mesures topologiques des mosaïques polycristallines.* PhD thesis. Paris: Ecole des Mines de Paris, 1978.

[209] Lee, T. S. (2003). "Computations in the early visual cortex." In *Neurogeometry and Visual Perception*, J. Petitot and J. Lorenceau eds., *Journal of Physiology-Paris*, 97, 2–3 (2003): 121–139.

[210] Levinson, S.C. (2003). *Space in language and cognition.* Cambridge, UK: Cambridge University Press, 2003.

[211] Levinson, S.C., Wilkins, D. (2006). *Grammars of space.* Cambridge, UK: Cambridge University Press, 2006.

[212] Leyton, M. (1992). *Symmetry, Causality, Mind.* Cambridge, MA: MIT Press, 1992.

[213] Logan, G. D., Sadler, D. D. (1996). "A computational analysis of the apprehension of spatial relations." In *Language and Space*, P. Bloom, M. A. Peterson, L. Nadel, and M. F. Garrett eds. Cambridge, MA: MIT Press, 1996, 493–529.

[214] LTC (1989). *Logos et Théorie des Catastrophes*, Proceedings of René Thom's Cerisy Conference, J. Petitot ed. Genève: Editions Patiño, 1989.

[215] Lumer, E. D., Huberman, B. A. (1992). "Binding Hierarchies: A Basis for Dynamic Perceptual Grouping." *Neural Computation*, 4 (1992): 341–355.

[216] Luria, A. R. (1975). " Scientific Perspectives and Philosophical Dead Ends in Modern Linguistics." *Cognition*, 3, 4 (1975): 377.

[217] Mac Lane, S., Moerdijk, I. (1992). *Sheaves in Geometry and Logic*. New York: Springer, 1992.

[218] Mallat, S. (1989). "A theory for multiresolution signal decomposition: the wavelet representation." *IEEE Transactions on Pattern Analysis and Machine Intelligence*, 11, 7 (1989): 674–693.

[219] ——— (1998). *A Wavelet Tour of Signal Processing*. New York: Academic Press, 1998.

[220] Manjali, F. (1991). *Nuclear Semantics*. New Dehli: Bahri Publications, 1991.

[221] ——— (1994). "Dynamic Semiotics or the Case for Actantial Case." *Linguistique cognitive et Modèles dynamiques*, J. Petitot ed. *Sémiotiques*, 6–7 (1994): 85–98.

[222] Mandler, J. (2004). *The Foundations of Mind: Origins of Conceptual Thought*. Oxford: Oxford University Press.

[223] ——— (2004). "Thought before language." *Trends in Cognitive Sciences*, 8, 11 (2004): 508–513.

[224] Marr, D. (1982). *Vision*. San Francisco, CA: Freeman Publishers, 1982.

[225] Mather, J., (1968/71). "Stability of C^∞ Mappings. I: the Division Theorem." *Ann. of Math.*, 87 (1968): 89-104. "II: Infinitesimal Stability Implies Stability." *Ann. of Math.*, 89 (1969): 254–291. "III: Finitely Determined Map Germs." *Publications Mathématiques de l'IHES*, 35 (1968): 127–156 (Paris: Presses Universitaires de France). "IV: Classification of Stable Germs by $\mathbb{R}$-algebras." *Publications Mathématiques de l'IHES*, 37 (1969): 223–248 (Paris: Presses Universitaires de France). "V: Transversality." *Advances in Math.*, 4 (1970): 301–336. "VI: the Nice Dimensions." *Proceedings of Liverpool Singularities Symposium*, C.T.C. Wall ed. *Lecture Notes in Mathematics*, 192. New-York, Heidelberg, Berlin: Springer, 1971, 207–253.

[226] Mavridis, N., Roy, D. (2005). "Grounded Situation Models for Robots: Bridging Language, Perception, and Action." *AAAI-05 Workshop on Modular Construction of Human like Intelligence*, Seattle, WA, 2005.

[227] McDonough, L., Choi, S., Mandler, J. M. (2003). "Understanding spatial relations: Flexible infants, lexical adults". *Cognitive Psychology*, 46, 3 (2003): 229-259.

[228] Mc Dowell, J. (1994). *Mind and World*. Cambridge, MA: Harvard University Press, 1994.

[229] McIntyre, R. (1999). "Naturalizing Phenomenology? Dretske on Qualia." In *Naturalizing Phenomenology: Issues in Contemporary Phenomenology and Cognitive Science*, J. Petitot, F. J. Varela, J.-M. Roy, B. Pachoud eds. Stanford: Stanford University Press, 429-439.

[230] McKevitt, P., ed. (1994). *Proceedings of AAAI-94 Workshop on Integration of Natural Language and Vision Processing*, Seattle, WA, 1994.

[231] Merleau-Ponty, M. (1968). *Résumés de Cours. Collège de France 1952–1960*. Paris: Gallimard, 1968.

[232] Meyer, Y. (1988). *Ondelettes et Opérateurs*. Paris: Hermann, 1988.

[233] Mézard, M., Parisi, G. Virasoro, M. A. (1987). *Spin glass theory and Beyond, Lecture Notes in Physics*, 9. Singapore: World Scientific, 1987.

[234] Michotte, A. (1946). *La perception de la causalité*. Louvain: *Études de Psychologie*, 6, 1946. *The perception of causality*, transl. T.R. Miles and E. Miles. London: Methuen, 1963.

[235] Michotte, A., Thinès, G., Crabbé, G. (1964). *Les compléments amodaux des structures perceptives (Amodal completions of perceptual structures)*. Louvain: Publications Universitaires, *Studia Psychologica*, 1964.

[236] Milton, J. G., Chu, P. H., Cowan, J. D. (1993). "Spiral waves in integrate-and-fire neural networks." In *Advances in Neural Information Processing Systems*, 5. San Francisco: Morgan Kaufmann, 1001–1007.

[237] Mirollo, R. E., Strogatz, S. H. (1990). "Synchronization of pulse-coupled biological oscillators." *SIAM J. Appl. Math.*, 50 (1990): 1645–1662.

[238] Moerdijk, I., Reyes, G. (1991). *Models for Smooth Infinitesimal Analysis*. Berlin: Springer, 1991.

[239] Morel, J-M., Solimini, S. (1995). *Variational Methods in Image Segmentation*. Berlin: Birkhäuser, 1995.

[240] Mulligan, K. (1995). "Perception." In *The Cambridge Companion to Husserl*, B. Smith and D. W. Smith eds. Cambridge, UK: Cambridge University Press, 1995, 168–238.

[241] Mulligan, K., Simons, P., Smith, B. (1984). "Truth-Makers." *Philosophy and Phenomenological Research*, XLIV, 3 (1984): 287–321.

[242] Mumford, D. (1994). "Bayesian rationale for the variational formulation." In *Geometry-driven Diffusion in Computer Vision*, B. M. ter Haar Romeny ed. Dordrecht: Kluwer Academic Publishers, 1994, 135–146.

[243] Mumford, D., Shah J. (1989). "Optimal Approximations by Piecewise Smooth Functions and Associated Variational Problems." *Communications on Pure and Applied Mathematics*, XLII (1989): 577–685.

[244] Munnich, E., Landau, B., Dosher, B.A. (2001). "Spatial language and spatial representation: a cross-linguistic comparison." *Cognition*, 81 (2001): 171–207.

[245] NP (1999). *Naturalizing Phenomenology: Issues in Contemporary Phenomenology and Cognitive Science*, J. Petitot, F. J. Varela, J.-M. Roy, B. Pachoud eds. Stanford: Stanford University Press.

[246] Ninio, J. (1989). *L'Empreinte des Sens*. Paris: Odile Jacob.

[247] O'Keefe, J., Recce, M. L. (1993). "Phase relationship between hippocampal place units and the EEG θ rhythm." *Hippocampus*, 3 (1993): 317–330.

[248] Osgood, C. E. (1971). "Where Do Sentences Come From?" *Semantics*, D. Steinberg, L. A. Jacobovitz eds. Cambridge, UK: Cambridge University Press.

[249] Osher, S., Sethian, J. A. (1988). "Fronts Propagating with Curvature-Dependent Speed: Algorithms Based on Hamilton-Jacobi Formulations." *Journal of Computational Physics*, 79, 1 (1988): 12–49.

[250] Ouellet, P., (1987). "Une Physique du Sens". *Critique*, 481/482: 577–597.

[251] ——— (1994). "La Sémiotique cognitive: les sciences de l'esprit entre la nature et la culture." *Linguistique cognitive et Modèles dynamiques*, J. Petitot ed. *Sémiotiques*, 6–7 (1994): 137–160.

[252] Parisi, D. (1991). "Emerging Compositionality in Neural Nets." *COMPCOG I*, 90–98.

[253] P.D.P., (1986). *Parallel Distributed Processing*, D.E. Rumelhart, J.L. McClelland eds. Cambridge MA: MIT Press.

[254] Parret, H. (1976). "Sémantique structurale et sémantique générative." *Sémantique et Logique*, B. Pottier ed. Paris: J.P. Delarge, 1976, 85–108.

[255] Pentland, A. P., ed. (1986). *From Pixels to Predicates: Recent Advances in Computational Vision*. Norwood, NJ: AblexCorp., 1986.

[256] Perlmutter, D., ed. (1983/1984). *Studies in Relationnal Grammars, I/II*. Chicago: University of Chicago Press.

[257] Perona, P., Malik, J. (1990). "Scale-space and edge detection using anisotropic diffusion." *IEEE Trans. on PAMI*, 12, 7 (1990): 629–639.

[258] Petitot, J. (1979). "Hypothèse localiste et Théorie des Catastrophes." *Théories du langage, théories de l'apprentissage*, M. Piattelli ed. Paris: Le Seuil, 1979.

[259] ——— (1979). "Infinitesimale." *Enciclopedia Einaudi*, VII. Torino: Einaudi, 1979, 443–521.

[260] ——— (1982). *Pour un Schématisme de la Structure*, Thèse de Doctorat d'Etat. Paris: École des Hautes Études en Sciences Sociales, 1982.

[261] ——— (1985). *Morphogenèse du Sens*. Paris: Presses Universitaires de France. *Morphogenesis of Meaning* , transl. F. Manjali. Bern: Peter Lang, 2003.

[262] ——— (1986). "Structure." *Encyclopedic Dictionary of Semiotics*, T. Sebeok ed., Vol. 2. New-York: de Gruyter, 1986, 991–1022.

[263] ——— (1987). "Sur le réalisme ontologique des universaux sémio-linguistiques." In *Sémiotique en jeu*, M. Arrivé and J.C. Coquet eds. Paris-Amsterdam: Hadès-Benjamins, 1987, 43–63.

[264] ——— (1988). "Approche morphodynamique de la formule canonique du mythe." *L'Homme*, 106–107, XXVIII, 2–3, (1988): 24–50.

[265] ——— (1989a). "On the Linguistic Import of Catastrophe Theory." *Semiotica*, 74, 3–4 (1989): 179–209.

[266] ——— (1989b). "Hypothèse localiste, Modèles morphodynamiques et Théories cognitives : Remarques sur une note de 1975." *Semiotica*, 77, 1–3 (1989): 65–119.

[267] ——— (1989c). "Catastrophe Theory and Semio-Narrative Structures." In *Paris School of Semiotics*, P. Perron and F. Collins eds. Amsterdam: John Benjamins, 1989, 177–212.

[268] ——— (1989d). "Modèles morphodynamiques pour la Grammaire cognitive et la Sémiotique modale." *RSSI (Canadian Semiotic Association)*, 9, 1–2–3 (1989): 17–51.

[269] ——— (1989e). "Morphodynamics and the Categorical Perception of Phonological Units." *Theoretical Linguistics*, 15, 1–2 (1989): 25–71.

[270] ——— (1989f). "Eléments de dynamique modale." *Poetica et Analytica*, 6 (1989): 44–79, Aarhus University.

[271] ——— (1989g). "Forme." *Encyclopaedia Universalis*, XI (1989): 712–728. Paris.

[272] ——— (1989h). "La modélisation : formalisation ou mathématisation? L'exemple de l'approche morphodynamique du langage." In *Perspectives méthodologiques et épistémologiques dans les sciences du langage*, M.J. Reichler-Bégulin ed. Bern: Peter Lang, 1989, 205–220.

[273] ——— (1990a). "Le Physique, le Morphologique, le Symbolique. Remarques sur la Vision." *Revue de Synthèse*, 1–2 (1990): 139–183.

[274] ——— (1990b). "Semiotics and Cognitive Science: the Morphological Turn." *The Semiotic Review of Books*, 1, 1 (1990): 2–4.

[275] ———, ed. (1990). "Sciences cognitives : quelques aspects problématiques." *Revue de Synthèse*, IV (1990): 1–2.

[276] ——— (1991a). "Why Connectionism is such a Good Thing. A Criticism of Fodor's and Pylyshyn's Criticism of Smolensky." *Philosophica*, 47, 1 (1991): 49–79.

[277] ——— (1991b). "Syntaxe topologique et Grammaire cognitive." *Langages*, 103 (1991): 97–128.

[278] ——— (1991c). "Le schématisme morphodynamique de l'aspectualité." In *Le discours aspectualisé*, J. Fontanille ed. Limoges-Amsterdam: Pulim-Benjamins, 1991, 177–193.

[279] ——— (1992). *Physique du Sens*. Paris: Editions du CNRS, 1992.

[280] ——— (1993a). "Modeling: Formalization or Mathematization? The example of a morphodynamical approach to language." In *Structures of Signification*, vol. III, H.S. Gill ed. New Dehli: Wiley Eastern, 1993, 700–709.

[281] ——— (1993b). "Natural Dynamical Models for Cognitive Grammars." In *Language and Cognition*, F. Manjali ed. New Dehli: Bahri Publications, 1993.

[282] ——— (1994a). "Phenomenology of Perception, Qualitative Physics and Sheaf Mereology." *Philosophy and the Cognitive Sciences, Proceedings of the 16th International Wittgenstein Symposium*, R. Casati, B. Smith and G. White eds. Vienna: Verlag Hölder-Pichler-Tempsky, 1994, 387–408.

[283] ——— (1994b). "La sémiophysique : de la physique qualitative aux sciences cognitives." In *Passion des Formes, à René Thom*, M. Porte ed. École Normale Supérieure, Éditions Fontenay-Saint Cloud, 1994, 499–545.

[284] ——— (1994c). "Attractor Syntax: Morphodynamics and Cognitive Grammars." In *Continuity in Linguistic Semantics*, C. Fuchs and B. Victorri eds. Amsterdam: John Benjamins, 1994, 167–187.

[285] ——— (1994d). "Natural Dynamical Models for Visual Morphology, Topological Syntax and Cognitive Grammar." In *Representation: Relationship between Language and Image*, S. Levialdi and C. Bernardelli eds. Singapore: World Scientific, 1994, 187–208.

[286] ——— (1994e). "Dynamical Constituency: an Epistemological Approach." *Linguistique cognitive et Modèles dynamiques,* J. Petitot ed. *Sémiotiques*, 6–7 (1994): 187–225.

[287] ——— (1995a). "Morphodynamics and Attractor Syntax. Dynamical and morphological models for constituency in visual perception and cognitive grammar." In *Mind as Motion*, R. Port and T. van Gelder eds. Cambrige, MA: MIT Press, 1995, 227–281.

[288] ——— (1995b). "Sheaf Mereology and Husserl's Morphological Ontology." *International Journal of Human-Computer Studies*, 43 (1995): 741–763.

[289] ———, ed. (1995c). *Conference on Topology and Dynamics in Cognition and Perception.* Università di San Marino.

[290] ———, ed. (1995d). *Linguistique cognitive et Modèles Dynamiques.* Special issue of *Sémiotiques*, 6–7, Paris: Didier.

[291] ——— (1995e). "Approche morphodynamique de l'iconicité des stemmas." *Lucien Tesnière aujourd'hui*, F. Madray-Lesigne, J. Richard-Zapella eds. Louvain: Editions Peeters, 1995, 105–112.

[292] ——— (1997). "Sheaf Mereology and Space Cognition." In *Modelli della Cognizione e Teoria della Complessità, La Nuova Critica*, 29, 1 (1997): 49–74, Union Printing.

[293] ——— (1998). "Modèles morphodynamiques de catégorisations phon'etiques." In *The Roman Jakobson Centennial Symposium*, P. Aa. Brandt, F. Gregersen eds. *Acta Linguistica Hafniensia*, 29 (1998): 239–269.

[294] ——— (1999a). "Morphological Eidetics for Phenomenology of Perception." In *Naturalizing Phenomenology: Issues in Contemporary Phenomenology and Cognitive Science*, J. Petitot, F. J. Varela, J.-M. Roy, B. Pachoud eds. Stanford: Stanford University Press, 330–371.

[295] ——— (1999b). "Vers une Neurogéométrie. Fibrations corticales, structures de contact et contours subjectifs modaux." *Mathématiques, Informatique et Sciences Humaines*, Special Issue, 145, (1999): 5–101, Paris, École des Hautes Études en Sciences Sociales.

[296] ——— (2000). "Philosophie de la Nature et phénoménologie dans *Le monde comme perception et réalité* de Roger Chambon." *Science et philosophie de la Nature*, L. Boi ed. Berne: Peter Lang, 363–384.

[297] ——— (2001). "A Morphodynamical Schematization of the Canonical Formula for Myths." *The Double Twist. From Ethnography to Morphodynamics*, P. Maranda ed. Toronto: Univ. of Toronto Press, 267–311.

[298] ——— (2003a). "The Neurogeometry of pinwheels as a sub-Riemannian contact structure." In *Neurogeometry and Visual Perception*, J. Petitot and J. Lorenceau eds. *Journal of Physiology-Paris*, 97, 2–3, (2003): 265–309.

[299] ——— (2003b). "Modèles de structures émergentes dans les systèmes complexes." In *Complexity and Emergence*, E. Agazzi and L. Montecucco eds. *Proceedings of the Annual Meeting of the International Academy of the Philosophy of Science*. Singapore: World Scientific, 2003, 57–71.

[300] ——— (2003c). "Neurogeometry of V1 and Kanizsa contours." *Axiomathes*, 13, (2003): 347–363.

[301] ——— (2004). "Théorie des singularités et équations de diffusion." *Gazette des Mathématiciens*, Special issue on René Thom, 2004, 89–107.

[302] ——— (2006a). "Neurogéométrie des architectures fonctionnelles de la vision." In *Mathématiques et Vision, Journée annuelle de la Société Mathématiques de France*, 24 juin 2006, 69–128.

[303] ——— (2006b). "Les modèles morphogénétiques de René Thom." In *Morphogenèse. L'origine des formes*, P. Bourgine and A. Lesne eds. Paris: Belin, 2006.

[304] ——— (2008). *Neurogéométrie de la vision. Modèles mathématiques et physiques des architectures fonctionnelles.* Paris: Les Éditions de l'École Polytechnique, Distribution Ellipses.

[305] Petitot, J., Doursat, R. (1998). "Modèles dynamiques et linguistique cognitive : vers une sémantique morphologique active." Technical Report 9809, *Rapports et documents du CREA*. Paris: École Polytechnique, 1998.

[306] Petitot, J., Lorenceau, J., eds. (2003). *Neurogeometry and Visual Perception*. Special Issue of the *Journal of Physiology-Paris*, 97, 2–3 (2003).

[307] Petitot, J., Smith, B. (1991). "New Foundations for Qualitative Physics." In *Evolving Knowledge in Natural Science and Artificial Intelligence*, J.E. Tiles, G.J. McKee and G.C. Dean eds. London: Pitman, 1991, 231–249.

[308] Petitot, J., Varela, F. J., Roy, J.-M., Pachoud, B., eds. (1999). *Naturalizing Phenomenology: Issues in Contemporary Phenomenology and Cognitive Science*. Stanford: Stanford University Press, 1999.

[309] Piatelli-Palmarini, M. (1989). "Evolution, Selection and Cognition: From 'Learning' to Parameter Setting in Biology and the Study of Language." *Cognition*, 31 (1989): 1–44.

[310] Piotrowski, D. (1994). "Approche morphodynamique de deux régimes fondamentaux du langage." *Linguistique cognitive et Modèles dynamiques*, J. Petitot ed. *Sémiotiques*, 6–7 (1994): 175–186.

[311] Poli, R., Simons, P. (1996). *Formal Ontology*. Dordrecht: Kluwer, 1996.

[312] Port, R., van Gelder, T., eds. (1995). *Mind as Motion: Explorations in the dynamics of cognition*. Cambridge, MA: MIT Press, 1995.

[313] Pottier, B. (1992) *Théorie et Analyse en Linguistique*. Paris: Hachette, 1992.

[314] ——— (1992). *Sémantique générale*. Paris: Presses Universitaires de France, 1992.

[315] ——— (2000). *Représentations mentales et catégorisations linguistiques*. Bibliothèque de l'Information grammaticale, 47. Louvain-Paris: Peeters, 2000.

[316] Prewitt, J. M. S. (1970). "Object enhancement and extraction." In *Picture Processing and Psychopictorics*, B. S. Lipkin, A. Rosenfeld eds. New York: Academic Press, 1970.

[317] Psotka, J., (1978). "Perceptual Process that May Create Stick Figures and Balance." *Journal of Experimental Psychology, Human Perception and Performance*, 4, 1 (1978): 101–111.

[318] Pylyshyn, Z. (1986). *Computation and Cognition*. Cambridge, MA: MIT Press, 1986.

[319] ——— (2001). "Visual indexes, preconceptual objects, and situated vision." *Cognition*, 80 (2001): 127–158.

[320] Regier, T. (1988). "Recognizing Image-schemata Using Programmable Networks." In *Proceedings of the 1988 Connectionist Models Summer School*. San Francisco: Morgan Kaufman, 1988.

[321] ——— (1996). *The Human Semantic Potential*. Cambridge, MA: MIT Press, 1996.

[322] Regier, T., Carlson, L.A. (2001). "Grounding Spatial Language in Perception: An Empirical and Computational Investigation." *Journal of Experimental Psychology. General*, 130, 2 (2001): 273–298.

[323] Renals, S., Rohwer, R. (1990). "A Study of Network Dynamics." *Journal of Statistical Physics*, 58, 5–6 (1990): 825–848.

[324] Richter, J., Ullman, S. (1986). "Non-Linearities in Cortical Simple Cells and the Possible Detection of Zero Crossings." *Biol. Cybern.*, 53, (1986): 195–202.

[325] Riemann, B. (1867). "Über die Hypothesen, welche der Geometrie zu Grunde liegen. (On the Hypotheses which lie at the Bases of Geometry.)" *Abhandlungen der Königlichen Gesellschaft der Wissenschaften zu Göttingen*, 13, 1867.

[326] Rosch, E. (1975). "Cognitive representations of semantic categories." *Journal of Experimental Psychology. General*, 104, (1975): 192–233.

[327] Roskies, A. L. (1999). "The binding problem." *Neuron* 24 (1999): 7–9.

[328] Ross, J. R. (1972). "The Category Squish: Endstation Hauptwort." *Papers from the Eighth Regional Meeting, Chicago Linguistic Society (CLS 8)*, P. Peranteau, J. Levi, G. Phares eds. 8 (1972): 316–328.

[329] RS (1990). *Sciences cognitives : quelques aspects problématiques*, J. Petitot ed. Special Issue of *Revue de Synthèse*, IV, 1–2 (1990).

[330] Sarti, A., Citti, G., Manfredini, M. (2003). "From neural oscillations to variational problems in the visual cortex." In *Neurogeometry and Visual Perception*, J. Petitot and J. Lorenceau eds. *Journal of Physiology-Paris*, 97, 2–3 (2003): 87–385.

[331] Sato Y. D., Shiino, M. (2002). "Spiking neuron models with excitatory or inhibitory synaptic couplings and synchronization phenomena." *Phys. Rev. E*, 66, 4 (2002) id. 041903.

[332] Schank, R., Abelson, R.P. (1977). *Scripts, Plans, Goals and Understanding.* Hillsdale: Lawrence Erlbaum, 1977.

[333] Schank, R.C. (1991). *Tell Me a Story: A New Look at Real and Artificial Intelligence.* New York: Simon & Schuster, 1991.

[334] Schlesinger, I.M. (1971). "Production of Utterances and Language Acquisition." *The Ontogenesis of Grammar*, D. I. Slobin ed. Cambridge, MA: Academic Press, 1971, 63–101.

[335] Scholl, B. J., Pylyshyn, Z. W. (1999). "Tracking multiple items through occlusion: Clues to visual objecthood." *Cognitive Psychology*, 38, (1999): 259–290.

[336] Scholl, B. J., Tremoulet, P. D. (2000). "Perceptual causality and animacy." *Trends in Cognitive Science*, 4, 8 (2000): 299–309.

[337] Scholtz, E. (1992). "Riemann's Vision of a New Approach to Geometry." In *1830-1930: A Century of Geometry*, L. Boi, D. Flament, and J.-M. Salanskis eds. Berlin, New-York: Springer, 1992, 22–34.

[338] Seibert, M., Waxman, A. (1989). "Spreading Activation Layers, Visual Saccades, and Invariant Representations for Neural Pattern Recognition Systems." *Neural Networks*, 2, 1 (1989): 9–28.

[339] Seiler, H.J., ed. (1978). *Language Universals*, Papers from the Gummersbach Conference (1976). Tübingen: Gunter Narr, 1978.

[340] ——— (2000). *Language Universals Research: A Synthesis*, Language Universals Series, 8. Tübingen: Gunter Narr, 2000.

[341] Serra, J. (1982). *Image Analysis and Mathematical Morphology.* New York: Academic Press, 1982.

[342] Sethian, J. A. (1990). "Numerical Algorithms for Propagating Interfaces: Hamilton-Jacobi Equations and Conservation Laws." *J. Differential Geometry*, 31 (1990): 131–161.

[343] Shastri, L., Ajjanagadde, V. (1993). "From simple associations to systematic reasoning." *Behavioral and Brain Sciences* 16, 3 (1993): 417–451.

[344] Shepard, R.N., Cooper, L.A. (1982). *Mental Images and their Transformations.* Cambridge, MA: MIT Press.

[345] Siddiqi, K., Shokoufandeh, A., Dickinson, S. J., Zucker, S. W. (1999). "Shock graphs and shape matching." *International Journal of Computer Vision*, 35, 1 (1999): 13–32.

[346] Simon, H. (1994). "Literary Criticism: A Cognitive Approach", In *Bridging the Gap. Where Cognitive Science Meets Literary Criticism*, S. Franchi, G. Guzeldere eds. *Stanford Humanities Review*, 4, 1 (1994).

[347] Smale, S. (1967). "Differentiable Dynamical Systems." *Bull. Amer. Math. Soc.*, 73, 6 (1967): 747–817.

[348] Smith, B., ed. (1982). *Parts and Moments. Studies in Logic and Formal Ontology.* München, Wien: Philosophia Verlag, Analytica, 1982.

[349] Smith B., ed. (1988). *Foundations of Gestalt Theory.* München, Wien: Philosophia Verlag, 1988.

[350] Smith, B. (1993). "The Structures of the Common-Sense World." In *Gestalt Theory. Its Origins, Foundations and Influence*, A. Pagnini, S. Poggi eds. Florence: Olschky, 1993.

[351] Smith, B. (2000). "Truth and the Visual Field." In *Naturalizing Phenomenology. Issues in Contemporary Phenomenology and Cognitive Science,* J. Petitot, F. J. Varela, B. Pachoud, J. M. Roy eds. Stanford: Stanford University Press, 2000, 317–329.

[352] Smolensky, P. (1986). "Information processing in dynamical systems: Foundations of harmony theory." In *Parallel distributed processing: Explorations in the microstructure of cognition. Vol. 1, Foundations*, D. E. Rumelhart, James L. McClelland and the PDP Research Group eds. Cambridge MA: MIT Press, 1986, 194–281.

[353] ——— (1987a). *On Variable Binding and the Representation of Symbolic Structures in Connectionist Systems.* Technical Report CU-CS-355-87. University of Colorado at Boulder: Department of Computer Science.

[354] ——— (1987b). "The Constituent Structure of Mental States: A Reply to Fodor and Pylyshyn." *Southern Journal of Philosophy*, 26, (1987): 137–160.

[355] ——— (1988). "On the Proper Treatment of Connectionism." *The Behavioral and Brain Sciences*, 11, (1988): 1–74.

[356] ——— (1990). "Tensor Product Variable Binding and the Representation of Symbolic Structures in Connectionist Networks." *Artificial Intelligence*, 46, (1990): 159–216.

[357] ——— (2000). "Grammar-based Connectionist Approaches to Language." *Cognitive Science*, 23, (2000): 589–613. Reprinted in *Connectionist Psycholinguistics*, M.H. Christiansen and N. Chater eds. Westport, CT: Ablex, 2001.

[358] Smolensky, P., Legendre, G. (2006). "Harmony optimization and the computational architecture of the mind/brain." In *The Harmonic Mind: From Neural Computation to Optimality-Theoretic Grammar, Vol. 1*, P. Smolensky and G. Legendre eds. Cambridge, MA: MIT Press, 2006, 3–56.

[359] Smolensky, P., Legendre, G., Miyata, Y. (1992). *Principles for an Integrated Connectionist/Symbolic Theory of Higher Cognition.* Technical Report CU-CS-600-92. University of Colorado at Boulder: Department of Computer Science.

[360] Somers, D., Kopell, N. (1993). "Rapid synchronization through fast threshold modulation." *Biol. Cybern.*, 68 (1993): 393–407.

[361] ——— (1995). "Waves and synchrony in networks of oscillators of relaxation and non-relaxation type." *Physica D*, 89 (1995): 169–183.

[362] Sompolinski, H., Crisanti, A., Sommers, H.-J. (1988). "Chaos in Random Neural Networks." *Phys. Rev. Lett.*, 61 (1988): 259–262.

[363] Sporns, O., Tononi, G., Edelman, G. M. (1991). "Modeling perceptual grouping and figure-ground segregation by means of active reentrant connections." *Proc. Natl. Acad. Sci. USA*, 88 (1991): 129–133.

[364] Strogatz, S. H. (2000). "From Kuramoto to Crawford: Exploring the onset of synchronization in populations of coupled oscillators." *Physica D*, 14 (2000): 31–20.

[365] Sweetser, E. (1990). *From Etymology to Pragmatics.* Cambridge, UK: Cambridge University Press, 1990.

[366] Swinney, H. L., Krinsky, V. I., eds. (1991). *Waves and Patterns in Chemical and Biological Media.* Cambridge, MA: The MIT Press, 1991.

[367] Talmy, L. (1975a). "Figure and ground in complex sentences." In *Proceedings of the First Annual Meeting of the Berkeley Linguistics Society*, Berkeley, 1975.

[368] ——— (1975b). "Semantics and syntax of motion." In *Syntax and Semantics*, 4, P. J. Kimball ed. New York: Academic Press, 1971.

[369] ——— (1978). "Relation of Grammar to Cognition." In *Proceedings of TINLAP-2*, D.Waltz ed. Urbana: University of Illinois, 1978.

[370] ——— (1983). "How Language Structures Space." In *Spatial Orientation: Theory, Research and Applications*, H. Pick and L. Acredolo eds. New York: Plenum Press, 1983.

[371] ——— (1985). "Force Dynamics in Language and Thought." *Parasession on Causatives and Agentivity*, Chicago Linguistic Society, 1985.

[372] ——— (1990). "Fictive Motion in Language and Perception." Workshop *Motivation in Language*, University of San Marino: International Center for Semiotic and Cognitive Studies, 1990.

[373] ——— (1996). "Fictive motion in language and 'ception'." In *Language and Space*, P. Bloom, M. Peterson, L. Nadel, and M. Garrett eds. Cambridge, MA: The MIT Press, 1996.

[374] ——— (2000a). *Toward a Cognitive Semantics, Vol. I: Concept Structuring Systems.* Cambridge, MA: MIT Press, 2000.

[375] ——— (2000b). *Toward a Cognitive Semantics, Vol. II: Typology and Process in Concept Structuring.* Cambridge, MA: MIT Press, 2000.

[376] Taylor, J. R. (2002). *Cognitive Grammar.* Oxford: Oxford University Press, 2002.

[377] Terman D., Wang, D. L. (1995). "Global competition and local cooperation in a network of neural oscillators." *Physica D*, 81 (1995): 148–176.

[378] Tesnière, L. (1959). *Eléments de Syntaxe stucturale.* Paris: Klincksieck, 1969 (2nd ed.).

[379] Thom, R. (1972). *Stabilité structurelle et Morphogenèse.* New York: Benjamin, Paris: Ediscience, 1972.

[380] ——— (1973). "Sur la typologie des langues naturelles : essai d'interprétation psycho linguistique." In *L'analyse formelle des langues naturelles*, M. Gross, M. Halle, M. P. Schutzenberger eds. Mouton, 233–248. (See Thom, 1980.)

[381] ——— (1978). "La Double Dimension de la Grammaire Universelle." In *Language Universals*, H.J. Seiler ed. Tübingen: Gunter Narr Verlag, 1978, 78–89.

[382] ——— (1980). *Modèles mathématiques de la Morphogenèse.* Paris: Christian Bourgois, 1980.

[383] ——— (1984). "Classification des sciences et des techniques." *Séminaire de philosophie et mathématiques*, l'École Normale Supérieure, Paris. Also in *Apologie du Logos.* Paris: Hachette, 1990, 515–543.

[384] ——— (1988). *Esquisse d'une Sémiophysique.* Paris: InterEditions, 1988.

[385] ——— (1990). *Apologie du Logos.* Paris: Hachette, 1990.

[386] Tirozzi, B., Tsodyks, M. (1991). "Chaos in Highly Diluted Neural Networks." *Europhy. Lett.*, 14 (1991): 727–732.

[387] Tougeron, J.C. (1972). *Idéaux de Fonctions Différentiables.* New-York, Heidelberg, Berlin: Springer, 1972.

[388] Vandeloise, C. (1984). *Description of Space in French.* PhD thesis. San Diego, CA: Univ. of California, 1984.

[389] ——— (1986). *L'Espace en Français : Sémantique des prépositions spatiales.* Paris: Editions du Seuil, 1986. Transl. *Spatial Prepositions: a Case Study in French*, Chicago University Press, 1991.

[390] ——— (2006). "Are there spatial prepositions?" In *Space in Languages: Linguistic Systems and Cognitive Categories*, M. Hickmann, S. Robert eds. Amsterdam: John Benjamins (*Typological Studies in Language* 66), 2006, 139–154.

[391] ——— (2009). "The genesis of spatial terms." In *Language, Cognition and Space: the State of the Art and New Directions*, V. Evans, P. Chilton eds. London: Equinox (*Advances in Cognitive Linguistics*), 2009, 157–178.

[392] van Gelder, T. (1992). "Connectionism and Dynamical Explanation." In *Proceedings of the Thirteenth Annual Conference of the Cognitive Science Society.* Hillsdale: L. Erlbaum, 1992.

[393] ——— (1994). "Compositionality and Dynamics in Neural Network Representations." *Linguistique cognitive et Modèles dynamiques*, J. Petitot ed. *Sémiotiques*, 6–7 (1994): 49–68.

[394] Victorri, B. (2010). "Le localisme à l'épreuve du verbe *aller*." In *Espace, Préposition, Cognition. Hommage à Claude Vandeloise*, G. Col, C. Collin eds. Corela, http://edel.univ-poitiers.fr/corela/document.php?id=2657

[395] Visetti, Y. M. (1990). "Modèles connexionnistes et représentations structurées." In *Modèles Connexionnistes*, D. Memmi and Y.M. Visetti eds. *Intellectica*, 9–10 (1990): 167–212.

[396] ——— (1994). "Les Modèles connexionnistes entre perception et sémantique." *Linguistique cognitive et Modèles dynamiques*, J. Petitot ed. *Sémiotiques*, 6–7 (1994): 15–48.

[397] von der Malsburg, C. (1981). *The Correlation Theory of Brain Function.* Internal Report 81-2. Göttingen: Max Planck Institute for Biophysical Chemistry.

[398] von der Malsburg, C., Bienenstock, E. (1986). "Statistical Coding and Short-Term Synaptic Plasticity: A Scheme for Knowledge Representation in the Brain." In *Disordered Systems and Biological Organization*, E. Bienenstock, F. Fogelman, and G. Weisbuch eds. Berlin: Springer, 1986, 247–272.

[399] Wagemans J., van Lier R., Scholl B.J. (2006). "Introduction to Michotte's heritage in perception and cognition research." *Acta Psychologica*, 123 (2006): 1–19.

[400] Weir, S. (1978). "The perception of motion: Michotte revisited." *Perception*, 7 (1978): 247–260.

[401] Whitaker, R. (1993). "Geometry-limited diffusion in the characterization of geometric patches in images." *Computer Vision, Graphics, and Image Processing: Image Understanding*, 57, 1 (1993): 111–120.

[402] Whorf, B. (1956). *Language, Thought and Reality.* New York: Technology Press of MIT and Wiley, 1956.

[403] Wickerhauser, M. V. (1991). *Lectures on Wavelet Packet Algorithms.* Technical Report. Washington University: Department of Mathematics.

[404] Wiener, N. (1948). *Cybernetics: or control and communication in the animal and machine.* New York: Wiley, 1948.

[405] Wildgen, W. (1981). "Archetypal Dynamics in Word Semantics: an Application of Catastrophe Theory." In *Words, Worlds and Contexts*, H. J. Eikmeyer, H. Reiser eds. New York: Walter de Gruyter, 1981, 234–296.

[406] ——— (1982). *Catastrophe Theoretic Semantics.* Amsterdam: Benjamins, 1982.

[407] ——— (1994). "Les Fondements sémiotiques et philosophiques d'une grammaire morphodynamique." *Linguistique cognitive et Modèles dynamiques*, J. Petitot ed. *Sémiotiques*, 6–7 (1994): 161–174.

[408] ——— (1999). *De la grammaire au discours. Une approche morphodynamique.* Bern, Peter Lang, 1999.

[409] ——— (2008). *Kognitive Grammatik.* Berlin, New York: Walter de Gruyter, 2008.

[410] Willems, D. (1978). "A la recherche d'une grammaire des cas. Les rapports avec la syntaxe et le lexique." *Valence, Semantic Case and Grammatical Relations*, W. Abraham ed. *Studies in Language Companion Series*, 1. Amsterdam: John Benjamins, 1978, 243-260.

[411] Wilson H. R., Cowan, J. D. (1972). "Excitatory and inhibitory interactions in localized populations of model neurons." Biophys. J., 12 (1972): 1–24.

[412] Winfree, A. (1980). *The Geometry of Biological Time.* Berlin: Springer-Verlag, 1980.

[413] Witkin, A. (1983). "Scale-Space Filtering." In *Proc. Int. Joint Conf. on Artificial Intelligence.* Karlsruhe, 1983, 1019-1021.

[414] Wittgenstein, L. (1978). *Philosophical Investigations.* Oxford: Basil Blackwell, 1978.

[415] Yuille, A., Poggio, T. A. (1986). "Scaling theorems for zero crossings." *IEEE Trans. Patt. Anal. Mach. Intell. PAMI-8*, 1986: 15–25.

[416] Zeeman, C. (1965). "Topology of the Brain." In *Mathematics and Computer Science in Biology and Medicine.* Medical Research Council.

[417] ——— (1976). "Brain Modelling." In *Structural Stability, the Theory of Catastrophes and Applications in the Sciences*, Lecture Notes in Mathematics, 525. Berlin: Springer 1976, 367–372. (See also Zeeman (1977), chap. 8).

[418] ——— (1977). *Catastrophe Theory: Selected Papers 1972–1977.* Redwood City: Addison-Wesley, 1977.

[419] Zhu, S.C., Yuille, A.L. (1996). "FORMS: A flexible object recognition and modeling system." *Int. J. Comp. Vis.* 20, 3 (1996): 187–212.

[420] Zibetti, E. (2001). *Catégorisation contextuelle et compréhension d'événements visuellement perçus et interprétés comme des actions.* PhD thesis, University of Paris VIII.

[421] Zibetti, E., Tijus, C. (2003). "Perceiving Action from Static Images: the Role of Spatial Context." *CONTEXT* (2003): 397–410.

[422] Zucker, S.W., David, C., Dobbins, A., Iverson L. (1988). "The organization of curve detection: coarse tangent fields and fine spline covering." In *Proc. 2nd Intern. Conf. Comp. Vis.*, IEEE, 1988.

[423] Zurek, W. H., ed. (1990). *Complexity, Entropy, and Physics of Information.* Reading, MA: Addison-Wesley, 1990.

Index

List of abbreviations

nD n-dimensional
AC apparent contour
BCS boundary contour system
CG cognitive grammar
CL classic
CL cut locus
CLC classical cognitivism
CN connectionism
CNC connectionist cognitivism
CR cognitive representation
CS conceptual structure
CT cognitive topology
DO direct object
ES expanse scanning
FCS featural contour system
F-P (argument) Fodor-Pylyshyn
FS field scanning
GC ganglion cell
LH localist hypothesis
LM landmark
MM mathematical morphology
MT mathematical topology
NP noun phrase
OO oblique object
PDE partial differential equation
PhW phenomenal world
PrW projected world
PS periphery scanning
PTC proper treatement of connectionism
R field of real numbers
RAM random access memory
RF receptive field
RH relational hierarchy
RP receptive profile
RW real world
SKIZ skeleton by influence zones
TR trajector
WFT window Fourier transform
WT wavelet transform

European Semiotics: *Language, Cognition, and Culture*
Sémiotiques Européennes: *langage, cognition et culture*

Edited by / Série dirigée par
Per Aage Brandt (Cleveland), Wolfgang Wildgen (Bremen/Brême),
and/et Barend van Heusden (Groningen/Groningue)

European Semiotics originated from an initiative launched by a group of researchers in Semiotics from Denmark, Germany, Spain, France and Italy and was inspired by innovative impulses given by René Thom and his "semiophysics". The goal of the series is to provide a broad European forum for those interested in semiotic research focusing on *semiotic dynamics* and combining *cultural, linguistic and cognitive perspectives.*
This approach, which has its origins in Phenomenology, Gestalt Theory, Philosophy of Culture and Structuralism, views semiosis primarily as a cognitive process, which underlies and structures human culture. Semiotics is therefore considered to be the discipline suited *par excellence* to bridge the gap between the realms of the Cognitive Sciences and the Sciences of Culture.
The series publishes monographs, collected papers and conference proceedings of a high scholarly standard. Languages of publication are mainly English and French.

Sémiotiques européennes est le résultat d'une initiative prise par un groupe de chercheurs en sémiotique, originaires du Danemark, d'Allemagne, d'Espagne, de France et d'Italie, inspirée par l'impulsion innovatrice apportée par René Thom et sa "sémiophysique". Le but de cette collection est de fournir une tribune européenne large à tous ceux qui s'intéressent à la recherche sémiotique portant sur *les dynamiques sémiotiques*, et réunissant des *perspectives culturelles, linguistiques et cognitives.*
Cette approche, qui combine différentes sources, telle que la phénoménologie, le gestaltisme, la philosophie de la culture et le structuralisme, part du principe que la sémiosis est essentiellement un procès cognitif, qui sous-tend et structure toute culture humaine. Dans cette approche, la sémiotique est donc considérée comme la discipline par excellence capable de créer un pont entre les domaines des sciences cognitives et ceux des sciences de la culture.
Sémiotiques européennes accueille tant des monographies que des anthologies et des actes de colloques d'un haut niveau de recherche, rédigés de préférence en anglais et en français.

Volume 1 Wolfgang Wildgen: De la grammaire au discours
Une approche morphodynamique
xi + 351 pp. 1999. ISBN 3-906762-03-3 / US-ISBN 0-8204-4226-7

Volume 2 Lene Fogsgaard: Esquemas copulativos de SER y ESTAR
Ensayo de semiolingüística
382 pp. 2000. ISBN 3-906764-22-2

Volume 3 Jean Petitot: Morphogenesis of Meaning
xii + 279 pp. 2003. ISBN 3-03910-104-8 / US-ISBN 0-8204-6858-4

Volume 4 Per Aage Brandt: Spaces, Domains, and Meaning
Essays in Cognitive Semiotics
271 pp. 2004. ISBN 3-03910-227-3 / US-ISBN 0-8204-7006-6

Volume 5 Marcel Bax, Barend van Heusden & Wolfgang Wildgen:
Semiotic Evolution and the Dynamics of Culture
xx + 318 pp. 2004. ISBN 3-03910-394-6 / US-ISBN 0-8204-9

Volume 6 Ángel López-García: The Grammar of Genes
How the Genetic Code Resembles the Linguistic Code
182 pp. 2005. ISBN 3-03910-654-6 / US-ISBN 3-03910-7171-2

Volume 7 Ole Kühl: Musical Semantics
261 pp. 2nd unmodified printing. 2007, 2008. ISBN 978-3-03911-718-5

Volume 8 Todd Oakley: From Attention to Meaning
Explorations in Semiotics, Linguistics, and Rhetoric
268 pp. 2009. ISBN 978-3-03911-442-9

Volume 9 Wolfgang Wildgen & Barend van Heusden (eds.):
Metarepresentation, Self-Organization and Art
298 pp. 2009. ISBN 978-3-03911-684-3

Volume 10 Wolfgang Wildgen & Per Aage Brandt (eds.):
Semiosis and Catastrophes
René Thom's Semiotic Heritage
185 pp. 2010. ISBN 978-3-0343-0467-2

Volume 11 Jean Petitot:
Cognitive Morphodynamics
Dynamical Morphological Models of Constituency
in Perception and Syntax
In collaboration with René Doursat
305 pp. 2011. ISBN 978-3-0343-0475-7